Volumes already published

Volume 1: Trees I (1986)
Volume 2: Crops I (1986)
Volume 3: Potato (1987)
Volume 4: Medicinal and Aromatic Plants I (1988)
Volume 5: Trees II (1989)
Volume 6: Crops II (1988)
Volume 7: Medicinal and Aromatic Plants II (1989)
Volume 8: Plant Protoplasts and Genetic Engineering I (1989)
Volume 9: Plant Protoplasts and Genetic Engineering II (1989)
Volume 10: Legumes and Oilseed Crops I (1990)
Volume 11: Somaclonal Variation in Crop Improvement I (1990)
Volume 12: Haploids in Crop Improvement I (1990)
Volume 13: Wheat (1990)
Volume 14: Rice (1991)
Volume 15: Medicinal and Aromatic Plants III (1991)
Volume 16: Trees III (1991)
Volume 17: High-Tech and Micropropagation I (1991)
Volume 18: High-Tech and Micropropagation II (1992)
Volume 19: High-Tech and Micropropagation III (1992)
Volume 20: High-Tech and Micropropagation IV (1992)
Volume 21: Medicinal and Aromatic Plants IV (1993)
Volume 22: Plant Protoplasts and Genetic Engineering III (1993)
Volume 23: Plant Protoplasts and Genetic Engineering IV (1993)
Volume 24: Medicinal and Aromatic Plants V (1993)

Volumes in preparation

Volume 25: Maize (1994)
Volume 26: Medicinal and Aromatic Plants VI (1994)
Volume 27: Somatic Hybridization in Crop Improvement
Volume 28: Medicinal and Aromatic Plants VII
Volume 29: Plant Protoplasts and Genetic Engineering V
Volume 30: Somatic Embryogenesis and Synthetic Seed I

Biotechnology in Agriculture and Forestry 22

Plant Protoplasts and Genetic Engineering III

Edited by Y. P. S. Bajaj

With 103 Figures

Springer-Verlag
Berlin Heidelberg New York
London Paris Tokyo
Hong Kong Barcelona Budapest

Professor Dr. Y.P.S. BAJAJ
A-137
New Friends Colony
New Delhi 110065, India

ISBN-13: 978-3-642-78008-0 **e-ISBN-13: 978-3-642-78006-6**
DOI: 10.1007/ 978-3-642-78006-6

Library of Congress Cataloging-in-Publication Data. (Revised for volume 3). Plant protoplasts and genetic engineering. (Biotechnology in agriculture and forestry ; 8–9, 22) Includes bibliographies and indexes. 1. Plant protoplasts. 2. Plant genetic engineering. I. Bajaj, Y.P.S., 1936– . II. Series. QK725.P578 1989 581.87'3 88-31197

Softcover reprint of the hardcover 1st edition 1993

Production Editor: Ingrid Samide, Heidelberg
Typesetting: Macmillan India Limited, Bangalore
31/3145-5 4 3 2 1 0 – Printed on acid-free paper

Dedicated to Professor Dr. Martin Bopp with whom I had the privilege of working at the University of Heidelberg, Germany, from 1970–1972

Preface

Since the publication of *Plant Protoplasts and Genetic Engineering I and II* in 1989, there has been a steady inflow of new information in the area of genetic engineering of plants. The number of species in which complete plants have been regenerated from isolated protoplasts has increased considerably. Likewise, numerous genetic transformation studies have been conducted successfully on a number of crop species. This is the most compelling reason for compiling the following two new volumes which contain 58 chapters contributed by international experts.

Plant Protoplasts and Genetic Engineering III comprises 29 chapters dealing with the regeneration of plants from protoplasts, and genetic transformation in various species of *Agrostis*, *Arabidopsis*, *Atropa*, *Brassica*, *Catharanthus*, *Cucumis*, *Datura*, *Daucus*, *Digitalis*, *Duboisia*, *Eustoma*, *Festuca*, *Helianthus*, *Hordeum*, *Kalanchoe*, *Linum*, *Lobelia*, *Lolium*, *Lotus*, *Lycium*, *Lycopersicon*, *Mentha*, *Nicotiana*, *Pelargonium*, *Pisum*, *Pyrus*, *Salvia*, *Scopolia* and *Solanum*.

Plant Protoplasts and Genetic Engineering IV comprises 29 chapters on various species of *Actinidia*, *Armoracia*, *Beta*, *Brassica*, *Cicer*, *Citrus*, *Cucumis*, *Duboisia*, *Fragaria*, *Glycine*, *Ipomoea*, *Lactuca*, *Lotus*, *Lycopersicon*, *Manihot*, *Medicago*, *Nicotiana*, *Petunia*, *Phaseolus*, *Picea*, *Pisum*, *Prunus*, *Psophocarpus*, *Saccharum*, *Solanum*, *Sorghum*, *Stylosanthes*, *Vitis* and *Zea mays*.

These studies reflect the far-reaching implications of protoplast technology in agricultural and forest biotechnology. Therefore, these books may be of special interest to advanced students, teachers, and research scientists in the field of plant tissue culture, molecular biology, genetic engineering, plant breeding, and general biotechnology.

New Delhi, September 1993

Professor Y. P. S. Bajaj
Series Editor

Contents

Section I Regeneration of Plants from Protoplasts

Section II Genetic Transformation

II.1 Transformation in *Arabidopsis thaliana*
R. Bilang and I. Potrykus (With 3 Figures)

II.2 Transformation in *Atropa belladonna*
Y. Suzuki, Y. Kurioka, T. Ogasawara, and H. Kamada (With 4 Figures)

II.3 Transformation in *Catharanthus* Species (Madagascar Periwinkle)
C. David and J. Tempé (With 3 Figures)

II.4 Transformation in *Datura* Species
P. Christen and M. F. Roberts (With 1 Figure)

II.18 Transformation in *Scopolia*
Y. L. Zhang (With 4 Figures)

II.19 Transformation in *Solanum melongena* L. (Eggplant)
M. Leone, E. Filippone, and P. F. Lurquin (With 2 Figures)

List of Contributors

ASANO, Y., Faculty of Horticulture, Chiba University, 6-2-1 Kashiwa-no-ha, Kashiwa city, Chiba Pref., Japan 277

BAJAJ, Y. P. S., Former Professor of Tissue Culture, Punjab Agricultural University, Ludhiana, India. (Present address: A-137 New Friends Colony, New Delhi 110065, India)

BELLINI, C., Laboratoire de Biologie Cellulaire, I.N.R.A., Route de Saint-Cyr, 78026 Versailles Cédex, France

BILANG, R., Institute of Plant Sciences, Swiss Federal Institute of Technology, ETH Zentrum, 8092 Zürich, Switzerland

BIRCH, R. G., Department of Botany, The University of Queensland, Brisbane, 4072, Australia

BOWER, R., Department of Botany, The University of Queensland, Brisbane, 4072, Australia

CHRISTEN, P., Dépt. de Pharmacognosie, Ecole de Pharmacie, Université de Genève, 30, Quai E.-Ansermet, 1211 Genève 4, Switzerland

DALTON, S. J., Institute of Grassland and Environmental Research, WPBS, Plas Gogerddan, Aberystwyth, Dyfed, SY23 3EB, United Kingdom

DAVID, C., Institut des Sciences Végétales, CNRS, Avenue de la Terrasse, 91198 Gif sur Yvette, France

DAVIES, K. M., Levin Research Centre, Crop and Food Research Ltd, Private Bag, Levin, New Zealand

DEROLES, S. C., Levin Research Centre, Crop and Food Research Ltd, Private Bag, Levin, New Zealand

DUNBAR, K. B., Division of Natural Science and Mathematics, Dalton College, 213 North College Drive, Sequoya 127A,Dalton, GA 30720, USA

EDNO, TOHRU, Research Institute for Molecular Genetics, Tsumura & Co., 3586 Yoshiwara, Ami-machi, Inashiki, Ibaraki 300-11, Japan

FILIPPONE, E., Dipartimento Scienze Agronomiche e Genetica Vegetale, Universita di Napoli, 80055 Portici, Italy

GIVEN, N. K., died in June 1991

GOLDBERG, K.-B., Department of Plant Pathology, University of Kentucky, Lexington, KY 40546, USA

GOWDA, S., Department of Plant Pathology, University of Kentucky, Lexington, KY 40546, USA

HAMILL, J. D., Department of Genetics and Developmental Biology, Monash University, Melbourne, Australia

HOEKEMA, A., MOGEN Int. nv, Einsteinweg 97, 2333 CB Leiden, The Netherlands

MCHUGHEN, A., National Research Council of Canada, Plant Biotechnology Institute, Saskatoon, Sask. S7N 0W9, and Crop Development Centre, University of Saskatchewan, Saskatoon, Sask. S7N 0W0, Canada

ISHIMARU, KANJI, Genetic Engineering Laboratory, Faculty of Agriculture, Saga University, Honjo, Saga 840 Japan

JACOBSEN, H.-J., Lehrgebiet Molekulargenetik, Universität Hannover, Herrenhäuserstr. 2, 30419 Hannover, Germany

JIA, SHI-RONG, Biotechnology Research Center, Chinese Academy of Agricultural Sciences, 30 Baishiqiao Road, Beijing 100081, China

JORDAN, M. C., National Research Council of Canada, Plant Biotechnology Institute, Saskatoon, Sask. S7N 0W9, and Crop Development Centre, University of Saskatchewan, Saskatoon, Sask. S7N 0W0, Canada

KAJI, AKIRA, Department of Microbiology, School of Medicine, University of Pennsylvania, Philadephia, PA 19104-6076, USA

KAMADA, HIROSHI, Gene Experiment Center, University of Tsukuba, Tsukuba, Ibaraki 305, Japan

KIERNAN, J. M., Department of Plant Pathology, University of Kentucky, Lexington, KY 40546, USA

KURIOKA, YURIKO, Gene Experiment Center, University of Tsukuba, Tsukuba, Ibaraki 305, Japan

LEDGER, S. E., Levin Research Centre, Crop and Food Research Ltd, Private Bag, Levin, New Zealand

LEHMINGER-MERTENS, R., Institut f. Genetik, Universität Bonn, Kirschallee 1, 53115 Bonn, Germany. (Present address: Reiherweg 4, 50374 Erftstadt-Liblar, Germany)

LEONE, M., Centro di Studio per il Miglioramento Genetico degli Ortaggi, CNR, 80055 Portici, Italy

LURQUIN, P. F., Department of Genetics and Cell Biology, Washington State University, Pullman, WA 99164-4234, USA

MANO, YOSHIHIRO, Department of Biological Science and Technology, Tokai University, 317 Nishino, Numazu, Shizuoka 410-03, Japan

MILLER, R. M., Levin Research Centre, Crop and Food Research Ltd, Private Bag, Levin, New Zealand

MOCHIDA, SHUNJI, Research Institute for Molecular Genetics, Tsumura & Co., 3586 Yoshiwara, Ami-machi, Inashiki, Ibaraki 300-11, Japan

MURAKOSHI, ISAMU, Faculty of Pharmaceutical Sciences, Chiba University, Yayoi-cho 1-33, Inage-ku, Chiba 263, Japan

NIIZEKI, M., Plant Breeding Laboratory, Faculty of Agriculture, Hirosaki University, Hirosaki, Aomori-ken 036, Japan

OCHATT, S. J., I.N.R.A., Station d'Amélioration des Espèces Fruitières et Ornementales, Centre d'Angers, B.P. 57, 49071 Beaucouzé Cédex, France

OGASAWARA, TAKESHI, Gene Experiment Center, University of Tsukuba, Tsukuba, Ibaraki 305, Japan

PIVEN, N. M., Institute of Cell Biology and Genetic Engineering, Ukrainian Academy of Sciences, Acad. Lebedev Str., 1, 252650, Kiev-GSP-22, Ukraine

POTRYKUS, I., Institute of Plant Sciences, Swiss Federal Institute of Technology, ETH Zentrum, 8092 Zürich, Switzerland

PUA, ENG-CHONG, Institute of Molecular and Cell Biology, National University of Singapore, 10 Kent Ridge Crescent, Singapore 0511, Republic of Singapore. (Present address: Department of Botany, National University of Singapore, Kent Ridge, Singapore 0511)

PUNJA, Z. K., Department of Biological Sciences, Simon Fraser University, Burnaby, British Columbia, Canada V5A 1S6

RAHARJO, S. H. T., Department of Biological Sciences, Simon Fraser University, Burnaby, British Columbia, Canada V5A 1S6

RATUSHNYAK, Y. I., Institute of Cell Biology and Genetic Engineering, Ukrainian Academy of Sciences, Acad. Lebedev Str., 1, 252650, Kiev-GSP-22, Ukraine

RHODES, M. J. C., Genetics and Microbiology Department, AFRC Institute of Food Research, Norwich Laboratory, Norwich, United Kingdom

ROBERTS, M. F., Department of Pharmacognosy, The School of Pharmacy, University of London, London WC1N 1AX, United Kingdom

RUDAS, V. A., Institute of Cell Biology and Genetic Engineering, Ukrainian Academy of Sciences, Acad. Lebedev Str., 1, 252650, Kiev-GSP-22, Ukraine

SAITO, KAZUKI, Faculty of Pharmaceutical Sciences, Chiba University, Yayoi-cho 1-33, Inage-ku, Chiba 263, Japan

SCHRAMMEIJER, B., MOGEN Int. nv, Einsteinweg 97, 2333 CB Leiden, The Netherlands

SHEPHERD, R. J., Department of Plant Pathology, University of Kentucky, Lexington, KY 40546, USA

SHIMOMURA, KOICHIRO, Tsukuba Medicinal Plant Research Station, National Institute of Hygienic Sciences, 1 Hachimandai, Tsukuba, Ibaraki 305, Japan

SIJMONS, P. C., MOGEN Int. nv, Einsteinweg 97, 2333 CB Leiden, The Netherlands

SPENCER, A., Department of Biological Sciences, University of East Anglia, Norwich, Norfolk, United Kingdom

SUZUKI, YOSHIHITO, Faculty of Agriculture, University of Tokyo, Yayoi 1-1-1, Bunkyo-ku, Tokyo 113, Japan

TANAKA, MASAHIKO, Research Institute for Molecular Genetics, Tsumura & Co., 3586 Yoshiwara, Ami-machi, Inashiki, Ibaraki 300-11, Japan

TEMPÉ, J., Institut des Sciences Végétales, CNRS, Avenue de la Terrasse, 91198 Gif sur Yvette, France, and Institut National Agronomique Paris Grignon, 75231 Paris, France

TOYODA, HIDEYOSHI, Faculty of Agriculture, Kinki University, Nakamachi 3327-204, Nara 631, Japan

WU, F. G., Department of Plant Pathology, University of Kentucky, Lexington, KY 40546, USA

XU, ZHI-HONG, Shanghai Institute of Plant Physiology, Academia Sinica, Shanghai 200032, China

YAMAZAKI, MAMI, Faculty of Pharmaceutical Sciences, Chiba University, Yayoi-cho 1-33, Inage-ku, Chiba 263, Japan

YANG, LI-JUN, Shanghai Institute of Plant Physiology, Academia Sinica, Shanghai 200032, China

YONEMITSU, HIROSHI, Research Institute for Molecular Genetics, Tsumura & Co., 3586 Yoshiwara, Ami-machi, Inashiki, Ibaraki 300-11, Japan

YOSHIMATSU, KAYO, Tsukuba Medicinal Plant Research Station, National Institute of Hygienic Sciences, 1 Hachimandai, Tsukuba, Ibaraki 305, Japan

ZHANG, Y. L., Institute of Medicinal Plant Development, Chinese Academy of Medical Sciences, Hai Dian District, Dong Bei Wang, Beijing 100094, China

Section I
Regeneration of Plants from Protoplasts

I.1 Regeneration of Plants from Protoplasts of *Agrostis alba* (Redtop)

Y. Asano[1]

1 Introduction

The genus *Agrostis*, of the Gramineae, includes about 125 species that occur in temperate and subarctic climates as well as at high altitudes in tropical and subtropical zones. *A. alba* (Redtop) (Fig. 1) is native to Europe, Asia, and N. America and adapts to a wide range of soil conditions in these areas. *A. alba* is an economically important grass cultivated as a pasture grass and as a hay crop in temperate regions around the world. It is also used as a component of cool-season turf.

The improvement of grasses by genetic manipulation would be facilitated by the development of an efficient protoplast-to-plant regeneration system. In *Agrostis*, plant regeneration via organogenesis from seed-derived callus culture has been reported only in *A. palustris* (Krans 1981; Krans et al. 1982; Blanche et al. 1986). In this chapter the establishment of protoplast-to-plant regeneration system of *A. alba* using the embryogenic suspension culture is described.

2 Induction of Embryogenic Callus and Establishment of Embryogenic Suspension Culture

Explants such as immature embryos and young inflorescence tissues have been used most frequently as the best sources to initiate regenerable cultures in gramineous species (Lorz et al. 1988; Ozias-Akins and Vasil 1988; Vasil 1988; Bhaskaran and Smith 1990). These materials are, however, difficult to use in this species because of its very small size, and we use the mature seeds as explants to induce callus. By culturing on MS medium (Murashige and Skoog 1962) supplemented with 5 mg/l 2,4-D for 45–60 days, 80–90% of the seeds plated produce calli, of which 3–8% are embryogenic, indicated by the differentiation of a whitish, nodular embryogenic sector on the watery, nonembryogenic callus (Fig. 2A).

Recently, it was shown that more than two to three times more embryogenic calli are induced by the addition of L-proline (5–20 mM) to the induction medium.

[1] Faculty of Horticulture, Chiba University, 6-2-1 Kashiwa-no-ha, Kashiwa city, Chiba Pref., Japan 277

Biotechnology in Agriculture and Forestry, Vol. 22
Plant Protoplasts and Genetic Engineering III (ed. by Y.P.S. Bajaj)

Fig. 1. An *Agrostis alba* plant regenerated from a protoplast

Similarly, L-glutamine has the same effect, but to a lesser extent (Shetty and Asano 1991).

The somatic embryos (Fig. 2B) can be readily germinated to develop plantlets when the calli are transferred onto hormone-free MS medium. The embryogenic sector developed on the callus is excised and transferred to a liquid culture medium to initiate suspension culture. The composition of the medium (= suspension

Table 1. Composition of suspension maintenance medium for *Agrostis*; N6AA medium. (Asano and Sugiura 1990)

Major elements	(mg/l)	Organic components	(mg/l)
$(NH_4)_2SO_4$	463	Nicotinic acid	0.5
KNO_3	2830	Pyridoxine HCl	0.5
$MgSO_4 \cdot 7H_2O$	185	Thiamine HCl	1.0
$CaCl_2 \cdot 2H_2O$	166	Proline	1150
KH_2PO_4	400	Glutamine	876
Minor elements	(mg/l)	Asparatic acid	266
$MnSO_4 \cdot 4H_2O$	4.4	Arginine	174
$ZnSO_4 \cdot 7H_2O$	1.5	Glycine	7.5
H_3BO_3	1.6		
KI	0.8	2,4-D	2.0
$FeSO_4 \cdot 7H_2O$	27.8	Sucrose	20 (g/l)
$Na_2 \cdot EDTA$	37.3	Sorbitol	30 (g/l)
$NaMoO_4 \cdot 2H_2O$	0.25		
$CuSO_4 \cdot 5H_2O$	0.025	pH: 5.8	
$CoCl_2 \cdot 6H_2O$	0.025		

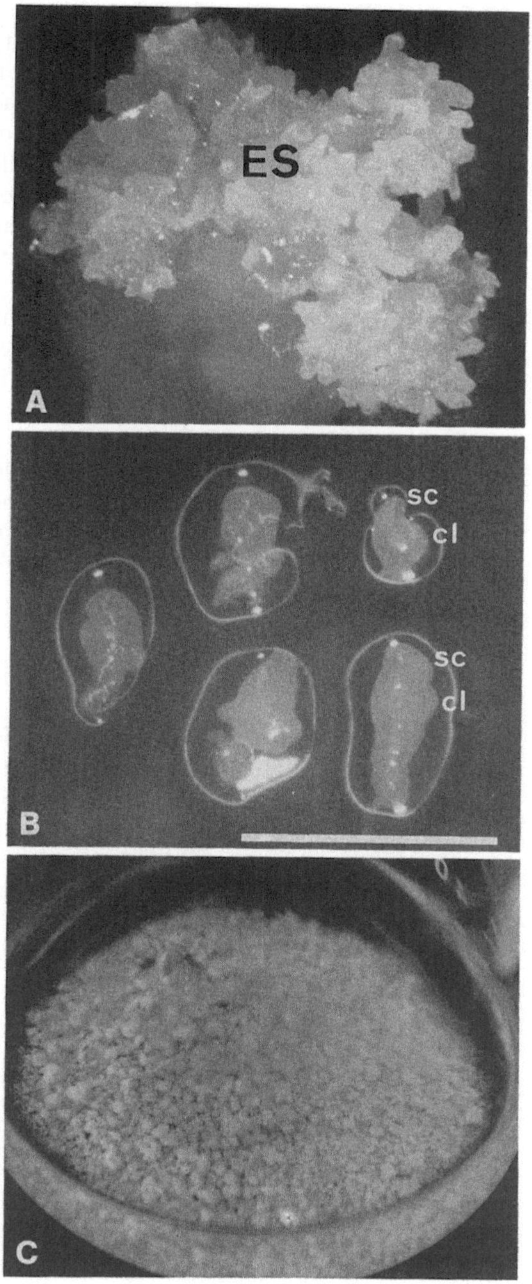

Fig. 2A–C. Induction of embryogenic suspension culture of *A. alba*. **A** Seed-derived callus showing differentiation of an embryogenic sector (ES) on MS + 5 mg/l 2,4-D medium: 50 days after plating. **B** Somatic embryos. *sc* scutellum; *cl* coleoptile *Bar* 1 mm. **C** Established embryogenic suspension culture in N6AA medium: 7 days after subculture. (Asano and Sugiura 1990)

maintenance medium) developed for *Agrostis* is listed in Table 1. The N6AA medium, including L-proline and four other kinds of amino acids, appears to be superior to the media for rice (Toriyama and Hinata 1985; Ozawa and Komamine 1989), with respect to promoting the growth and retaining the morphogenic potential of *Agrostis*.

Suspension cultures derived from the embryogenic calli vary considerably in their growth response and aggregation ability. During repeated subcultures for a few months, a few percent of fast-growing, finely dispersed cell lines with plant regeneration ability are selected. Established regenerable suspension cultures (Fig. 2C) consist of compact, organized cell colonies with proembryonic structures. *A. alba* is an outbreeding species and the difference among embryogenic cell lines with respect to suspension formation could be genotype-dependent. It is therefore important for the initiation of suspension cultures of *A. alba* to screen for more responsive genotypes with a high amenability to culture in liquid medium, as in other outbreeding species.

The embryogenic suspension culture maintained in N6AA medium recovers plantlets when transferred onto the regeneration medium following the steps described below. Plant regeneration ability is retained for at least 2 years in this medium.

Embryogenic callus can be induced from the basal segments of young leaves of the regenerated plants, and genotypically identical suspension culture can be re-established. This serves as a substitute for the original suspension if the latter has been lost by contamination or has lost its morphogenic potential.

3 Isolation of Protoplasts

The embryogenic suspension culture is used for protoplast isolation. However, very few protoplasts are isolated by enzyme treatment of the suspension that has been maintained in the maintenance medium. To obtain a higher yield of protoplasts, 4–5 days of subculture step using MS medium prior to protoplast isolation are essential.

For isolation of protoplasts, suspension culture is added to an enzyme solution containing 0.5–1% Cellulase Onozuka RS (Yakult Co. Ltd., Japan) and 0.05% Pectolyase Y-23 (Seishin Pharmaceutical Co. Ltd., Japan). Using this combination of enzymes, a protoplast yield after mesh filtration of $1.5\text{–}2 \times 10^7$/g fresh wt.

Table 2. The effect of enzyme concentrations on protoplast yields from *A. alba* suspension culture: after 2 h of incubation at 28 °C

Pectolyase Y-23	Cellulase Onozuka RS (%)			
(%)	0.25	0.5	1	2
0.012	0.8[a]	1.2	1.1	1.5
0.025	1.3	1.3	1.8	1.5
0.05	1.7	2.1	1.7	1.7
0.1	1.7	1.8	1.7	1.5

[a] $\times\ 10^7$/g fresh wt. suspension.
All enzyme solutions contain 0.1% potassium dextran sulfate, 0.01% $CaCl_2$, and 9% mannitol, pH 5.8.

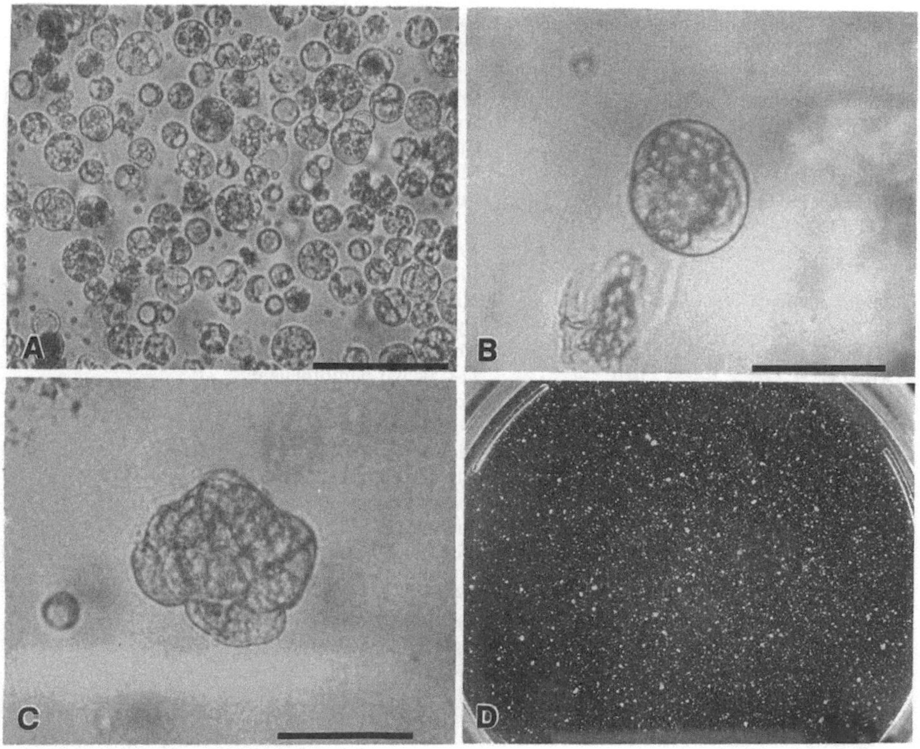

Fig. 3A–D. Colony development from protoplasts of *A. alba*. **A** Freshly isolated protoplasts. (*Bar* 100 μm). **B, C** The fresh division and colony formation from a protoplast in agarose-solidified mod. K8p medium: 5 and 11 days after plating, respectively. *Bar* 50 μm. **D** Visible cell colonies recovered after 3 weeks of culture. (Asano and Sugiura 1990)

suspension is obtained (Table 2). Embryogenic protoplasts obtained are small and richly cytoplasmic (Fig. 3A).

4 Culture of Protoplasts

Protoplasts are plated at a density of $2–5 \times 10^5$/ml in an agarose-solidified culture medium. Below this density, no cell division occurs. Modified K8p medium (Kao 1977) is used to culture protoplasts (Table 3). Glucose concentration of 10–11% in the medium is optimum (Fig. 4).

Cell division is initiated 5–6 days after plating. The plating efficiency based on the percentage of plated protoplasts that form substantial cell colonies is 0.5–0.6% (Figs. 3B–D, 4).

After a 3-week culture period, agarose blocks, including protoplast-derived cell colonies, are resuspended into the liquid suspension culture medium to ensure the growth of embryogenic cell colonies.

Table 3. Composition of protoplast culture medium for *Agrostis*; modified K8p medium (Kao 1977)

Major elements	(mg/l)	*Organic acids*	
NH_4NO_3	600	Sodium pyruvate	5
KNO_3	1900	Citric acid	10
$CaCl_2 \cdot 2H_2O$	600	Malic acid	10
$MgSO_4 \cdot 7H_2O$	300	Fumaric acid	10
KH_2PO_4	170		
KCl	300	*Vitamins*	
		Inositol	100
Minor elements	(mg/l)	Nicotinamide	1
KI	0.75	Pyridoxine HCl	1
H_3BO_3	3.0	Thiamine HCl	10
$MnSO_4 \cdot H_2O$	10.0	D-Calcium pantothenate	0.5
$ZnSO_4 \cdot 7H_2O$	2.0	Folic acid	0.2
$FeSO_4 \cdot 7H_2O$	27.8	p-Aminobenzoic acid	0.01
$Na_2 \cdot EDTA$	37.3	Biotin	0.005
$Na_2MoO_4 \cdot 2H_2O$	0.25	Choline chloride	0.5
$CuSO_4 \cdot 5H_2O$	0.025	Ascorbic acid	1
$CoCl_2 \cdot 6H_2O$	0.025	Vitamin A	0.005
Organic components	(mg/l)	Vitamin D_3	0.005
Sugar		2,4-D	0.5–2
Glucose	100–110 (g/l)	Casein hydrolysate	125
Sucrose	125	Agarose (Sea Plaque)	12 (g/l)
Fructose	125		
Ribose	125	pH: 5.8	
Xylose	125		
Mannose	125		
Rhamnose	125		
Cellobiose	125		
Sorbitol	125		
Mannitol	125		

Sugar, organic acid, and vitamin supplements are filter sterilized.

5 Regeneration of Plants

After culturing cell colonies in the suspension maintenance medium for 2 weeks, the colonies are transferred to identical liquid medium excluding hormones, with reduced concentration of sucrose, and incubated for a further 7–10 days. Somatic embryos are induced to develop to the germinable stage during these subculture steps. Following these steps, whitish and compact embryogenic cell colonies are selected from among nonmorphogenic, translucent colonies, and transferred onto agarose-solidified regeneration medium. Somatic embryos on the colonies germinate within 1–2 weeks on this medium.

Although the colonies can regenerate plants on the standard regeneration medium, it appears that addition of sorbitol as osmoticum as well as higher concentration of agarose enhance plant regeneration: the frequency of shoot-regenerating colonies increases from ca. 80% to more than 90% with increase of agarose concentration from 0.8 to 1.6%. Also, addition of 2–3% sorbitol significantly improves the shoot growth (Fig. 5, Table 4). Normal plants are readily established by transplanting the plantlets developed to soil (Fig. 1).

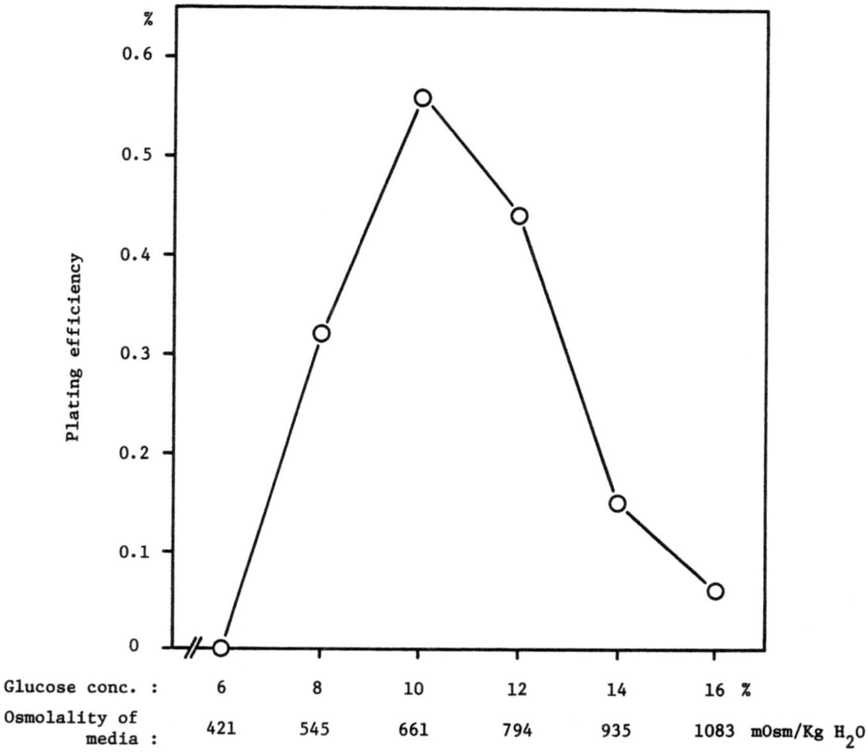

Fig. 4. Effect of glucose concentrations in culture media on plating efficiency (colony formation) from *A. alba* protoplasts. Protoplasts were cultured at a density of 5×10^5/ml in an agarose-solidified mod. K8p medium. Data were collected 3 weeks after plating

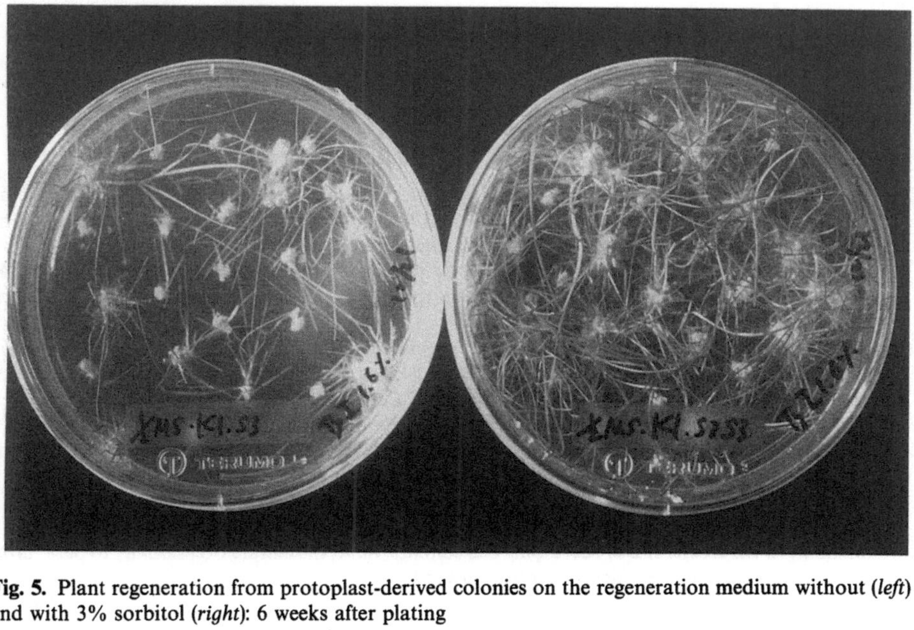

Fig. 5. Plant regeneration from protoplast-derived colonies on the regeneration medium without (*left*) and with 3% sorbitol (*right*): 6 weeks after plating

Table 4. The effect of sorbitol concentrations in regeneration media on plant regeneration from protoplast-derived colonies: after 6 weeks of culture

Additive[a]	% Colonies with shoots[b]	Shoot growth
3% sucrose	96	+
3% sucrose + 1% sorbitol	93	+ +
3% surcose + 2% sorbitol	96	+ + +
3% sucrose + 3% sorbitol	98	+ + +
3% sucrose + 4% sorbitol	97	+ +

[a] Basal medium is 0.5 × MS + 1 mg/l kinetin, solidified with 1.6% agarose Sigma Type I, pH 5.8.
[b] Out of 125 colonies/treatment.

6 Conclusions

Keys to our successful system for *Agrostis* are the following: (1) Development of a suspension culture medium including an amino acid mixture that allows for retaining the morphogenic potential, (2) Subculturing the suspension in MS medium prior to protoplast isolation to maximize protoplast yield, (3) Resuspending protoplast-derived cell colonies that have grown in agarose-solidified medium into liquid suspension medium to ensure embryo development; otherwise the colonies tend to dedifferentiate, resulting in very poor plant regeneration. This culture system has proved applicable to protoplasts of the related species *A. palustris* and *A. tenuis*.

7 Protocol

7.1 Establishment of Embryogenic Suspension Culture

1. Sterilize mature seeds of *A. alba* in 70% ethanol for 1 min followed by immersion in a sodium hypochlorite solution (1% available chlorine, 0.1% Tween 20) for 20 min with vigorous agitation. Wash the seeds three times in sterile distilled water.
2. Place seeds on an agar-solidified callus induction medium consisting of MS medium, 5 mg/l 2,4-D, 3% sucrose, and 10 mM proline, pH 5.8, in 90 mm plastic Petri dishes and incubate for 45–60 days at 25 °C in the dark.
3. Excise embryogenic sector of the callus under a stereomicroscope and suspend in a 1–2 ml N6AA liquid culture medium (Table 1) in a 22-mm glass tube.
4. Incubate the tubes on a rotary culture (2 rpm) at 25 °C in the dark. As the volume of cultures increase, add the same volume of fresh medium weekly up to a total volume of 20 ml.
5. Transfer the cultures to 100 ml culture flasks and maintain by weekly subculture of approx. 0.3 ml of culture into 20 ml fresh medium. The flasks are incubated on a rotary shaker (120 rpm) at 25 °C in the dark.
6. Screen for fast-growing, finely dispersed cell lines with plant regeneration ability during a few months. To reduce the size of developing cell colonies, grind the culture occasionally to pass through a 1 mm diameter stainless sieve when subcultured.

7.2 Isolation of Protoplasts

1. Transfer approx. 1/3 volume of the suspension culture to a 20 ml of MS medium with 2 mg/l 2,4-D and 6% sucrose and incubate at 120 rpm for 4–5 days at 30 °C in the dark.
2. Add approx. 1.5 g fresh wt. suspension culture to 10 ml of a filter-sterilized enzyme solution consisting of 0.5–1% Cellulase Onozuka RS (Yakult Co. Ltd., Japan), 0.05% Pectolyase Y-23 (Seishin Pharmaceutical Co. Ltd., Japan), 0.1% potassium dextran sulfate (KDS), 0.01% $CaCl_2$, and 9% mannitol, pH 5.8, in a Petri dish and incubate for 30 min on a rotary shaker (40 rpm), followed by 1.5 h without shaking at 28 °C.
3. Pass the protoplast suspension through a 32-μm nylon sieve to remove undigested cell aggregates.
4. Centrifuge the suspension at 90 g for 3 min.
5. Dispose supernatant and wash protoplasts three times in a washing solution containing 0.1% KDS, 0.01% $CaCl_2$ and 9% mannitol, pH 5.8, by centrifugation at 90 g for 3 min.

7.3 Protoplast Culture and Plant Regeneration

1. Resuspend protoplasts in 1 ml of protoplast culture medium at twice the final density: $2 \times (2\text{–}5 \times 10^5)$/ml. K8p medium (Kao 1977) with vitamins as modified by Vasil and Vasil (1980) and 10–11% glucose is used as the culture medium (Table 3).
2. Mix the protoplast suspensin with an equal volume of 2 × agarose (2.4% Sea Plaque, FMC) including culture medium in a 55 mm plastic Petri dish and cool immediately on ice to solidify. Incubate the dishes at 25 °C in the dark for 3 weeks.
3. Cut the agarose, including protoplast-derived cell colonies, into small blocks, transfer to 20 ml of suspension culture medium in a 100-ml culture flask and incubate on a rotary shaker (120 rpm) for 2 weeks at 25 °C in the dark.
4. Transfer approx. 1/3 volume of the culture to identical liquid medium excluding 2,4-D, with reduced conc. of sucrose (= N6AA + 1% sucrose, 3% sorbitol) and incubate for 7–10 days at 30 °C in the dark.
5. Select whitish and compact embryogenic colonies and transfer onto agarose-solidified (1.6% Type I, Sigma) regeneration medium consisting of 0.5 × MS medium, 1 mg/l kinetin, 3% sucrose, and 2–3% sorbitol, pH 5.8, in plastic Petri dishes. Incubate the dishes at 25 °C in light (approx. 3000 lx) for 4–6 wks for plant regeneration.
6. Separate each plantlet and transplant into plastic containers containing 30 ml of agar-solidified 0.5 × MS medium with 2% sucrose, pH 5.8.
7. Transplant the 5–10-cm-tall plantlets to soil.

Acknowledgment. The author is grateful to Dr. Y. Otsuki of the National Agriculture Research Center, for his valuable technical advice in this research.

References

Asano Y, Sugiura K (1990) Plant regeneration from suspension culture-derived protoplasts of *Agrostis alba* L. (Redtop). Plant Sci. 72:267–273

Bhaskaran S, Smith RH (1990) Regeneration in cereal tissue culture: a review. Crop Sci 30:1328–1337

Blanche FeC, Krans JV, Coats GE (1986) Improvement in callus growth and plantlet formation in creeping bentgrass. Crop Sci 26:1245–1248

Chu CC, Wang CC, Sun CS, Hsu C, Yin KC, Chu CY, Bi FY (1975) Establishment of an effective medium for anther culture of rice through comparative experiments on the nitrogen sources. Sci Sin 18:659–668

Kao KN (1977) Chromosomal behaviour in somatic hybrids of soybean-*Nicotiana glauca*. Mol Gen Genet 150:225–230

Krans JV (1981) Cell culture of turfgrasses. In: Sheard RW (ed) Proc Fourth Int Turfgrass Res Conf pp 27–33 Huddleston and Barney Limited, Woodstock, Ontario

Krans JV, Henning VT, Torres KC (1982) Callus induction, maintenance and plantlets regeneration in creeping bentgrass. Crop Sci 22:1193–1197

Lorz H, Gobel E, Brown P (1988) Advances in tissue culture and progress towards genetic transformation of cereals. Plant Breed 100:1–25

Murashige T, Skoog F (1962) A revised medium for rapid growth and bioassays with tobacco tissue cultures. Physiol Plant 15:473–497

Ozawa K, Komamine K (1989) Establishment of a system of high-frequency embryogenesis from long-term cell suspension cultures of rice (*Oryza sativa* L.). Theor Appl Genet 77:205–211

Ozias-Akins P, Vasil IK (1988) In vitro regeneration and genetic manipulation of grasses, Physiol Plant 73:565–569

Shetty K, Asano Y (1991) The influence of organic nitrogen sources on the induction of embryogenic callus in *Agrostis alba* L. J. Plant Physiol. 139:82–85

Toriyama K, Hinata K (1985) Cell suspension and protoplast culture in rice. Plant Sci 41:179–183

Vasil IK (1988) Progress in the regeneration and genetic manipulation of cereal crops. Bio/Technol 6:397–402

I.2 Regeneration of Plants from Protoplasts of *Atropa belladonna* L. (Deadly Nightshade)

Y.P.S. Bajaj[1]

1 Introduction

Atropa belladonna (deadly nightshade), a member of the family Solanaceae, is a rich source of tropane alkaloids, which are antimuscaric drugs and have been used for inducing sleep, as a cure for toothache, also having hallusinigenic effects. Hyoscyamine and scopolamine are the two main tropanes. Because of the pharmaceutical interest in this plant, a considerable amount of work has been done on the in vitro culture, morphogenesis, and the production of tropane alkaloids, and the subject has been recently reviewed by Bajaj and Simola (1991). Recently, genetic transformation of *Atropa belladonna* through *Agrobacterium rhizogenes* has attracted much attention to increase production of medicinal compounds (Kamada et al. 1986; Jung and Tepffer 1987).

Isolated protoplasts have been shown to be an excellent system for morphogenetic studies (Bajaj 1989), and complete plants of *Atropa belladonna* have been regenerated from cell suspension-derived protoplasts (Gosch et al. 1975), mesophyll protoplasts (Bajaj et al. 1978), and somatic hybrids obtained from them (Krumbiegel and Schieder 1979; Gleba et al. 1988; Kushnir et al. 1991). These studies are reviewed here and summarized in Table 1.

2 Isolation and Culture of Protoplasts

2.1 Protoplasts from Cell Suspensions (Gosch et al. 1975)

Callus raised from stem segments of *A. belladonna* cv. *lutea* Döll on Wood and Braun's medium (1961) supplemented with NAA (2 mg/l) and kinetin (0.5 mg/l) was transferred to liquid medium of the same composition except for the lower concentration (0.1 mg/l) of kinetin. The cell suspensions were subcultured at weekly intervals and incubated in the dark at 24 ± 1 °C on a rotary shaker. The protoplasts were then isolated from actively growing cell suspensions (3–4 days after subculture) by incubating them at 30 °C in an enzyme mixture containing Cellulase R10 (1%), Macerozyme (0.5%), and 0.6 M sorbitol. In 4–5 h, about 80% of the cells

[1] Former Professor of Tissue Culture, Punjab Agricultural University, Ludhiana, India. (Present Address: A-137 New Friends Colony, New Delhi 110065, India)

Biotechnology in Agriculture and Forestry, Vol. 22
Plant Protoplasts and Genetic Engineering III (ed. by Y.P.S. Bajaj)

Table 1. Summary of protoplast work done on *Atropa belladonna*

Nature of study	Reference	Observations/Results
Isolation and culture	Gosch et al. (1975)	Protoplasts isolated from cell suspension, embryogenesis and plants from protoplast-derived callus
	Bajaj et al. (1978)	Protoplasts isolated from mesophyll cells of anther-derived haploid plants. Plants of various ploidy from callus
Fusion and somatic hybridization	Gosch and Reinert (1976)	Fusion with *Petunia* protoplasts, heterokaryocytes
	Krumbiegel and Schieder (1979)	Somatic hybrids between *Datura innoxia*-*Atropa belladonna*. The chromosome number of hybrids varied from 84 to 175
	Krumbiegel and Schieder (1981)	Comparison of somatic and sexual incompatibility between *Datura innoxia* and *A. belladonna*
	Gleba et al. (1988) Kushnir et al. (1991)	Somatic hybrids between *Nicotiana plumbaginifolia* and *A. belladonna*
Cryopreservation	Bajaj (1988)	Plants from protoplasts frozen in liquid nitrogen (− 196 °C)

Table 2. Effect of protoplast density on cell wall regeneration and division of *Atropa belladonna* protoplasts after one week of culture in liquid medium. (Gosch et al. 1975)

Protoplast density	Dead protoplasts	Viable protoplasts	Cells	Dividing cells	Colonies
		Percentage			
3×10^4	61.6	13.3	20.0	5.0	–
6×10^4	44.1	4.9	31.0	16.0	4.0
1×10^5	43.0	6.4	34.6	12.9	3.0
1.2×10^5	51.3	7.1	26.6	11.0	3.3

yielded protoplasts. To ensure the purity of the protoplasts, they were first filtered through a 40-μm nylon mesh and then floated on 15% sucrose in CPW salts solution (see Cocking and Peberdy 1974), followed by washing four times with the appropriate medium. The filtration through a fine 40-μm mesh also encouraged the collection of relatively highly cytoplasmic protoplasts, while the large and vacuolated protoplasts which were normally obtained from the older cells were discarded. The protoplasts were cultured in liquid as well as in agar medium and incubated in the dark at 28 °C.

The freshly isolated protoplasts (Fig. IA) when cultured in a modified liquid MS with NAA (2 mg/l) and kinetin (0.1 mg/l) regenerated walls within 2 days. The reconstituted cells (Fig. 1B) underwent first division in 2–3 days and exhibited both unequal (Fig. 1C) and equal divisions (Fig. 1D, E). In 1 weeks's time, some of them divided two or three times, and about 4% of them formed small colonies consisting of 4–8 cells (Fig. 1F, G). The viability and growth of the cultured protoplasts was strongly influenced by their density per unit volume of the medium (Table 2). Density in the range between 6 × 10 and 1×10^5/ml of the liquid medium was

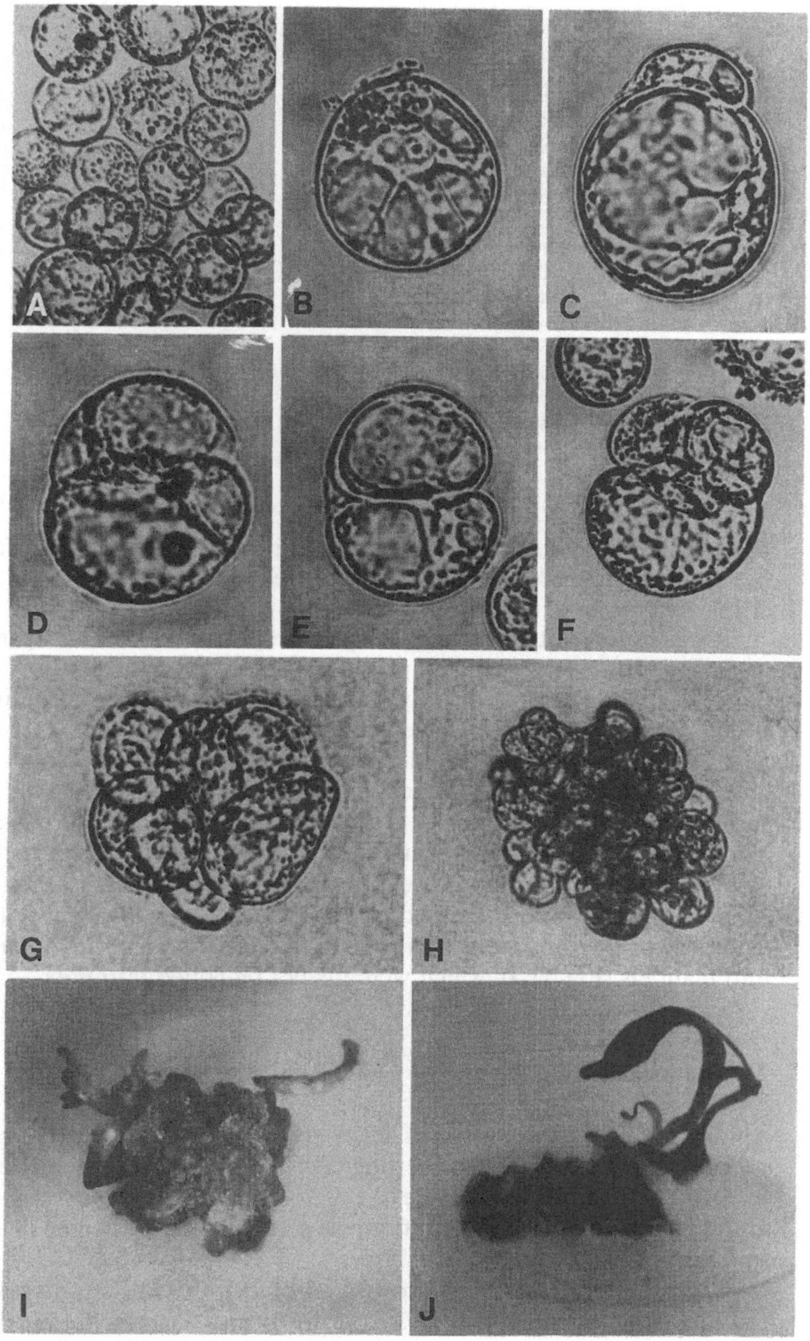

Fig. 1A–J. Isolation and culture of protoplasts from cell suspensions of *Atropa belladonna*. **A** Freshly isolated protoplasts from 4-day-old cell suspensions after 4 h treatment with a mixture of Cellulase, Macerozyme, and sorbitol in a modified MS medium. **B** A protoplast after 48 h of culture in liquid medium; the wall has already been regenerated. **C–F** Three- to 4-day-old cultures in liquid MS medium; note the unequal division in **C**, equal division in **D** and **E**, and a second division in **F**. **G** One-week-old colony of cells obtained after plating 5-day-old suspension of protoplasts in agar medium, and **H** showing a small mass of callus 1 week later. **I, J** Ten- and 12-week-old cultures showing the regeneration of plantlets from calli obtained from isolated protoplasts. (Gosch et al. 1975)

Table 3. Effect of protoplast density on division of cells obtained from isolated protoplasts 2 weeks after culture. The protoplasts were first cultured in liquid medium for a week, and then plated on agar medium of the same composition. (Gosch et al. 1975)

Protoplast density	Percentage of dividing cells
5×10^4	52.9
1×10^5	56.0
1.2×10^5	51.0
2×10^5	46.6

observed to be optimal and yielded 16% division within 1 week (Table 2), which increased up to 56% during the next week when transferred to agar medium (Table 3).

It was observed that direct plating of the protoplasts on the agar medium was inefficient. However, when the protoplasts were cultured first for a week in the liquid medium and subsequently transferred to agar (0.6%) medium of the same composition, they divided actively and formed small colonies (Fig. 1H). After the next 2–3 weeks when the colonies had attained a sufficiently large size, they were once again transferred to a medium with relatively low osmotic value (0.2 M sorbitol), which stimulated the growth of colonies into callus masses. The calli, when grown on a NAA-free liquid medium (Thomas and Street 1970), underwent embryogenesis, and within 2 weeks globular proembryos and heart- and torpedo-shaped embryos were observed in the same culture (Gosch et al. 1975). The embryos, when transferred to an auxin-free agar medium, eventually developed into plantlets (Fig. 1I, J), which were further replanted into pots.

2.2 Protoplasts from Mesophyll Cells (Bajaj et al. 1978)

About 5- and 6-cm-long leaves from greenhouse-reared plants were used. The peeled leaves were incubated overnight (14–17 h) in an enzyme mixture of Cellulase R-10 (0.5–1%) and Macerozyme (0.1–0.3%) in 0.5 M mannitol at a room temperature. To obtain a clean preparation, the protoplasts were floated in a 20% sucrose solution.

The protoplasts were first grown in drop cultures in Petri dishes in liquid MS medium supplemented with NAA (2 mg/l) and kinetin (0.1 mg/l). They were incubated in the dark, or at a low intensity light for 2–3 weeks. When colonies of cells were formed, they were poured onto the top of an agar medium in Petri dishes and stored at 27 °C.

The protoplasts isolated in low enzyme level (Fig. 2B) showed better growth response. First division (Fig. 2C–E) was noticed after 7 days of culture, and within 2 weeks small clumps of cells (Fig. 2F, G) were formed. At this stage, dividing protoplasts and cell colonies were plated on agar medium of the same composition,

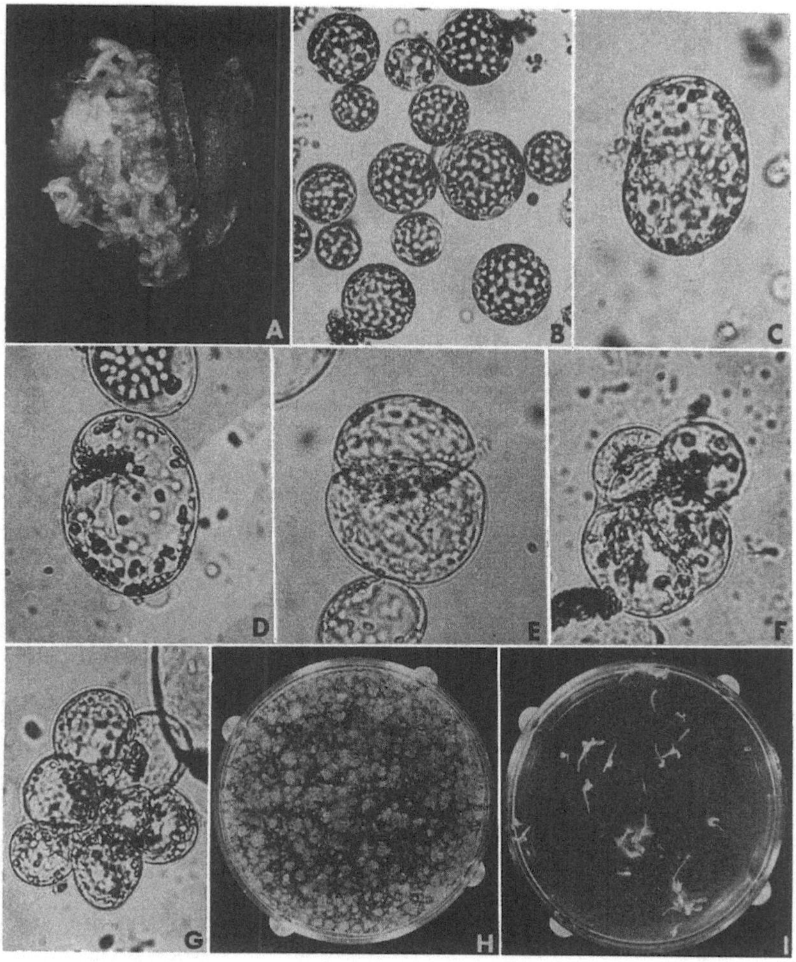

Fig. 2A–I. Isolation and culture of haploid mesophyll protoplasts. **A** 5-week-old culture of an excised anther undergoing androgenesis (the leaves from the haploid plants thus raised were employed for the isolation of protoplasts). **B** Mesophyll protoplasts obtained from the peeled leaves incubated in an enzyme mixture containing Cellulase (1.5%) and Macerozyme (0.3%) in mannitol. **C–E** Various stages in the elongation and division of an isolated protoplast cultured in a liquid nutrient medium 5 days after culture. **F, G** One- and 2-week-old cluster and a colony of callus cells obtained from protoplasts. **H** Masses of calli obtained from cell colony (F, G) 3 weeks after plating on agar-solidified medium containing NAA. **I** Differentiation of embryoids and plantlets from calli after transfer to an auxin-free medium. (Bajaj et al. 1978)

and after 2–3 weeks colonies of callus were formed (Fig. 2H). When transferred to an auxin-free medium, within 2 weeks the callus started to differentiate, i.e., about 7–8 weeks from the time of incubation of protoplasts to the differentiation of plantlets (Fig. 2I).

The cultures exhibited three modes of morphogenesis, i.e., rhizogenesis, shoot-bud differentiation, and embryogenesis (Bajaj et al. 1978). Callus mostly differentiated to form only roots. In such cultures where roots once started to initiate,

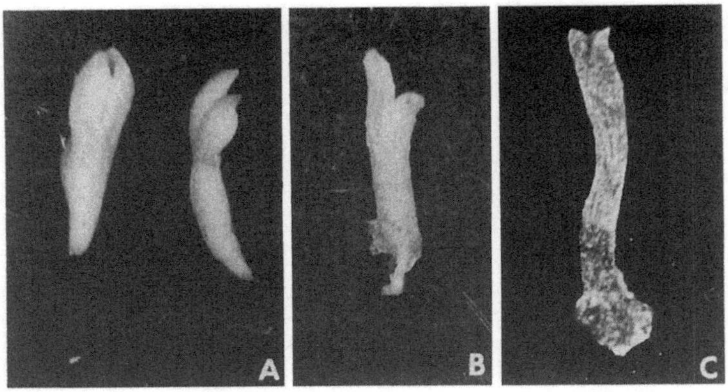

Fig. 3A–C. Various stages in the differentiation of embryos originated from mesophyll protoplast-derived callus grown on auxin-free medium. (Bajaj et al. 1978)

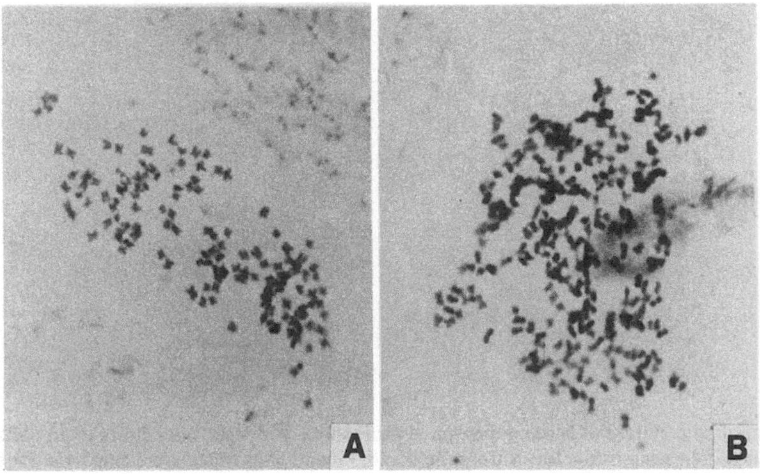

Fig. 4A, B. Root tip squashes from protoplast-derived polyploid plants of *Atropa belladonna*; note 132 chromosomes in **A** (n = 36) and about 200 chromosomes in **B** (6n = 216). (Bajaj et al. 1978)

the growth of callus was slowed down, and no shoot initiation was observed, while in others, on the agar medium shoots started to appear without forming any roots. In liquid cultures, sometimes, in addition to rhizogenesis, the formation of embryos, was also observed (Fig. 3A–C). The removal of cell wall did not in any way impair the process of embryogenesis.

The plants transferred to the pots continued to grow and flowered. These plants showed various levels of chromosome number ranging from haploid (n = 36), diploid, triploid, and hexaploid, along with aneuploids of various types, were obtained (Fig. 4A, B).

3 Protoplast Fusion and Somatic Hybridization

Successful regeneration of plants from isolated protoplasts led to fusion studies and the production of intergeneric somatic hybrids. Heterokaryons were first obtained between *Atropa belladonna* and *Petunia hybrida* (Gosch and Reinert 1976), but no hybrids were formed. This was followed by fusion of protoplasts of two chlorophyll-deficient mutants of *Datura innoxia* (2n = 24) with green protoplasts of *A. belladonna* (2n = 72) (Krumbiegel and Schieder 1979). Thirteen somatic hybrids were selected. Hybrid calli were recognizable by the production of hairs, typical of *Datura*, and the green color derived from *Atropa*. The chromosome number of hybrids varied from 84 to 175. In a subsequent study (Krumbiegel and Schieder 1981), somatic sexual incompatibility between *D. innoxia* and *A. belladonna* was compared. Experiments were carried out to test the feasibility of producing sexual hybrids through in vivo and in vitro methods by cross-pollination; however, no embryos, seeds or plantlets were obtained, thus demonstrating that protoplast fusion is the only possibility for obtaining hybrids between *Datura* and *Atropa*. Recently, somatic hybrids between *A. belladonna* and *Nicotiana plumbaginifolia* have also been reported (Gleba et al. 1988; Kushnir et al. 1991). Hybrid formation frequency was 0.7–3.7%. Some propagated lines flowered and were partially fertile. These studies on somatic hybridization would be more meaningful if hybrid cell lines/plants capable of producing higher amounts of tropanes are isolated.

4 Cryopreservation of Protoplasts (Bajaj 1988)

The protoplasts differ from cell suspensions in two important aspects, (1) absence of cell wall, and (2) lack of cytoplasmic connections with the adjacent cells. Thus protoplasts, in not being interdependent, are in fact a better system to study the mechanism of cryopreservation at the cellular level. The protoplasts, like the freeze preservation of cell suspension, can be employed for the storage and retention of biosynthetic potential and for genetic manipulation studies. There is no substantial literature on the freeze preservation of protoplasts of medicinal plants. Of the three alkaloid-producing plants (*Atropa, Datura, Nicotiana*) studied by the author (Bajaj 1988), *Atropa* gave the best response. The enzymatically isolated mesophyll protoplasts were subjected to freezing in the presence of mannitol, sucrose, and dimethyl sulfoxide, and subsequently thawed at 37 °C. Initially, the protoplasts lay quiescent for 3–5 days as compared to controls. After that they started to elongate, the chloroplasts became scanty and yellowish, and reorganized themselves at the periphery. The first division and the subsequent colony formation were invariably delayed; however, once they regenerated walls they looked like the control protoplasts. A maximum of 44% cultures showed differentiation of plantlets in callus. The morphogenetic events were not impaired and cultures showed, as in control, both the patterns of differentiation, embryogenesis and organogenesis. The plants thus regenerated, when transferred to pots, continued to grow to form normal flowers and fruits (Fig. 5). The seeds obtained from these plants were fertile, and in turn showed normal germination (Bajaj 1988).

Fig. 5A, B. Plants regenerated from frozen (−196 °C) protoplasts of *Atropa belladonna* 4 weeks after transfer to a pot (**A**), and showing normal flowering (**B**) (Bajaj 1988)

5 Summary

Protoplasts isolated from cell suspensions and mesophyll cells of *Atropa belladonna* in cultures underwent three modes of morphogenesis, i.e., rhizogenesis, somatic embryogenesis, and shoot/plant regeneration. The plants showed a wide range of chromosome numbers varying from haploids to hexaploids with various types of aneuploids. Also protoplasts frozen at −196 °C retained their morphogenetic potential and regenerated complete plants capable of normal flowering and seed set. Intergeneric somatic hybrids through protoplast fusion between *A. belladonna* + *Datura innoxia*, and *A. belladonna* + *Nicotiana plumbaginifolia* have been reported. These studies have opened up the possibility of generating genetic variability for increased production of alkaloids, and of cryopreservation of such cultures.

6 Protocol

Following are the references for protocol on various aspects of the work done on *Atropa belladonna* protoplasts:

Cell-suspension protoplasts (Gosch et al. 1975)
Mesophyll protoplasts (Bajaj et al. 1978)
Somatic hybridization (Krumbiegel and Schieder 1979; Gleba et al. 1988)
Cryopreservation of protoplasts (Bajaj 1988)

References

Bajaj YPS (1988) Regeneration of plants from frozen (– 196 °C) protoplasts of *Atropa belladonna* L., *Datura innoxia* Mill and *Nicotiana tabacum* L. Indian J Exp Biol 26:289–292

Bajaj YPS (1989) Recent advances in the isolation and culture of protoplasts and their implications in crop improvement. In: Bajaj YPS (ed) Biotechnology in agriculture and forestry, vol 8, Plant protoplasts and genetic engineering I. Springer, Berlin Heidelberg New York pp 3–22

Bajaj YPS, Simola LK (1991) *Atropa belladonna* L.: in vitro culture, regeneration of plants, cryopreservation, and the production of tropane alkalods. In: Bajaj YPS (ed) Biotechnology in agriculture and forestry, vol 15, Medicinal and aromatic plants III. Springer, Berlin Heidelberg New York, pp 1–23

Bajaj YPS, Gosch G, Ottma M, Weber A, Gröbler A (1978) Production of polyploid and aneuploid plants from anthers and mesophyll protoplasts of *Atropa belladonna* and *Nicotiana tabacum*. Indian J Exp Biol 16:947–953

Cocking EC, Peberdy JF (eds) (1974) The use of protoplasts from fungi and higher plants as genetic systems – a practical handbook. Nottingham Univ

Gleba YY, Hinnisdacis S, Sidorov VA, Kaleda VA, Parokonny AS, Boryshuk NV (1988) Intergeneric asymmetric hybrids between *Nicotiana plumbaginifolia* and *Atropa belladonna* obtained by gamma-fusion. Theor Appl Genet 76:760–766

Gosch G, Reinert J (1976) Nuclear fusion in intergeneric heterokaryocytes and subsequent mitosis of hybrid nuclei. Naturwissenschaften 63:534–535

Gosch G, Bajaj YPS, Reinert J (1975) Isolation, culture, and induction of embryogenesis in protoplasts from cell suspensions of *Atropa belladonna*. Protoplasma 86:405–410

Jung G, Tepffer D (1987) Use of genetic transformation by the Ri T-DNA of *Agrobacterium rhizogenes* to stimulate biomass and tropane alkaloid production in *Atropa belladonna* and *Calystegia sepium* roots grown in vitro. Plant Sci 50:145–151

Kamada H, Okamura N, Satake M, Harada H, Shimomura K (1986) Alkaloid production by hairy root cultures in *Atropa belladonna*. Plant Cell Rep 5:239–242

Krumbiegel G, Schieder O (1979) Selection of somatic hybrids after fusion of protoplasts from *Datura innoxia* Mill. and *Atropa belladonna* L. Planta 145:371–375

Krumbiegel G, Schieder O (1981) Comparison of somatic and sexual incompatibility between *Datura innoxia* and *Atropa belladonna*. Planta 153:466–470

Kushnir S, Babiychuk E, Bannikova M, Momot V, Komarnitsky I, Cherep N, Gleba N (1991) Nucleo-cytoplasmic incompatibility in cybrid plants possessing an *Atropa* genome and *Nicotiana* plastome. Mol Gen Genet 225:225–230

Thomas E, Street HE (1970) Organogenesis in cell suspension cultures of *Atropa belladonna* L. and *Atropa belladonna* cultivar lutea Döll. Ann Bot 34:657–669

Wood HN, Braun AC (1961) Studies on the regulation of certain essential biosynthetic systems in normal and crown-gall tumor. Proc Natl Acad Sci USA 47:1907–1913

I.3 Regeneration of Plants from Protoplasts of *Brassica alboglabra* Bailey (Chinese Kale)

Eng-Chong Pua[1]

1 Introduction

The genus *Brassica* comprises a wide range of crop plants with great economic value worldwide. *Brassica alboglabra* Bailey (Chinese kale or Kailan, 2n = 18) (Fig. 1), which is believed to have originated from the southern part of China, is an important vegetable crop in Southeast Asia. In recent years, major efforts have focused on the utilization of protoplast technology for *Brassica* improvement via

Fig. 1. Six-week-old white-flowered *Brassica alboglabra* plant

[1] Institute of Molecular and Cell Biology, National University of Singapore, 10 Kent Ridge Crescent, Singapore 0511, Republic of Singapore. (Present address: Department of Botany, National University of Singapore, Kent Ridge, Singapore 0511)

Biotechnology in Agriculture and Forestry, Vol. 22
Plant Protoplasts and Genetic Engineering III (ed. by Y.P.S. Bajaj)

somatic hybridization or cell transformation, and considerable progress has been made in some species including *B. napus*, *B. oleracea*, and *B. juncea* (Schenck and Hoffmann 1984; Pua 1990, Thomzik 1993). However, work on protoplast culture of *B. alboglabra* has been limited, although plant regeneration from various cultured explants, e.g., cotyledon, hypocotyl, and stem, has been reported (Zee and Hui 1977; Wong and Loh 1987; Pua et al. 1989).

2 Isolation of Protoplasts

Various plant parts, e.g., cotyledon, hypocotyl, leaf, and stem, can be used as a source of protoplasts. While both cotyledon and hypocotyl are usually excised from 5–6-day-old aseptically germinated seedlings, stem and leaf can be prepared from in vitro-grown plants. Explants are cut into small pieces, then digested for a few hours with the appropriate amount of enzyme solution (Table 1) (Pua 1987). The duration of tissue digestion depends on the enzyme concentration used; the higher the enzyme concentration, the shorter the time required. After incubation, the digestion mixture is filtered through a 100-μm nylon or brass sieve. If the sieve is clogged by the tissue debris, remove the debris by centrifuging the mixture once prior to filtration. The protoplasts in the filtrate are purified by centrifugation at 800 rpm for 5 min; viable protoplasts usually float on the surface of the filtrate, and tissue debris and broken protoplasts stay at the bottom. After washing two or three times with CPW9G (Table 1), the protoplast density is determined using a haemacytometer. In general, the protoplast yield of $1-4 \times 10^6$/g cotyledons or leaves and $0.6-2.5 \times 10^6$/g hypocotyls or stems can be obtained after purification.

Table 1. Composition of various solutions[a] used for protoplast isolation of *B. alboglabra*

Constituent	mg/l	Enzyme solution[b]	CPW21S	CPW9G
$CaCl_2 \cdot 2H_2O$	1500	+	+	+
$MgSO_4 \cdot 7H_2O$	246	+	+	+
KNO_3	101	+	+	+
KH_2PO_4	27.2	+	+	+
KI	0.16	+	+	+
$CuSO_4 \cdot 5H_2O$	0.025	+	+	+
MES[c]	1086	+	+	+
Cellulase R10	10000	+		
Macerozyme R10	2000	+		
Sucrose	210000	+	+	
Glucose	90000			+

[a] The solution is adjusted to pH 5.8 prior to sterilization by filtration or autoclaving.
[b] The solution is filter-sterilized. Both cellulase R10 and macerozyme R10 purchased from Yakult Pharmaceutical Industry Co., Nishinimiya, Japan.
[c] 2-(N-morphino)ethanesulfonic acid, sodium salt.

3 Culture of Protoplasts

The freshly isolated protoplasts (Fig. 2a) were cultured at a density of 0.2–1 $\times 10^6$/ml PC medium (Table 2) (Pua 1987) in the dark for 5–7 days, then grown under dim light for the rest of the culture period prior to transfer to SI medium (Table 2) for shoot regeneration. Cell division generally begins 36–48 h after isolation (Fig. 2b). However, protoplast growth is markedly affected by the explant type and the growth regulator in the medium. 2, 4–D, particularly at 0.2 mg/l, is generally more effective than NAA in the presence or absence of BA (Table 3) (Pua 1987). With respect to explant type, hypocotyls and stems are a better source of protoplasts than cotyledons and leaves, since browning usually occurs in protoplast cultures originating from cotyledons or leaves, and consequently cell growth

Table 2. Composition of various media used for protoplast culture and plant regeneration in *Brassica alboglabra*

Constituent	mg/l	Medium				
		SG	PC	SR	SI	R
MS salts	–	+	+	+	+	+
myo-Inositol	100	+	+	+	+	+
Thiamine·HCl	0.4	+	+	+	+	+
Difco-Bacto agar	8000	+		+	+	+
Sucrose	10000				+	
	30000	+				+
Glucose	20000				+	
	60000		+			
NAA	1			+		
IBA	0.1					+
2,4-D	0.25		+			
BA	2			+		
	0.25		+			

Table 3. Effect of growth regulator and explant source on cell division of *Brassica alboglabra*

Growth regulator[a]	Source of protoplasts			
	Cotyledon	Hypocotyl	Leaf	Stem[b]
	Cell division frequency[c]			
NAA	4	8	4	13
NAA + BA	4	9	4	13
2,4–D	24	36	19	46
2,4–D + BA	22	42	18	47

[a] The concentrations used were 0.2–1 mg/l for NAA and 2,4-D and 0.2–0.4 mg/l for BA.
[b] Stem segments were pre-cultured on SR medium for 10–14 days.
[c] The frequency was calculated by counting the number of cells showing division in a sample of 200 cells scored 7 days after isolation.

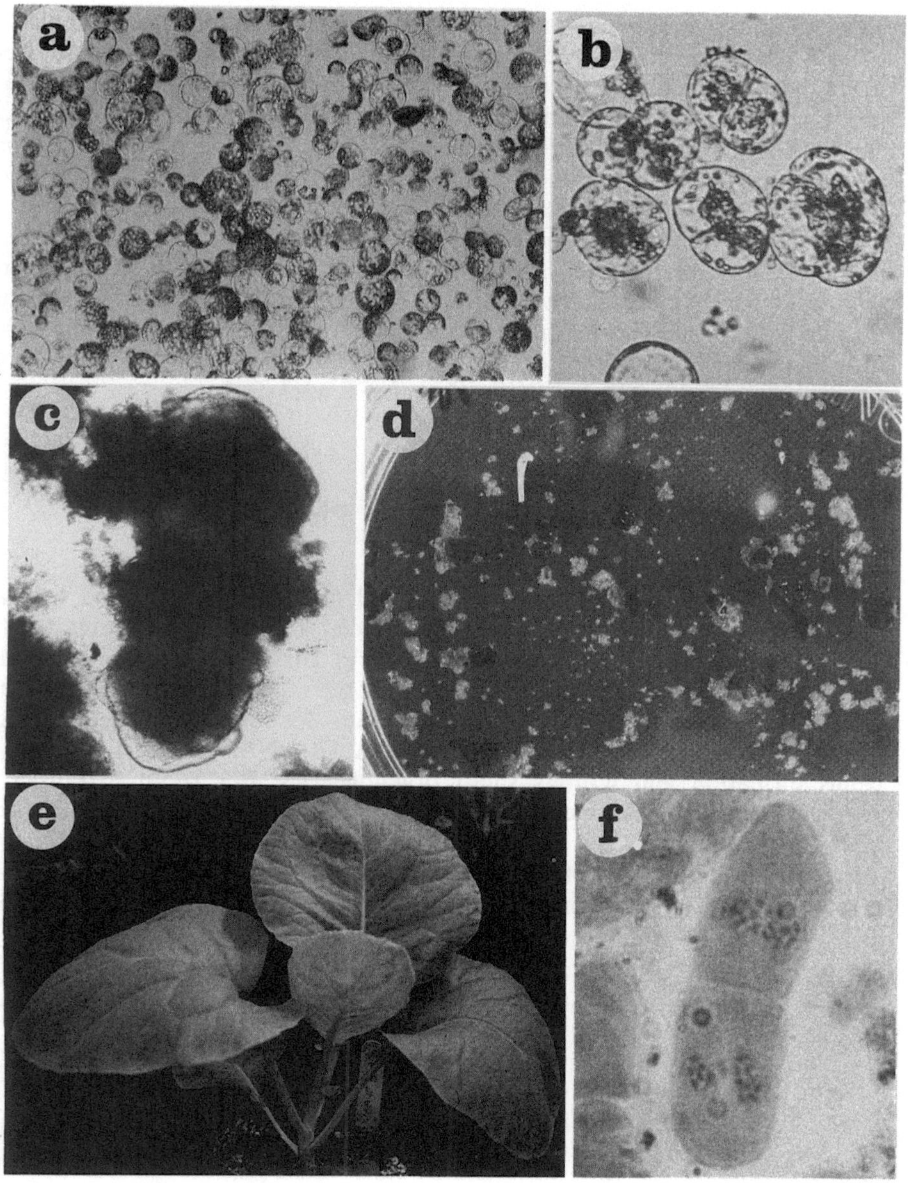

Fig. 2a–f. Protoplast culture and plant regeneration of *Brassica alboglabra* (see also Pua 1987). **a** Freshly isolated protoplasts in PC medium. **b** First and second cell division in PC medium 36–48 h after isolation. **c** Microcalli with organized structures grown in PC medium after 5–6 weeks. **d** Shoot regeneration from microcalli grown on SR medium after 2 weeks. **e** Four-week-old acclimatized plant. **f** Chromosome number (2n = 18) in root tip cells of a 3-week-old acclimatized protoplast-derived plant

becomes arrested. Browning has also been reported in mesophyll protoplasts of other *Brassica* species (Schenck and Robbelen 1982; Glimelius 1984). It has been suggested that browning is due mainly to the presence of phenolic compounds produced by protoplasts, and these compounds may be toxic to cultured cells (Schenck and Hoffmann 1979; Ulrich et al. 1980). The use of 0.05% activated charcoal has been shown to promote formation of protoplast-derived microcolonies of *B. napus* (Kohlenbach et al. 1982).

Hypocotyl- or stem-derived protoplasts of *B. alboglabra* divide rapidly in PC medium, and after 5–6 weeks, protoplasts grown into yellowish and light green microcalli which is usually accompanied by multicellular organized structures (Fig. 2c). Microcalli float on the surface of the medium and sometimes give rise to root hair. Prolonged culture of microcalli in PC medium is inhibitory to shoot formation and cultures begin to discolor after 8 weeks.

4 Regeneration and Acclimatization of Plants

Shoot organogenesis from protoplast-derived microcalli are induced by transferring 5–6-week-old microcalli (> 1 mm diameter) to SR medium (Table 2), where shoot formation usually occurs after 2 weeks (Fig. 2d). However, the organogenetic pathway is affected by the growth regulator in the medium. BA is generally better than other cytokinins, e.g., 2ip, kinetin, and zeatin, for shoot organogenesis (Fig. 3) (Pua 1987). The presence of an auxin, e.g., IAA, IBA, NAA, and 2, 4–D, promotes callus formation and inhibits shoot organogenesis. For rooting, transfer the regenerated shoots to R medium (Table 2), where root formation usually occurs after 10 days.

In vitro-grown plants of *B. alboglabra* are quite hardy. A 100% survival rate can be readily obtained using a simple plant acclimatization process. In general,

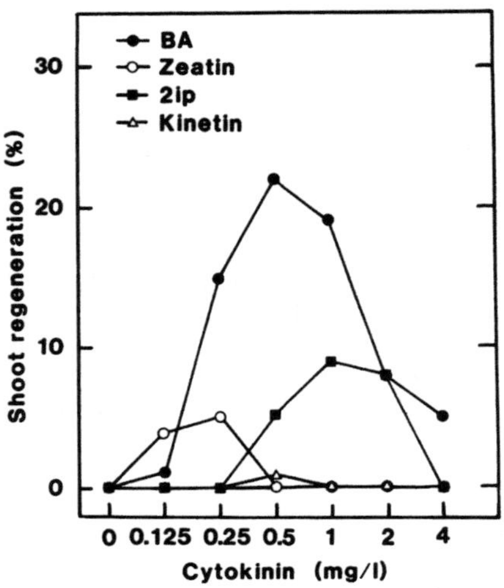

Fig. 3. Effect of cytokinin on shoot regeneration from protoplast-derived microcalli of *Brassica alboglabra* grown on SR medium after 3 weeks

agar from in vitro-grown plants is removed prior to transferring to soil. Plants are initially maintained at high humidity for 2–3 days and then the humidity is gradually reduced for 1–2 days. The acclimatized plants (Fig. 2e) grow vigorously under greenhouse conditions with the day and night temperatures at 20 and 15 °C, respectively, and with a weekly fertilization of 20N:20P:20K.

5 Characterization of Protoplast-Derived Plants

In *Brassica* spp., protoplast-derived plants and their progeny have shown considerable variation in morphology and ploidy level (Glimelius 1984; Newell et al. 1984). However, 19 out of 20 protoplast-derived plants of *B. alboglabra* were phenotypically normal, and chromosome number determination using root tips of 3-week-old acclimatized plants also showed the diploid chromosome number of 18 (Fig. 2f). One plant which showed reduced growth rate and smaller and distorted leaves was aneuploid ($2n = 15$).

6 Gene Transfer into Protoplasts Using Electroporation

To date, the method of *Agrobacterium tumefaciens*-mediated transformation has been most commonly used for transferring foreign genes into plants, including *B. napus* (Pua et al. 1987), *B. oleracea* (de Block et al. 1989) and *B. juncea* (Barfield and Pua 1991). Although various *Brassica* species have been shown to be highly susceptible to *A. tumefaciens* (Holbrook and Miki 1985; Charest et al. 1989), in our hands, *B. alboglabra* is refractory to transformation using *A. tumefaciens*. This is due primarily to the hypersensitive response of the inoculated explants, which exhibit necrosis at the wound area. As an alternative approach, direct DNA transfer into protoplasts, e.g., electroporation, can be used to by-pass the hypersensivity of the plant tissue and host range limitation in *A. tumefaciens*-mediated transformation. Electroporation has been successfully employed to produce transgenic plants of several plant species including *B. napus* (Guerche et al. 1987).

Successful transfer of foreign genes into protoplasts of *B. alboglabra* using electroporation can be achieved as follows: mix $1–2 \times 10^6$ freshly isolated protoplasts with 10–40 µg/ml plasmid DNA in 1 ml electroporation buffer in a cuvet with or without calf thymus DNA as a carrier. The plasmid DNA usually carries chloramphenicol acetyltransferase (CAT) or β-glucuronidase as a reporter and/or neomycin phosphotransferase as a selectable marker. Prior to electroporation, cool the DNA and protoplast mixture on ice for 5 min, since electroporation generates heat, which may be detrimental to protoplasts. Place the stainless steel electrodes at the two opposite sides of the cuvet containing the mixture. A desired voltage (100–400 V) can be discharged from a power supply (ISCO model 494) to a capacitor which is subsequently discharged to the cuvet. After electroporation, place the protoplasts on ice for 10 min, and then dilute the protoplasts with 2 ml fresh PC medium in a 60 × 15-mm Petri dish. After 2 days of incubation in the

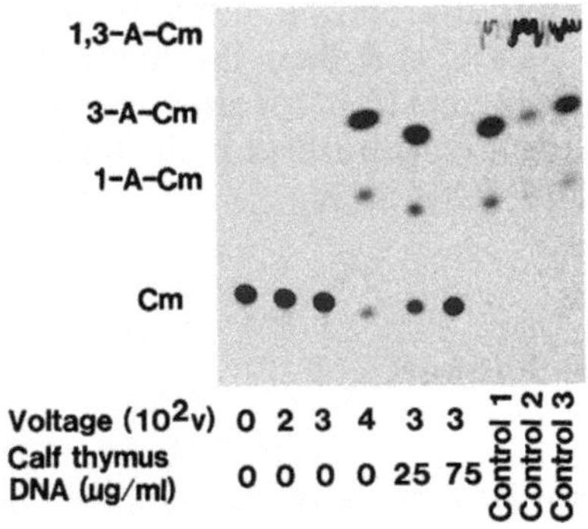

Fig. 4. CAT activity in *Brassica alboglabra* protoplasts after electroporation with 40 µg/ml pUC35SCAT and cultured for 2 days in PC medium. pUC35SCAT (5.2 kb) is a pUC derivative carrying a chimeric gene containing the coding sequence of a bacterial CAT gene and polyadenylation region of the tobacco small subunit of ribulose-1,5-biphosphate carboxylase gene driven by the cauliflower mosaic virus 35S promoter. CAT assay was conducted according to Nagy et al. (1988). *Control 1* pUC35SCAT; *Control 2 E. coli* carrying the CAT gene; *control 3* tissue from the transgenic tobacco plant carrying the CAT gene. *Cm* [^{14}C]-chloramphenicol; *1–A–Cm* 1-acetyl-[^{14}C]-chloramphenicol; *3-A-Cm* 3-acetyl-[^{14}C]-chloramphenicol; *1,3–A–Cm* 1,3-diacetyl-[^{14}C]-chloramphenicol

dark, protoplasts can be assayed for transient expression of the transgene, e.g., CAT, to determine the transformation efficiency.

Various factors, e.g., voltage (field strength), polyethylene glycol, and DNA carrier concentration and buffer, have been shown to affect protoplast transformation using electroporation (Shillito et al. 1985; Tsukada et al. 1989). In *B. alboglabra*, the chimeric CAT gene can be successfully transferred to and expressed in protoplasts by the use of 400 V alone or 300 V in combination with 25 µg calf thymus DNA (Fig. 4). However, regeneration of transgenic plants from transformed protoplasts of *B. alboglabra* has not been attempted.

7 Conclusions and Prospects

Phenotypically normal plants of *B. alboglabra* have been regenerated from protoplasts within 12 weeks. However, the capacity of plant regeneration varied considerably among genotypes, the yellow-flowered being/generally more regenerative than the white-flowered genotypes. In *Brassica*, while genotypic variability in plant regeneration from protoplasts has been well documented (Jourdan and Earle 1989), our findings suggest that cell recalcitrance in regeneration is attributed mainly to ethylene produced by cultured cells and tissues (Chi et al. 1990, 1991; Pua et al. 1990). It is probable that ethylene may also play a regulatory role in cell differentiation of recalcitrant white-flowered *B. alboglabra* in vitro. Future study to investigate this possibility is warranted.

The system for high frequency plant regeneration from protoplasts is essential for protoplast fusion, to overcome the species barrier to sexual hybridization. After fusion, a large number of somatic hybrids can be selected for the desired characteristics for genetic improvement of *B. alboglabra*. The regeneration system coupled with the ability to transfer foreign genes to and expressed in protoplasts of

B. alboglabra also offers a great potential for producing transgenic plants with improved agronomic traits, e.g., herbicide tolerance and insect and virus resistance. However, further study is needed to develop an efficient system for regeneration of transgenic plants from transformed protoplasts before it can be utilized for crop improvement.

8 Protocol

8.1 Plant Regeneration from Stem Protoplasts

(Pua 1987) (see Tables 1 and 2 for solutions and media)

1. Excise 3–5 mm internodal stem segments from 3–6-week-old in vitro-grown plants maintained on R medium, and then transfer to SR medium for 10–14 days.
2. Cut the stem segments into smaller pieces with a sterile scalpel.
3. Digest 1 g tissue with 15 ml enzyme solution in a 100 × 20-mm Petri dish for 4–5 h at room temperature in the dark on an orbital shaker at 40 rpm.
4. Filter the digestion mixture twice with a 100 μm brass or nylon sieve.
5. Gently transfer 10–12 ml of the filtrate to a 15-ml disposable tube, layer 1 ml CPW9G onto the filtrate and centrifuge for 5 min at 800 rpm.
6. Collect the protoplasts at the interface between the filtrate and CPW9G with a plastic pipet and wash twice with CPW9G and once with PC medium by centrifugation.
7. Determine the protoplast density by counting with a haemacytometer.
8. Culture the protoplasts at a density of $0.2–0.5 \times 10^6$/ml in a 60 × 15-mm Petri dish containing 3 ml PC medium.
9. Incubate the culture in the dark for 5–7 days and then transfer to dim light for 5–7 weeks. Replace half of the culture medium with fresh medium at weekly intervals.
10. After 5–6 weeks, transfer the microcalli to culture tubes and stand for 30–60 min.
11. Retain the pellet, and those microcalli with > 1 mm diameter can be transferred to SI medium for shoot regeneration.
12. After 3–4 weeks on SI medium, excise the regenerated shoots (> 10 mm) and transfer to R medium for rooting.
13. For acclimatization, remove the plants from culture containers and rinse the roots with water to remove the agar, and then plant in a pot containing greenhouse compost moistened with water.
14. Cover the plants with a lid to maintain high humidity for 2–3 days after which reduce the humidity by gradually opening the lid. Fertilize the plants with 20N : 20P : 20K at weekly intervals.

8.2 Plant Regeneration from Hypocotyl Protoplasts

1. Surface-sterilize the seeds by dipping into 95% ethanol (v/v) for 30 s, followed by 30 min in 20% Clorox® solution, and rinse twice with sterile distilled water.
2. Germinate the seeds on SG medium.
3. Excise the hypocotyls from 5–6-day-old seedlings and cut into small pieces with a sterile scalpel.
4. Digest 0.5 g tissue with 15 ml enzyme solution and incubate as described in (3) of Sect. 8.1.
5. Follow the procedure from (4) to (14) in Sect. 8.1.

References

Barfield DG, Pua E-C (1991) Gene transfer in plants of *Brassica juncea* using *Agrobacterium tumefaciens*-mediated transformation. Plant Cell Rep 10:308–314 de Block M, de Brouwer D, Tenning P (1989) Transformation of *Brassica napus* and *Brassica oleracea* using *Agrobacterium tumefaciens* and the expression of the *bar* and *neo* genes in the transgenic plants. Plant Physiol 91:694–701

Charest PJ, Iyer VN, Miki BL (1989) Virulence of *Agrobacterium tumefaciens* strains with *Brassica napus* and *Brassica juncea*. Plant Cell Rep 8:303–306

Chi G-L, Barfield DG, Sim G-E, Pua E-C (1990) Effect of $AgNO_3$ and aminoethoxyvinylglycine on in vitro shoot and root organogenesis from seedling explants of recalcitrant *Brassica* genotypes. Plant Cell Rep 9:195–198

Chi G-L, Pua E-C, Goh C-J (1991) Role of ethylene on de novo shoot regeneration from cotyledonary explants of *Brassica campestris* ssp. *pekinensis* (Lour) Olsson in vitro. Plant Physiol 96:178–183

Glimelius K (1984) High growth rate and regeneration capacity of hypocotyl protoplasts in some Brassicaceae. Physiol Plant 61:38–44

Guerche P, Charbonnier M, Jouanin L, Tourneur C, Paszkowski J, Pelletier G (1987) Direct gene transfer by electroporation in *Brassica napus*. Plant Sci 52:111–116

Holbrook LA, Miki BL (1985) *Brassica* crown gall tumerigenesis and in vitro of transformed tissue. Plant Cell Rep 4:329–332

Jourdan PS, Earle ED (1989) Genotypic variability in the frequency of plant regeneration from leaf protoplasts of four *Brassica* spp. and *Raphanus sativus*. J Am Soc Hortic Sci 114:343–349

Kohlenbach HW, Wenzel G, Hoffmann F (1982) Regeneration of *Brassica napus* plantlets in cultures from isolated protoplasts. Z Planzenphysiol 105:131–142

Nagy F, Kay SA, Chua NH (1988) Analysis of gene expression in transgenic plants. In: Gelvin SB, Schilperoorf RA (eds) Plant molecular biology manual. Kluwer Academic Publ, Dordrecht, ptB4, pp 1–29

Newell CA, Rhoads ML, Bidney DL (1984) Cytogenetic analysis of plants regenerated from tissue explants and mesophyll protoplasts of winter rape, *Brassica napus* L. Can J Genet Cytol 26:752–761

Pua E-C (1987) Plant regeneration from stem-derived protoplasts of *Brassica alboglabra* Bailey. Plant Sci 50:153–160

Pua E-C (1990) Somatic embryogenesis and plant regeneration from hypocotyl protoplasts of *Brassica juncea* (L.) Czern & Coss. Plant Sci 68:231–238

Pua E-C, Mehra-Palta A, Nagy F, Chua N-H (1987) Transgenic plants of *Brassica napus* L. Bio/Technol 5:815–817

Pua E-C, Trinh TH, Chua N-H (1989) High frequency plant regeneration from stem explants of *Brassica alboglabra* Bailey in vitro. Plant Cell Tissue Organ Cult 17:143–152

Pua E-C, Chi G-L, Barfield DG (1990) Effect of ethylene inhibitors on plant regeneration from seedling explants and somatic embryos of recalcitrant genotypes in Cruciferae. In: Abst VIIth Int Congr Plant Tissue and Cell Culture, IAPTC Amsterdam, p 248

Schenck HR, Hoffmann F (1979) Callus and root regeneration from mesophyll protoplasts of basic *Brassica* species: *B. campestris*, *B. oleracea* and *B. nigra*. Z Pflanzenzuecht 82:354–360

Schenck HR, Hoffmann F (1984) Isolation and culture of protoplasts: *Brassica*. In: Vasil IK (ed) Cell culture and somatic cell genetics of plants, vol 1. Academic Press, New York, pp 356–369

Schenck HR, Robbelen G (1982) Somatic hybrids by fusion of protoplasts from *Brassica oleracea* and *B. campestris*. Z Pflanzenzuecht 82:354–360

Shillito RD, Paszkowski SJ, Muller M, Potrykus I (1985) High efficiency direct gene transfer to plants. Bio/Technol 3:1099–1103

Thomzik JE (1993) Transformation in oilseed rape (*Brassica napus* L.) In: Bajaj YPS (ed) Biotechnology in agriculture and forestry, vol 23. Plant protoplasts and genetic engineering IV. Springer, Berlin Heidelberg New York pp 170–182

Tsukada M, Kusano T, Kitagawa Y (1989) Introduction of foreign genes into tomato protoplasts by electroporation. Plant Cell Physiol 30:599–603

Ulrich TH, Chowdhury JB, Widholm JW (1980) Callus and root formation from mesophyll protoplasts of *Brassica rapa*. Plant Sci Lett 19:347–354

Wong KW, Loh CS (1987) In vitro regeneration of plantlets in *Brassica alboglabra*. Plant Cell Tissue Organ Cult 10:143–148

Zee SY, Hui LH (1977) In vitro plant regeneration from hypocotyl and cotyledons of Chinese kale (*Brassica alboglabra* Bailey). Z Pflanzenphysiol 82:440–445

I.4 Regeneration of Plants from Protoplasts of *Cucumis sativus* L. (Cucumber)

Z.K. Punja and S.H.T. Raharjo[1]

1. Introduction

1.1 Importance and Distribution

Cucumber (*Cucumis sativus* L.) is one of about 25 species in the genus *Cucumis* (family Cucurbitaceae) and is one of only a few species which is grown for human consumption. The muskmelon, *C. melo* L., gherkin, *C. anguria* L. and jelly melon (kiwano), *C. metuliferus* Naud. are other edible species which are cultivated to various extents in different regions of the world. Most of the wild species of *Cucumis* are indigenous to Africa, while *C. sativus* and *C. sativus* var. *hardwickii* are native to Asia; cucumber is believed to have been first domesticated in India. Cucumber fruits are used primarily for fresh market consumption or for pickling. Commercial production of this crop can be found throughout different regions of the world, both tropical and temperate, under either field conditions or in controlled environments under glasshouses. Cultivars have been developed whose agronomic characteristics can differ markedly, reflecting the different growing environments and cultural requirements for this crop. Growth and development of cucumber are generally favored by temperatures above 20 °C and the plant has little to no frost tolerance.

Cucumber and a close relative, *C. sativus* var. *hardwickii* (Royle) Alef are unique among the *Cucumis* spp. in having a haploid chromosome number of n = 7 (2n = 14), and are for the most part sexually incompatible with a majority of the other species, in which n = 12 (2n = 24 or = 4x = 48) (Raamsdonk et al. 1989; den Nijs and Custers 1990; Singh 1990). Within many of these less extensively cultivated or wild species, however, useful agronomic traits, including resistance to diseases and insect pests, have been identified (Fassulliotis 1967; Provvidenti and Robinson 1974), and some of the species have increased vigor and fruit-bearing capacity. Research efforts therefore have been conducted to attempt to transfer these useful traits into cucumber using various breeding strategies and techniques in tissue culture and biotechnology. To date, however, success has been minimal (des Nijs and Custers 1990).

[1] Department of Biological Sciences, Simon Fraser University, Burnaby, British Columbia, Canada V5A 1S6

Biotechnology in Agriculture and Forestry, Vol. 22
Plant Protoplasts and Genetic Engineering III (ed. by Y.P.S. Bajaj)

Table 1. Summary of published procedures for protoplast isolation and culture in cucumber (1975–1990)

Tissue	Pretreatment	Enzymes	Culture medium	Plating procedure
Cotyledons and leaves of two cultivars, Ashley and China	Surface sterilized, epidermis peeled or abraded. Preplasmolyzed in 10% mannitol	Driselase + pectinase for cotyledons and/or Macerozyme R-10 + Cellulase R-10 for leaves	Harada Medium (half- and full-strength), with 10% mannitol	10^4–10^5/ml in liquid culture for 2–3 weeks, followed by 0.6% agar embedded for 2 weeks then onto agar medium
Leaves of one cultivar, Gy-13	Epidermis removed; preplasmolysed in 0.1 M glycine + 0.15 M mannitol	Cellulase R-10 Macerozyme R-10 Driselase	MS Medium with 0.3 M mannitol	1×10^5/ml in 0.4% agarose hanging drop (Potrykus et al. 1979) for 1–2 weeks, followed by bead culture (Shillito et al. 1983)
Cotyledons of one cultivar, Straight Eight	In PET solution (Shahin 1985) for 6–8 h at 4 °C	Cellulysin Macerozyme	MS Medium	1×10^6/ml in liquid (two types), followed by semi-solid (0.7% agar) medium
Cotyledons of 11 Chinese cultivars	None	Cellulase R-10 Macerozyme R-10	Modified DPD medium (Durand et al. 1973), then MS Medium	10^4–10^5/ml in DPD medium for 5 weeks, then onto MS agar
Cotyledons of two cultivars, Perfection Ridge and Hokus	Epidermis abraded; preplasmolyzed in 0.15 M mannitol + 0.1 M glycine	Cellulase R-10 Macerozyme R-10	MS medium with 0.25 M mannitol	10^4–10^5/ml in 0.6% agarose embedded for 2 weeks, then transfer to agarose bead culture (Davey 1983), or agarose disc (droplet) culture
Cotyledons and leaves of one cultivar, Hokus	Epidermis abraded	Cellulase R-10 Macerozyme R-10	Modified MS (half-strength macroelements, full-strenth minor)	1×10^5/ml in either liquid, 0.6% agarose bead, or agarose disc (droplet) culture
Cotyledons and leaves of four cultivars: Hokus, Salty, Gy-3, Borszczagowski	Epidermis abraded	Cellulase R-10 Macerozyme R-10	Modified MS (half-strength macroelements, full-strength minor)	1×10^5/ml in either liquid, 0.6% agarose bead (Shillito et al. 1983) or agarose disc culture
Cotyledons and leaves of two genotypes, Gy-14 and an inbred line	None	Cellulysin Macerozyme R-10	Modified MS (half-strength macroelements, full-strength minor)	4–6 $\times 10^4$/ml in either liquid, 0.4% agarose embedded, or agarose droplets

Growth regulators	Subcultures	Results	Reference
2,4-D/kin at 1/1 mg/l	Four subcultures over 8–10 weeks	Callus from leaf protoplasts, with root formation. No plantlets recovered	Coutts and Wood (1977)
NAA/2-iP at 5/3 mg/l in agarose, NAA/2-iP at 1/1 mg/l in bead culture, 2,4-D/2-iP at 0.8/0.8 mg/l for regeneration	Three-four subcultures over 8–11 weeks	Plantlets obtained	Orczyk and Malepszy (1985); Malepszy (1988)
NAA/Z/2,4-D/BAP/GA_3 at 0.2/0.25/0.5/0.25/0.4 mg/l for 7 days, 2,4-D/NAA/BAP at 1/1/0.5 for 3–5 days; NAA/BAP at 1/5 mg/l for 2–3 weeks	Three subcultures over 6–8 weeks	Eleven plantlets obtained	Trulson and Shahin (1986)
2,4-D/kin at 0.5/1 mg/l for 10 days; 0.25/0.5 mg/l for 25 days; 2,4-D/BA at 0.01/1 or IAA/BA at 0.2/0.5 mg/l, then IAA/kin at 0.2/5 mg/l	Four subcultures	Plantlets obtained from one cultivar	Jia et al. (1986)
NAA/2-iP at 5/15 μM	One subculture after 5–6 weeks	Callus, with no plantlets recovered	Dons and Bouwer (1986)
NAA/IPA at 25/15 μM, then 2,4-D/BA at 4/4 μM; then to 0.5 μM kin	Several over 3-month period	Plantlets from leaf protoplasts	Colijn-Hooymans et al. (1988)
NAA/IPA at 25/15 μM or 5/15 μM for 2–3 weeks; NAA/IPA at 5/5 μM or 2,4-D/IPA at 4/4 μM for 1–2 weeks; then 2,4-D/BA at 4/4 μM	Several over 4-month period	Plantlets obtained from leaf protoplasts of three cultivars	Colijn-Hooymans et al. (1988)
2,4-D/BA at 1/2.5 μM or 5/5 μM, or NAA/BA at 5/5 μM; then to 0.01–0.05 μM	Two subcultures over 8–12 weeks	Plantlets obtained from cotyledon and leaf protoplasts of two genotypes	Punja et al. (1990b)

1.2 Utility of Protoplasts in Research Studies

The development of general tissue culture methods for cucumber has been reported from many laboratories, and information is now available on various aspects, such as micropropagation, regeneration from callused explants via organogenesis and somatic embryogenesis, induction of flowers in vitro, direct regeneration from explants, and embryo, ovule, or anther culture (for reviews see Malepszy 1988; Punja et al. 1990a; Wehner et al. 1990; Raharjo and Punja, 1993). Efforts to regenerate plantlets from protoplasts were initiated in 1975 (Coutts and Wood 1975). The availability of a protoplast regeneration system for cucumber should make it possible to conduct studies on genetic transformation by introduction of foreign DNA through PEG-mediated uptake or electroporation, or for transient gene expression studies. In addition, the potential for use in somatic hybridization studies to introgress useful traits from wild *Cucumis* spp. is considerable. For example, the species *C. metuliferus* has received interest due to its tolerance to virus diseases and nematodes (Tang and Punja 1989; den Nijs and Custers 1990). Numerous techniques which have been developed for other important crop species, including electrofusion, PEG mediated fusion, or high pH/Ca^{2+} fusion (Kao and Michayluk 1989), could be applicable to cucumber, to generate interspecific hybrids. In addition, cucumber protoplasts may be used to study plant virus infection and replication (Maule et al. 1980) and potentially for selection of novel phenotypes through somaclonal variation.

1.3 Previous Work on Cucumber Protoplasts

Previous research studies on procedures for isolation and regeneration of protoplasts from cucumber have shown that the degree of success is quite variable and is dependent upon numerous factors: the plant genotype, source of donor tissue (cotyledons or leaves), enzyme concentration and combination (cellulase, pectinase), protoplast plating procedure (liquid, semi-solid, solid) and growth regulator combinations and concentrations. A summary of the different protocols that have been used is given in Table 1. In many of these previous studies, regeneration of plantlets from protoplast-derived calli has proved to be the most difficult step. In cases where this was accomplished, the frequencies of regeneration may not have been indicated, and plating efficiencies have been reported only in recent studies. Comparisons between these studies are made more difficult due to differences in the various methodologies used (Table 1). An overview of the procedures used that may be applicable to future research studies on cucumber is presented below.

2 Isolation of Protoplasts

2.1 Plant Material

Both fresh market and pickling cucumber, representing a broad range of cultivars, have been used as the starting material for protoplast isolation in previous studies (Table 1). Plant materials can be grown in the greenhouse (Coutts and Wood 1975), growth chamber, or under aseptic conditions on tissue culture medium such as MS (Murashige and Skoog (1962) basal salts medium with 20–30 g/l sucrose (Dons and Bouwer 1986; Punja et al. 1990b). The use of in-vitro-grown seedlings is preferred, since no tissue surface sterilization steps are needed. Seeds are commonly surface disinfested in 10–20% Clorox (containing 5.25% sodium hypochlorite) for 20–30 min, which may be preceded by a dip in 70% ethanol for 10–20s. Contamination with seed-borne bacteria in cucumber is frequent, however, which may necessitate the addition of 100–150 mg/l ampicillin, carbenicillin, or chloramphenicol to the culture medium (Trulson and Shahin 1986; Punja et al. 1990a).

To date, only cotyledons and true leaves have been used as the donor tissues for protoplast isolation. Cotyledons have been excised from 5- to 12-day-old seedlings, while true leaves (first to third) are obtained from 2- to 4-week-old seedlings. Alternatively, stem nodal cuttings can be propagated from the original seedling to provide an extended source of true leaves (Orczyk and Malepszy 1985; Colijn-Hooymans et al. 1988a, b). Uniformity of seed germination is enhanced by incubation in the dark for 24–96 h at 27 °C before transfer to culture media and by excising the seed coat and transfer of the embryos directly. Plants are grown under high light intensity at 14–16 h/day illumination. Prior to protoplast isolation, the lower epidermis has in some studies been peeled or abraded with a stiff hair brush, or used without treatment (Table 1). A preplasmolysis treatment in 10% mannitol for 1–3 h (Coutts and Wood 1977) or in 0.1 M glycine + 0.15 M mannitol (Dons and Bouwer 1986) was shown to enhance protoplast yields and integrity in some studies; in other studies, a preplasmolysis treatment was not used (Punja et al. 1990b).

2.2 Enzymes

In most research studies, a combination of Cellulase Onozuka R-10 (range of 0.6–2%) and Macerozyme R-10 (range of 0.3-1.25%) at specified concentrations provided optimal protoplast yields of cucumber. Alternatively, Driselase and Cellulysin have been substituted for Cellulase R-10 (Table 1). The choice of starting material, i.e., cotyledons versus true leaves, may require the use of different enzyme concentrations. For example, Coutts and Wood (1977) used 1.5% Driselase + 0.5% pectinase for cotyledon tissue and 2% Driselase + 0.5% pectinase for leaf tissue, while Punja et al. (1990b) found 0.5% Cellulysin + 1.25% pectinase to be optimal for cotyledons and 1% Cellulysin + 0.5% pectinase for leaves. Colijin-Hooymans et al. (1988a, b) have also shown the differential response of these two tissue types to enzymatic degradation. Length of exposure to the enzymes in various studies ranges from 3 to 18 h, with gentle agitation at 23–26 °C; the most commonly used procedure involves an overnight (16 h) digestion. The osmoticum used is mannitol (0.15-0.5 M) sometimes in combination with 0.1 M glycine (Orczyk and Malepszy 1985; Colijn-Hooymans et al. 1988a, b). The tissues are sliced or segmented prior to exposure. Protoplast yields have ranged from 0.5×10^5 to 8×10^6 per g (fr. wt.) of tissue.

2.3 Factors Influencing Yield

The yields of protoplasts from cucumber tissues have been shown to be enhanced by the following factors: pretreatment by peeling or abrasion of the epidermis

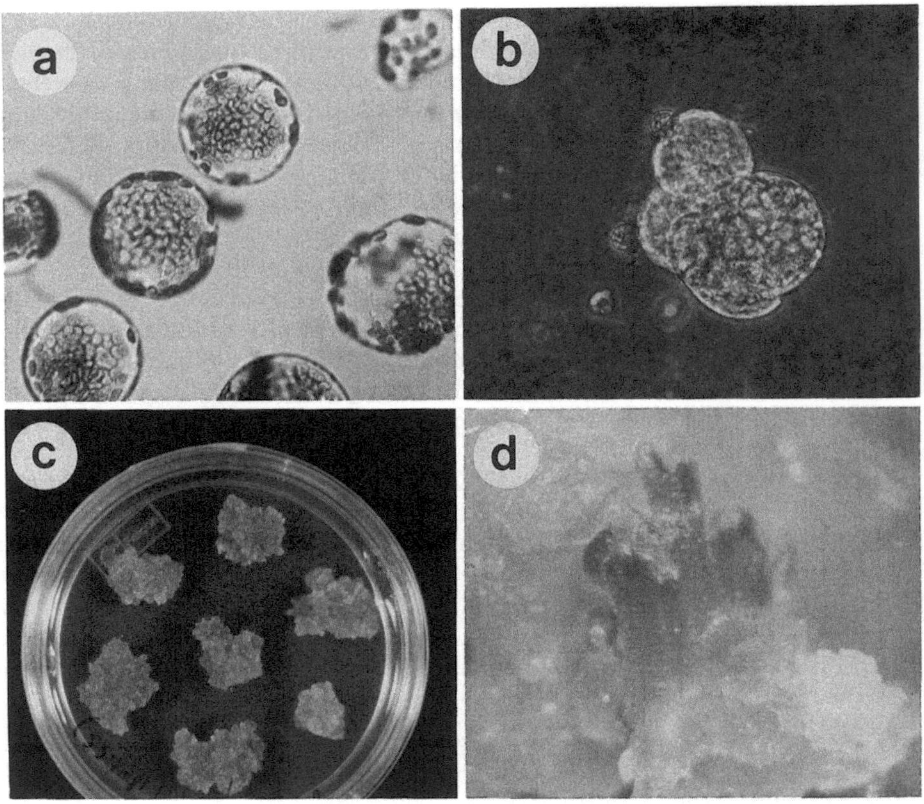

Fig. 1 a–d. Regeneration from cotyledon protoplasts of cucumber. **a** Freshly isolated protoplasts (× 525). **b** Second-third cell divisions (× 525). **c** Callus development from protoplast-derived minicalli on medium containing NAA/BA at 5/5 μM. **d** Shoot development on growth regulator-free medium

(Coutts and Wood 1977), preplasmolysis treatment (Coutts and Wood 1975), including presence of glycine (Orczyk and Malepszy 1985), use of young tissue (cotyledons not to exceed 10–12 days old and leaves not to exceed 4 weeks old), appropriate enzyme concentrations (as described in Sect. 2.2) depending on the tissues and digestion time. Placement of plants in the dark for 24 h prior to digestion may enhance protoplast yields (unpubl.). Yields from both cotyledons and leaves are generally comparable (Punja et al. 1990b), with cotyledonary protoplasts (Fig. 1a) being much larger in size than mesophyll protoplasts.

3 Culture of Protoplasts

Following isolation of cucumber protoplasts, several different methods for culture have resulted in sustained divisions and callus development. These include liquid culture (Coutts and Wood 1977; Trulson and Shahin 1986), agarose embedded (in

0.4–0.6% SeaPlaque agarose) followed by hanging-drop culture (Orcyzk and Malepszy 1985), agarose bead culture (Dons and Bouwer 1986) or agarose disc (droplet) culture (Colijn-Hooymans et al. 1988a, b). The culture media utilized were those described by Harada (1973), Durand et al. (1973), Kao (1977), and Murashige and Skoog (1962) (Table 1). Modified MS medium (half-strength macroelements, full-strength minor elements) has been frequently and successfully used in recent studies (Colijn-Hooymans et al. 1988a, b; Punja et al. 1990b). The results from some of the previous studies have shown that plating density has a profound influence on rate and frequency of cell divisions and rate of minicalli development from cucumber protoplasts (Dons and Bouwer 1986; Punja et al. 1990b). For example, at a density of $2.2–2.5 \times 10^4$/ml of cotyledonary protoplasts, it required over 30 days for minicalli to develop, whereas at $4–6 \times 10^4$/ml, minicalli were observed after 12–13 days (Punja et al. 1990). Similarly, Dons and Bouwer (1986) reported much higher plating efficiencies (50–80%) at a density of 1×10^5 protoplasts/ml, while at a density of 0.1×10^5/ml, no minicalli developed. First cell divisions are usually observed after 6–7 days, second-third divisions in 8–9 days (Fig. 1b), and minicalli formation by 13 days (Punja et al. 1990b).

In a comparative study of the three culture procedures (liquid, agarose embedded, and droplet), Punja et al. (1990b) reported that the highest planting efficiencies (about 15–28%) were achieved with agarose-embedded protoplast from cotyledons and leaves of three cucumber genotypes. These results were an extension of an earlier report by Dons and Bouwer (1986), who described the utility and applicability of the agarose-disc procedure for achieving high frequencies of cell divisions and plating efficiencies of cucumber protoplasts. The latter procedure required only regular replacements of the culture medium without the need for transfer of the minicalli. In several earlier studies, at least three to four subcultures of developing calli were required, frequently involving a complex array of growth regulators and extended culture (Table 1). Development of calli from cucumber protoplasts can be achieved within 8–10 weeks after plating from both cotyledon and leaf sources (Fig. 1c). A comparison of the procedures and variables for protoplast isolation and regeneration in three pickling cucumber genotypes is given in Table 2. Significant genotypic differences can be seen with regard to growth regulator requirements and regeneration frequency. Modifications of the protocol presented may be necessary for different cucumber genotypes.

4 Regeneration of Plants

Successful development of protoplast-derived calli of cucumber has been accomplished following a series of different procedures in previous studies (Table 1). The difficulty in inducing subsequent differentiation and achieving plantlet regeneration from these calli has been noted by several investigators (Orczyk and Malepszy 1985; Jia et al. 1986; Trulson and Shahin 1986; Colijn-Hooymans et al. 1988a, b). In those instances where plantlets were recovered, calli were subcultured onto lower growth regulator combinations, including 2,4-D/2 iP at 0.8/0.8 mg/l (Orcyzk and Malepszy 1985), NAA/BAP at 1/5 mg/l (Trulson and Shahin 1986), IAA/kin at 0.2/5 mg/l (Jia et al. 1986), kin at 0.5 μM or 2,4-D/BA at 4/4 μM (Colijin-Hooymans

Table 2. Comparison of procedures and variables for protoplast isolation and regeneration in pickling cucumber (Punja et al. 1990b)

Plant materials	In-vitro grown seedlings								
Tissues source	Cotyledons			True leaves					
Enzyme digestion	1.25% Pectinase 0.5% Cellulysin			0.5% Pectinase 1.0% Cellulysin					
Protoplast yields	1.8–2.0 × 10^6/0.5 g			4.0–5.0 × 10^6/g					
Genotype	3672			3672			3676		
Plating density[a]	3.5–4.0 × 10^4/ml			Same			Same		
Growth regulators[b]	2,4-D/BA at 1.0/2.5 μM			2,4-D/BA at 5.0/5.0 μM			NAA/BA 5.0/5.0 μM		
Plating procedure[c]	D	L	A	D	L	A	D	L	A
Days to minicalli formation	19	21	16	19	16	13	12	12	10
Plating efficiency (%)[d]	10	8	13	19	16	28	12	10	15
Callus development	2,4-D/BA or NAA/BA at 1.0/2.5 μM			NAA/BA or 2,4-D/BA at 2.5/5.0 or 5.0/5.0 μM					
Differentiation[e]	2,4-D/BA at 0.01–0.05 μM			2,4-D/BA at 0.01–0.05 μM			NAA/BA at 0.01–0.05 μM		
Regeneration frequency from callus (%)	12			0			17		
Regeneration frequency from protoplasts (%)	1.5			0			2.5		

[a] Plating density of 6.0–6.5 × 10^4/ml gave similar results. Lowever plating density (2–2.5 × 10^4/ml) delayed minicalli formation to > 30 days.

et al. 1988a, b) and 2,4-D/BA or NAA/BA at 0.01–0.05 μM (Punja et al. 1990b), prior to transfer onto MS medium without growth regulators. Plantlets have been recovered from both cotyledonary and mesophyll protoplasts (Punja et al. 1990b) (Fig. 1d). However, the preponderance of higher ploidy levels in cucumber protoplasts (Colijn-Hooymans et al. 1988b) and a significant proportion of tetraploid plantlets derived from cotyledon explants (Kim et al. 1988) may preclude cucumber cotyledonary tissues from being routinely used as starting material for genetic transformation or protoplast fusion studies.

5 Fusion of Protoplasts

The applicability of protoplast fusion studies for the genetic improvement of cucumber has potential, since sexually incompatible wild species are known to possess traits of interest (see Sect. 1.1). Breeding strategies have not to date yielded any success with regard to introgression of useful traits from the wild relatives, since crossing barriers have not yet been overcome (den Nijs and Custers 1990). Described below is one approach to the utilization of protoplast fusion to access traits from one species, *C. metuliferus*. The results are still preliminary, and considerable work remains before somatic hybridization produces utilizable results in cucumber.

5.1 Procedure

Plant materials of *C. sativus* Gy14 and *C. metuliferus* PI 292190 were grown under the conditions described below in Sect. 7. Protoplasts were isolated from true leaves of both species according to the protocol described (see Sect. 7). Cellulysin (1%) and pectinase (0.5%) were used for both species, which provided yields of about 5–6 $\times 10^6$/g tissue. Following the purification steps, a suspension containing 2.5–3 $\times 10^4$/ml was used for fusion studies. One-half ml of the suspension from each species was mixed in a 60 × 15-mm Petri dish and 1 ml of the following added: PEG (MW 8000) at a concentration of 15% (in final volume), with or without 1% DMSO, for 20 min, or high pH/Ca^{2+} (solution containing 80 g/l mannitol, 7.35 g/l $CaCl_2 \cdot 2H_2O$, 3.75 g/l glycine, pH 10) for 15–20 min. Following treatment, the protoplast mixture was washed two to three times with basal medium to remove the fusogenic agents, and the protoplasts were concentrated to a density of 2.5–3 $\times 10^4$/ml.

[b] Other genotypes may respond to 2,4-D/Z or NAA/Z at 2.5/5 μM or NAA/BA of 5/10 μM (unpubl.).
[c] Plating procedures are D = droplet culture (100 μl protoplast suspension in agarose on agar medium); L = liquid culture (protoplasts suspended in 5 ml of liquid medium); A = agarose embedded (protoplasts embedded in 0.4% agarose).
[d] Plating efficiency = proportion of original protoplasts that formed minicalli.
[e] Development of embryoid or shoot-like structures.

5.2 Protoplast Culture

Protoplasts of *C. sativus* and *C. metuliferus* at a density of 2.5–3.0 × 10^4/ml, alone or in a mixture, were plated without a fusion treatment or following fusion with PEG 8000 (Fig. 2) or high pH/Ca^{2+} in MS medium with half-strength major salts, full complement of minor salts and vitamins, 2% sucrose, and 0.25 M mannitol. Growth regulators were provided as 2,4-D/BA at 5/5 μM or NAA/BA at 5.0/2.5 μM. The suspension (2 ml) was added to soft agarose (final concentration 0.4%) and plated in 35 × 10 mm dishes. The influence of a nurse culture system (Fig. 3) was also evaluated. A droplet of the suspension (at densities as low as 1 × 10^2/ml) was placed on agar medium in the center of the dish, and along the periphery of the dish, a 1–1.5 cm zone of mesophyll protoplasts of *C. sativus* at 2.5–3 × 10^4/ml or higher was prepared, leaving a cell-free zone agar of about 1 cm in width around the droplet (Fig. 3). The Petri dishes were incubated in the dark at 24–26 °C for 7

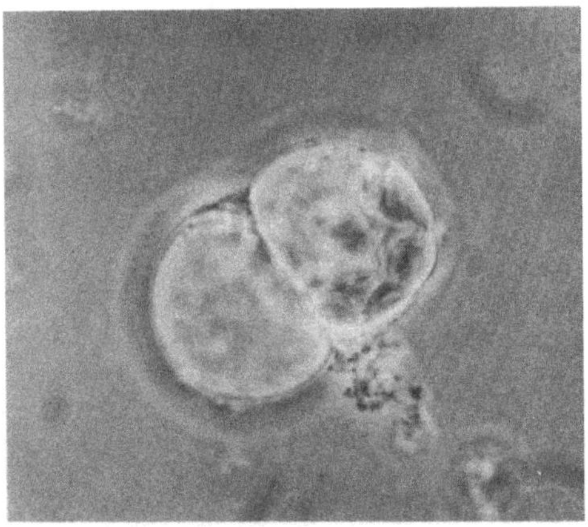

Fig. 2. Protoplast fusion between *Cucumis metuliferus* (*bottom*) and *Cucumis sativus* (*top*). Fusion mixture contained 15% PEG 8000, for 20 min

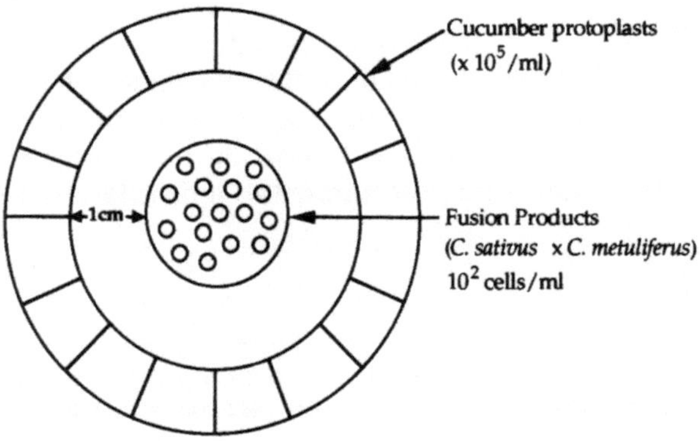

Fig. 3. Nurse culture system for enhancing division of protoplast fusion products which are plated in the center of the dish at a low density and are surrounded by protoplast nurse cells

Table 3. Response of protoplasts of *Cucumis sativus* and *Cucumis metuliferus* to a nurse culture and fusion treatments (Tang and Punja 1989)

Treatment	Days required for		
	First division	Second-Third division	Minicallus formation
C. sativus, with or without[a] nurse culture	4–5	6–7	13
C. metuliferus, with or without nurse culture	8–10	–	–
Mixture of both species[b]	10–20	>20	–
Mixture of both species,[b] with nurse culture	7–10	14	28
Mixture of both species,, following PEG 8000 or high pH/Ca^{2+} treatment, with nurse culture	7–8	12–16	28

[a] The nurse culture system is shown in Fig. 3.
[b] Without fusion treatment. Data were obtained from an average of the population of cells.

days, and then transferred to light (16 h/day, intensity of 100 μE/m²/s). The time required for first to third cell divisions and for minicallus development was noted.

5.3 Results

Protoplasts of *C. sativus* formed minicalli within 13 days in the absence or presence of the nurse culture (Table 3); at a lower plating density, however (0.5–0.6 × 10^4/ml), the nurse culture system was essential for sustained divisions. With *C. metuliferus*, only first cell divisions were observed. In a mixture of the two species, the rate of cell divisions was intermediate, and the presence of the nurse culture considerably enhanced cell divisions and minicallus development (Table 3). When a fusion treatment was applied, cell divisions and minicallus development were observed in the presence of the nurse culture. Fusion frequencies were estimated to be around 5–6% for PEG and 2–5% for high pH/Ca^{2+}. To date, only minicalli have been recovered following fusion treatments between the two *Cucumis* species. Additional studies are needed to establish whether these calli are of hybrid origin.

6 Conclusions and Prospects

The utility of protoplasts in basic research studies and for biotechnological improvements in cucumber warrant further research efforts. While *Agrobacterium*-mediated transformation has been accomplished with *A. tumefaciens* (Sarmento et al. 1992; Chee 1993) and *A. rhizogenes* (McInnes et al. 1991) using cucumber explants, the feasibility of using direct gene transfer methods via protoplasts of cucumber has not been investigated. The potential for somatic hybridization

Fig. 4. Leaves from selected greenhouse-grown plantlets derived from protoplasts showing the extent of variation in leaf morphology (parental leaf type shown in *top left corner*)

between cucumber and its wild relatives is known, yet little information is available regarding its practical utility. The inclusion of protoplast-derived calli in development of in-vitro screening assays for detecting resistance to plant pathogens, or the detection of novel traits, vis-a-vis somaclonal variation (Fig. 4) has not been extensively studied. The overview of the current information available for protoplast isolation and regeneration in cucumber presented here illustrates that a number of approaches may be successfully used, and that the frequency of regeneration can be affected by a variety of factors (Punja et al. 1990b).

7 Protocol for Isolation, Culture, and Regeneration of Cucumber Protoplasts

7.1 Plant Materials

1. Surface sterilize seeds by dipping in 70% ethanol for 20 s, followed by a 20-min soak in a 20% solution of commercial bleach (Clorox, 5.25% sodium hypochlorite) and three rinses in sterile distilled water.
2. To ensure uniform germination, place seeds in a shallow dish containing sterile water with 100 mg/l ampicillin (to inhibit bacterial growth) for 96 h at room temperature to permit the radicle to emerge. Then, remove seed coats with a scalpel and transfer the embryos to Magenta boxes containing 50 ml of growth regulator-free Murashige and Skoog basal medium (MS) with full-strength macroelements and microelements, myoinositol (100 mg/l), thiamine HCl (0.8 mg/l), 3% sucrose, and 0.65% Phytoagar. Adjust pH of the medium to 5.8 with 1 N KOH prior to autoclaving.
3. Incubate seedlings in a growth chamber set at 18/24 °C night/day temperature regime with a 16 h/day photoperiod provided by cool-white fluorescent lamps (intensity of 160 $\mu E/m^2/s$). Excise cotyledons from 10–12-day-old seedlings and first to third leaves from 19–21-day-old seedlings for protoplast isolation.

7.2 Isolation and Purification of Cotyledon and Leaf Protoplasts

1. Cut cotyledons (0.5 g fr. wt.) and true leaves (1 g) into 1–2 mm wide × 10 mm long strips under sterile conditions and place in 10 or 20 ml of the enzyme solution in a 100 × 25 mm Petri dish.
2. Make up enzyme combinations – 0.5% cellulysin + 1.25% pectinase for cotyledons and 1% cellulysin + 0.5% pectinase for true leaves – in modified MS medium containing half-strength major salts, full complement of minor salts and vitamins, 2% sucrose and 0.25 M mannitol, pH 5.8. Sterilize enzyme solutions by filtration through a syringe (B-D disposable) and Nalgene disposable filter unit (0.22 μm pore size).
3. Incubate tissues overnight (15–16 h) in the dark at 24 ± 2 °C on a reciprocating shaker set at 60 rpm. Pass the resulting suspension through multilayers of sieve cloth (pore sizes from 50 to 300 μm) to separate the protoplasts from undigested plant debris.
4. Rinse twice in MS basal medium containing 0.25 M mannitol with two centrifugation cycles at 70 *g* for 10 min at room temperature to remove the enzyme solution and concentrate the protoplasts (in Babcock bottles a dark green band is seen at the meniscus). Remove protoplasts with a Pasteur pipette and resuspend in basal medium. Estimate yields by counting on a hemocytometer. Dilute to the desired plating density.

7.3 Protoplast Culture and Regeneration

1. Embed the protoplasts in 0.4% Sea Plaque agarose in modified MS medium with 2% sucrose and 0.25 M mannitol, using 2 ml of the mixture in 35 × 10 mm Petri dishes at a density of 3.5–4 × 10^4 /ml. Growth regulator combinations are 2,4-D/BA at 1/2.5 μM (cotyledon) or 2.5/5 μM (leaf) or NAA/BA at 5/5 μM (leaf).
2. Incubate dishes in the dark at 24–26 °C in a growth chamber for the first 7 days and then transfer to a 16-h photoperiod provided by cool-white fluorescent lamps (intensity of 100 $\mu E/m^2/s$).
3. After 19–21 days, dilute medium (1 : 2, old : fresh) with modified MS basal salts to reduce mannitol to 0.08 M. As the calli develop, transfer to callus proliferation medium containing full strength MS salts, growth regulators as above, 3% sucrose, and 0.65% Phytoagar.
4. When calli are about 5 mm in size, subculture onto MS medium containing 2,4-D/BA or NAA/BA at similar concentrations as used above, or if signs of differentiation are evident, onto 0.05–0.01 μM of each. Make a final transfer to growth regulator-free MS basal medium. Allow embryos or buds to develop roots and transfer plantlets to MS medium or soil.

Acknowledgments. A major portion of the work described here was conducted at the Campbell Institute for Research and Technology, Davis, California, in conjuction with F.A. Tang and G.G. Sarmento. Their involvement in the project is gratefully acknowledged.

References

Chee PP (1993) Transformation in cucumber (*Cucumis sativus* L.) In: Bajaj YPS (ed) Biotechnology in agriculture and forestry, vol 23. Plant protoplasts and genetic engineering IV. Springer, Berlin Heidelberg New York pp 214–227

Colijn-Hooymans CM, Bouwer R, Dons JJM (1988a) Plant regeneration from cucumber protoplasts. Plant Cell Tissue Organ Cult 12 : 147–150

Colijn-Hooymans CM, Bouwer R, Orczyk W, Dons JJM (1988b) Plant regeneration from cucumber (*Cucumis sativus*) protoplasts. Plant Sci 57 : 63–71

Coutts RHA, Wood KR (1975) The isolation and culture of cucumber mesophyll protoplasts. Plant Sci Lett 4 : 189–193

Coutts RHA, Wood KR (1977) Improved isolation and culture methods for cucumber mesophyll protoplasts. Plant Sci Lett 9 : 45–51

Davey MR (1983) Recent developments in the culture and regeneration of plant protoplasts. In:Potrykus I, Harms CT, Hinnen A, Hutter R, King PJ, Shillito RD (eds) Protoplasts 1983. Birkhauser, Basel, pp 19–31

Dons JJM, Bouwer R (1986) Improving the culture of cucumber protoplasts by using an agarose-disc procedure. In: Symp Nuclear techniques and in vitro culture for plant improvement. IAEA, Vienna, pp 497–504

Durand J, Potrykus I, Donn G (1973) Plantes issues de protoplasts de *Petunia.* Z Pflanzenphysiol 69:26–34

Fassulliotis G (1967) Species of *Cucumis* resistant to the root-knot nematode *Meloidogyne incognita acrita.* Plant Dis Rep 51:720–723

Harada H (1973) A new method for obtaining protoplasts from mesophyll cells. Z Pflanzenphysiol 69:77–80

Jia S-R, Fu Y-Y, Lin Y (1986) Embryogenesis and plant regeneration from cotyledon protoplast culture of cucumber (*Cucumis sativus* L.). J Plant Physiol 124:393–398

Kao KN (1977) Chromosomal behaviour in somatic hybrids of soybean-*Nicotiana glauca.* Mol Gen Genet 150:225–230

Kao KN, Michayluk MR (1989) Fusion of plant protoplasts- Techniques. In: Bajaj YPS (ed) Biotechnology in agriculture and forestry, vol 8: Plant protoplasts and genetic engineering I. Springer, Berlin Heidelberg New York, pp 277–288

Kho YO, Nijs APM den, Franken J (1980) Interspecific hybridization in *Cucumis* L. II. The crossability of species, an investigation of in vivo pollen tube growth and seed set. Euphytica 29:661–672

Kim S-G, Chang J-R, Cha HC, Lee K-W (1988) Callus growth and plant regeneration in diverse cultivars of cucumber (*Cucumis sativus* L.). Plant Cell Tissue Organ Cult 12:67–74

Lu DY, Cocking EC, Davey MR, Pental D (1984) Cultural studies of protoplasts isolated from cotyledons of *Cucumis sativus, Lactuca sativa* and *Glycine max.* In: Symp Genetic manipulation in crops. IRRI, Beijing, pp 209–210

Malepszy S (1988) Cucumber (*Cucumis sativus* L.) In: Bajaj YPS (ed) Biotechnology in agriculture and forestry vol 6: Crops II. Springer, Berlin Heidelberg New York, pp 277–293

Maule AJ, Boulton MI, Wood KR (1980) An improved method for the infection of cucumber leaf protoplasts with cucumber mosaic virus. Phytopathol Z 97:118–126

McInnes E, Morgan AJ, Mulligan BJ, Davey MR (1991) Roots induced on cucumber cotyledons by the agropine Ri plasmid TR-DNA exhibit the transformed phenotype. Plant Cell Rep 9:647–650

Murashige T, Skoog F (1962) A revised medium for rapid growth and bioassays with tobacco tissue cultures. Physiol Plant 15:473–497

Nijs APM den, Custers JBM (1990) Introducing resistance into cucumbers by interspecific hybridization. In: Bates DM, Robinson RW, Jeffrey C (eds) Biology and utilization of the Cucurbitaceae. Cornell Univ Press, pp 382–396

Orczyk W, Malepszy S (1985) In vitro culture of *Cucumis sativus* L. *V.* Stabilizing effects of glycine on leaf protoplasts. Plant Cell Rep 4:269–273

Portykus I, Harms CT, Lörz H (1979) Multiple-drop assay (MDA) technique for the large-scale testing of culture media variations in hanging micro-drop cultures of single cell systems. I. The technique. Plant Sci Lett 9:231–235

Provvidenti R, Robinson RW (1974) Resistance to squash mosaic virus and watermelon mosaic virus 1 in *Cucumis metuliferus.* Plant Dis Rep 58:735–738

Punja ZK, Tang FA, Watkins LH (1988) Identification of resistance to root knot nematodes and virus diseases in *Cucumis metuliferus* and approaches to hybridization with *Cucumis sativus* by protoplast fusion. Phytopathology 78:1578 (Abstr)

Punja ZK, Abbas N, Sarmento GG, Tang FA (1990a) Regeneration of *Cucumis sativus* var. *sativus* and *C. sativus* var. *hardwickii, C. melo,* and *C. metuliferus* from explants through somatic embryogenesis and organogeneis. Plant Cell Tissue Organ Cult 21:93–102

Punja ZK, Tang FA, Sarmento GG (1990b) Isolation, culture and plantlet regeneration from cotyledon and mesophyll protoplasts of two pickling cucumber (*Cucumis sativus* L.) genotypes. Plant Cell Rep 9:61–64

Raamsdonk LWD van, Nijs APM den, Jongerius M C (1989) Meiotic analyses of *Cucumis* hybrids, with an evolutionary evaluation of the genus *Cucumis* (Cucurbitaceae). Plant Syst Evol 163:133–146

Raharjo SHT, Punja ZK (1993) Plantlet regeneration from petiole explants of the African horned cucumber, *Cucumis metuliferus.* Plant Cell Tissue Organ Cult. 32:169–174

Sarmento GG. Alpert K, Tang FA, Punja ZK (1992) Factors influencing *Agrobacterium tumefaciens-*

mediated transformation and expression of kanamycin resistance in pickling cucumber. Plant Cell Tissue Organ Cult 31:185–193.
Shahin EA (1985) Totipotency of tomato protoplasts. Theor Appl Genet 69:235–240
Shillito RD, Paszkowski J, Potrykus I (1983) Agarose plating and a bead type culture technique enable and stimulate development of protoplast-derived colonies in a number of plant species. Plant Cell Rep 2:244–247
Singh AK (1990) Cytogenetics and evolution in the Cucurbitaceae. In: Bates DM, Robinson RW, Jeffrey C (eds) Biology and utilization of the Cucurbitaceae. Cornell Univ Press, pp 10–28
Tang FA, Punja ZK (1989) Isolation and culture of protoplasts of *Cucumis sativus* and *Cucumis metuliferus* and methods for their fusion. Cucurb Gen Coop Rep 12:29–34
Trulson AJ, Shahin EA (1986) In vitro plant regeneration in the genus *Cucumis*. Plant Sci 47:35–43
Wehner TC, Cade RM, Locy RD (1990) Cell, tissue and organ culture techniques for genetic improvement of cucurbits. In: Bates DM, Robinson RW, Jeffrey C (eds) Biology and utilization of the Cucurbitaceae. Cornell Univ Press, pp 367–381

I.5 Regeneration of Plants from Protoplasts of *Lolium* (Ryegrasses) and *Festuca* (Fescues)

S.J. Dalton[1]

1 Introduction

Pasture grass is the natural food for herbivorous domestic animals and the single most important crop in temperate regions in terms of land area. Much of this is natural grassland but the Festuceae tribe, which includes the *Festuca-Lolium* complex, contains most important grasses sown in cultivated grassland and for amenity use (Table 1).

Plant regeneration from callus cultures is possible in certain genotypes of all the *Lolium* and fescue species and hybrids listed in Table 1. However, green plant regeneration from cell suspension cultures has been more difficult to achieve, while regeneration from protoplast cultures has been restricted so far to a few genotypes of *L. perenne*, *L. multiflorum* and *F. arundinacea*, (Table 2). Although *F. arundinacea* and *L. multiflorum* have been considered to be particularly amenable species (Dalton 1988b), this list probably reflects the greater economic importance of these species rather than any particular difficulty in the others, and Zaghmout and Torello (1990) have reported green leaves from *F. rubra* protoplast-derived colonies while albino shoots have been obtained from *L. temulentum* (Dalton, unpubl.).

The possibility of somatically hybridizing different species was a major reason for developing protoplast techniques. Many sexual hybridizations within the *Lolium-Festuca* complex have already been achieved (Lewis 1982) with varying degrees of chromosomal stability. However, protoplast fusion offers the chance to hybridize difficult crosses and to make asymmetric hybridizations. Somatic hybridizations with wild species and even species outside the *Lolium-Festuca* complex may also become possible; however, little has been published on this application.

Colonies developed in protoplast populations of electrofused *F. arundinacea* suspension protoplasts and alsike clover (*Trifolium hybridum*) mesophyll protoplasts (Webb et al. 1985), but these were not characterised. Takamizo et al. (1990) also reported colony formation after electrofusion of *F. arundinacea* suspension protoplasts with *L. multiflorum* mesophyll protoplasts. This group has since been able to regenerate hybrid plants with intermediate characteristics (1991) from electrofusions between iodoacetamide-inactivated embryogenic suspension protoplasts of *F. arundinacea* and non-morphogenic suspension protoplasts of *L. multiflorum*. Cybrid calli developing from fused iodoacetamide-treated and

[1] Institute of Grassland and Environmental Research, WPBS, Plas Gogerddan, Aberystwyth, Dyfed, SY23 3EB, United Kingdom

Biotechnology in Agriculture and Forestry, Vol. 22
Plant Protoplasts and Genetic Engineering III (ed. by Y.P.S. Bajaj)

Table 1. Uses, distribution and attributes of cultivated *Lolium* and *Festuca* species

Species and ploidy	Common name and attributes	Uses and distribution
Lolium perenne L. 2x = 14, 28 outbreeding perennial	Perennial ryegrass High yield, persistent, nutritious, palatable	Grassland, amenity Temperate regions
L. multiflorum Lam. 2x = 14, 28 outbreeding biennial	Italian ryegrass High early spring and late summer growth, palatable	Grassland Temperate and irrigated regions
L. rigidum Gaud. 2x = 14 outbreeding annual	Annual ryegrass Drought-resistant	Grassland Antipodes
L. multiflorum var *westerwoldicum* auct. 2x = 14 outbreeding annual	Westerwolds ryegrass Drought-resistant	Grassland Antipodes
L. temulentum L. 2x = 14 inbreeding annual	Darnel Drought-resistant	Grassland, research Antipodes
Festuca pratensis Huds. 2x = 14 outbreeding perennial	Meadow fescue Winter hardy, good aftermath grazing	Grassland Cold and temperate regions
F. arundinacea Schreb. 2x = 42 outbreeding perennial	Tall fescue Very winter hardy, heat, drought, water logging tolerance	Grassland, amenity Temperate and dry regions
F. rubra L. 2x = 42, 56 outbreeding perennial	Red fescue Winter hardy salinity and drought tolerance	Grassland, amenity reclamation Cold and temperate regions
F. ovina L. 2x = 28 outbreeding perennial	Sheep's fescue Winter hardy and persistent	Grassland, amenity Cold and temperate regions
L. perenne × *L. multiflorum* 2x = 28 *L. multiflorum* × *F. pratensis* 2x = 28 *L. perenne* × *F. pratensis* 2x = 28 outbreeding biennials or perennials	Tetraploid *Lolium* and *Festulolium* hybrids Combine yield, palatability, nutritional, establishment, tolerance, and hardiness characteristics	Grassland Temperate regions antipodes
F. gigantea (L.) Vill 2x = 42 *F. apennina* 2x = 28 *F. drymeja* Mert. & Koch. 2x = 14 *F. donax* 2x = 14 outbreeding perennials	Giant fescue No common name " "	Cytological research and hybridization with other species

Table 2. Sources and numbers of genotypes identified as being capable of green shoot or plant regeneration from protoplast cultures

F. arundinacea	*L. perenne*	*L. multiflorum*
cv. Nanryo 1 line, single genotype (Takamizo et al. 1990)	Lp 9a 1 line, 2–5 mixed genotypes (Creemers-Molenaar et al. 1989)	cv. Trident 1 line, 25–30 mixed genotypes (Dalton 1988b)
cv. S170 8 lines, 25–30 mixed genotypes (Dalton 1988a)	cv. Aurora 1 line, 25–30 mixed genotypes (Dalton 1988b)	cv. Trident 5 lines, single genotype (Dalton 1988c)
cv. S170 23 lines, 1 selected[a] genotype (Humphreys and Dalton 1991; Dalton 1988c)	cv. Aurora 6 lines, 1 selected[a] genotype (Dalton 1988c)	At least 2 responsive genotypes
cv. S170 2 lines, 1 selected[a] genotype (Dalton 1988c)	cv. Verna 1 line, single genotype (Dalton 1988c)	
cv. Yamani single genotype (Wang et al. 1992)	cv. S23 1 line, single genotype (Dalton unpubl.)	
cv. Fawn single genotype (Wang et al. 1992)	cv. Belfort (tetraploid) 1 line, single genotype (Dalton unpubl.)	*F. rubra*
cvs. Roa, Olga, Baral single genotypes (Spangenberg, pers. comm.)	Lp 22K 1 line, 40 mixed genotypes (Creemers-Molenaar et al. submitted b)	cv. Dawson 1 line, single genotype (Zaghmout and Torello 1990)
At least 14 responsive genotypes	At least 6 responsive genotypes	At least 1 responsive genotype

[a] Single cell suspension regenerent from a mixed genotype cell suspension.

gamma-irradiated *L. perenne* suspension protoplasts after polyethylene glycol (PEG)-treatment have also been reported by Creemers-Molenaar et al. (submitted a).

Regenerating protoplast cultures have hitherto been the only potential means of transferring foreign genes into *Lolium* and fescue species by genetic transformation. Nonregenerating *L. multiflorum* suspension protoplasts were transformed with a kanamycin resistance gene using PEG by Potrykus et al. in 1985, and transformation of the resultant calli was confirmed by Southern blot characterization. Unfortunately, natural resistance to many antibiotics makes selection of resistant colonies in grass cultures very difficult except at the plant level, when various antibiotics inhibit root growth (Lee et al. 1989). However, none of 137 *L. perenne*, 111 *F. arundinacea*, and 9 *L. multiflorum* plants tested for root growth after regenerating from similarly treated protoplasts grew roots in the presence of kanamycin (Dalton, unpubl.).

The development of nonselective reporter genes such as GUS (β-glucuronidase) (Jefferson et al. 1987) and the isolation of the ACT1 promotor from rice (McElroy et al. 1990), and ADH1 (Dennis et al. 1984) and pEmu (Last et al. 1991) from maize have led to improved transient expression in various grass cultures and cell suspension protoplasts, but stable integration and expression have only recently been reported. In this work, Wang et al. (1992) regenerated transgenic *F. arundinacea* plants from cell colonies expressing hygromycin (*hph* gene) or phosphinothricin resistance (*bar* gene) after PEG-mediated transformation of protoplasts isolated from young embryogenic cell suspensions. Transformation in the plants was confirmed by Southern hybridization analysis, HPH enzyme assays or phosphinothricin-herbicide spraying and in one case in situ hybridization of the labelled *hph* gene to metaphase chromosomes of a transformed plant.

A by-product of developing regeneration systems has been the production of variant plants (Humphreys and Dalton 1991). Somaclonal variation can be a problem or a benefit depending on the application. Certain causatory factors are becoming apparent (reviews: Stafford 1991, Karp 1991) so that efforts can be made to control and to use the variation. For instance, a *L. temulentum* cell suspension regenerent with a delayed greening response has been used in genetic and developmental regulation studies (Ougham et al. 1992). *Lolium-Festuca* sexual hybrid plants regenerated from cell suspension cultures show a higher frequency of multivalents at meiosis than uncultured plants, and this has applications in the introgression of useful fescue genes into the more economically important *Loliums* (Humphreys and Dalton 1992).

2 Isolation of Protoplasts

2.1 Sources of Protoplasts

Regenerating cell suspensions have been the source of protoplasts in all the published reports of plant regeneration from *Lolium* and *Festuca* protoplasts (Table 3). Other sources such as callus are generally too slow-growing to yield many protoplasts, while protoplasts isolated directly from meristematic explants such as immature leaves, shoot-tips, immature inflorescences, or from immature embryos are difficult to isolate in any number, are extremely small ($< 10\ \mu m$), and too delicate to manipulate (Dalton unpubl.). However, there is currently interest in isolating haploid protoplasts from sperm and egg cells of *Lolium perenne* for artificial fertilization (den Nijs et al. 1991), although no callus or plant regeneration has yet been reported.

2.2 Cell Suspension Initiation

As the source of protoplasts, the quality of the cell suspension is by far the most important factor for successful regeneration. A cell suspension which cannot regenerate will not yield regenerating protoplasts. Particular attention must therefore be paid to the cell suspension quality.

Cell suspension cultures have usually been initiated from embryogenic cultures induced from meristematic explants or embryos. A common feature of most

Table 3. Plant regeneration from protoplasts isolated from cell suspension cultures initiated from various sources either directly from explants or via callus

Species	Initial explant	Regeneration	Reference
F. arundinacea	Chopped mature embryos	Green plants	Dalton 1988a, b
F. arundinacea	Single shoot tips	Green plants	Dalton 1988c
F. arundinacea	Shoot-tip callus	Green plants	Dalton 1988c
F. arundinacea	Immature embryo callus	Albino shoots	Takamizo et al. 1990
F. arundinacea	Basal leaf callus	Green plants	Takamizo et al. 1990, Wang et al. 1992
F. rubra	Mature embryo callus	Green shoots	Zaghmout and Torello 1990
L. perenne	Chopped mature embryos	Green shoots	Dalton 1988a, b Creemers-Molenaar et al. submitted b
L. perenne	Single shoot tips	Green plants	Dalton 1988c
L. perenne	Shoot tip callus	Green plants	Dalton 1988c
L. perenne	Immature inflorescence callus	Green plants	Creemers-Molenaar et al. 1989
L. perenne	Mature embryo callus	Albino shoots	Zaghmout and Torello 1992b
L. multiflorum	Chopped mature embryos	Green plants	Dalton 1988a, b
L. multiflorum	Single shoot tips	Green plants	Dalton 1988c
L. multiflorum	Shoot tip callus	Green plants	Dalton 1988c
L. temulentum	Shoot tip callus	Albino shoots	Dalton unpubl.

reported methods of inducing embryogenic proliferation is the use of 1–3 mg/l 2,4-dichlorophenoxyacetic acid (2,4-D), with the exception of *L. perenne*, which generally requires a higher level (3–6 mg/l), although selected genotypes will proliferate on medium containing 2 mg/l 2,4-D. This has important implications for albino shoot formation, which is increased from calli cultured at high 2,4-D levels (Creemers-Molenaar et al. 1988).

Vegetative shoot tips are small and do not develop as rapidly as immature inflorescences. However, the latter must be induced and then cultured at the right size, while vegetative shoot tips can be dissected at any season from surface-sterilized in vivo plants or from sterile in-vitro plantlets.

With the exception of *L. temulentum*, all the species under discussion are outbreeding. An embryo therefore comprises a unique genotype, and stock plants of responsive genotypes must be regenerated from culture with the risk of somaclonal variation if embryos are used. However, initiation of cell suspensions with groups of immature or chopped mature embryos placed directly in liquid culture is a simple method of genotype selection, because few genotypes proliferate and regenerate. Plants that are regenerated are often subsequently very responsive in culture (Fig. 1).

Variation between species, cultivars, genotypes, and lines has been noted (Dalton 1988b, and unpubl.) both in the proportion of cultures which can regenerate and in the longevity of the regeneration response. A rapid loss in regeneration potential of cell suspension cultures is reported in both *Festuca*

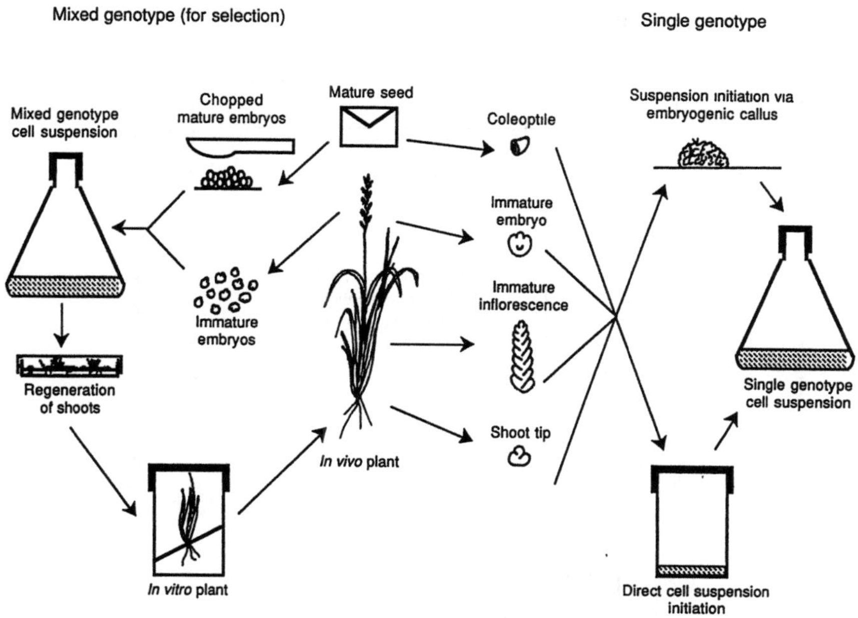

Fig. 1. Methods of cell suspension initiation: mixed and single genotype; direct and via callus

(Dalton 1988b; Takamizo et al. 1990; Radilofe-Rajoelina et al. 1990) and in *Lolium* (Dalton 1988b; Creemers-Molenaar et al. 1989), with loss of green shoot regeneration typically occurring within 30 weeks of cell suspension initiation (Fig. 2). It is therefore important to establish suspensions as quickly as possible, so that protoplasts can be isolated before regenerative ability is lost.

Criteria for the selection of suitable calli are difficult to define and even within a single genotype of *L. multiflorum*, relating embryogenic callus characteristics to subsequent cell suspension initiation and regeneration characteristics proved inconclusive. Only 7 out of 24 (29%) suspensions were capable of regenerating green shoots when all the tissue was plated 5 weeks after initiation, but the embryogenic calli from which these were initiated had no common feature. The two cell suspensions from which the highest numbers of green shoots were regenerated also produced albino shoots. The growth of these two suspensions was significantly slower than the others, even lines from which only green shoots were regenerated.

In a comparison of mixed embryo suspensions of different species, 51% of 53 *F. arundinacea* (cv. S170) suspensions, 70% of 10 *L. perenne* (cv. Aurora) suspensions, and 41% of 32 *L. multiflorum* (cv. Trident) suspensions were capable of green plant regeneration. However, all green regeneration ceased in the *L. perenne* suspensions plated after 30 weeks, although some *F. arundinacea* and *L. multiflorum* suspensions continued regenerating up to 50 weeks. In all three species the percentage of cell suspensions regenerating was halved between 15 and 20 weeks.

In another screen of mixed embryo suspensions from 30 *F. arundinacea* seed accessions, 13 (43%) were considered "good" (fast-growing, regenerative) after 4 weeks, while 45% were capable of green regeneration. However, only two lines (5%) were still capable of green shoot regeneration at 15 weeks.

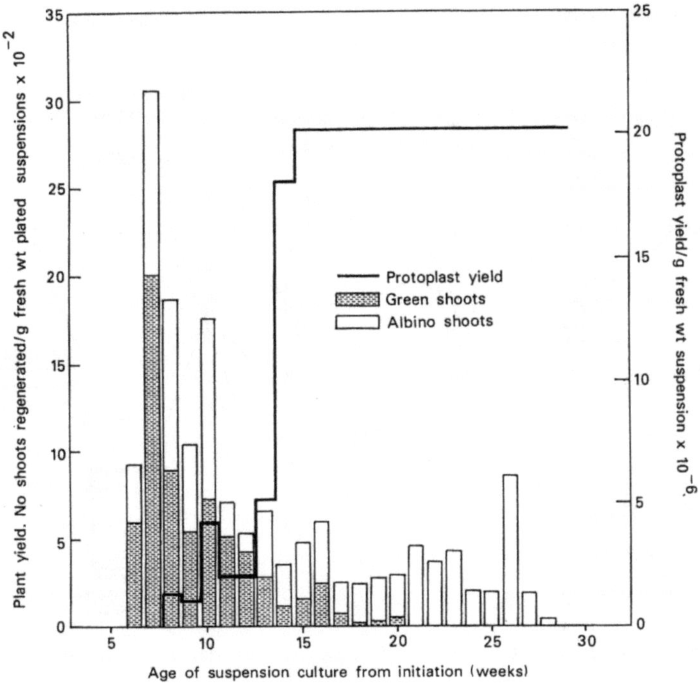

Fig. 2. Stability of plant regeneration and yield of isolated protoplasts from embryogenic cell suspension cultures of *Lolium multiflorum*

Table 2 lists the responsive cultivars of the species from which green shoots have been regenerated from cell suspension protoplasts. In a screen of ten *L. perenne* cultivars 20–80% of genotypes produced embryogenic callus depending on cultivar, but only 0–20% produced "good" suspensions after 3 weeks. A higher proportion of genotypes from four tetraploid cultivars screened were embryogenic and produced "good" suspensions than genotypes from six diploid cultivars. However, while five out of eight "good" diploid cell suspensions were capable of protoplast cell colony formation and one was capable of green plant regeneration, only one out of eight tetraploid suspensions produced protoplast colonies and green plants. Depending on the cultivar, 0–13% of genotypes were capable of colony formation and 0–8% of genotypes could regenerate green shoots. Overall, however, only 0.9% of the genotypes produced green plants from protoplasts (Fig. 3).

Within a genotype, regeneration loss generally occurs at about the same time (Table 4), whether shoot tips are first induced to form an embryogenic culture or are placed directly in liquid medium. However, although protoplasts can typically

Fig. 3. **A** Fresh *Lolium multiflorum* protoplasts cv. Trident, isolated from an 8-week-old cell suspension initiated with shoot-tip callus (×200). **B** Protoplast colonies of *L. perenne* cv. Aurora, 14 days after isolation from an 8-week-old cell suspension initiated with shoot-tip callus (×400). **C** Protoplast colonies of *L. multiflorum* cv. Trident, 14 days after isolation from an 8-week-old cell suspension initiated

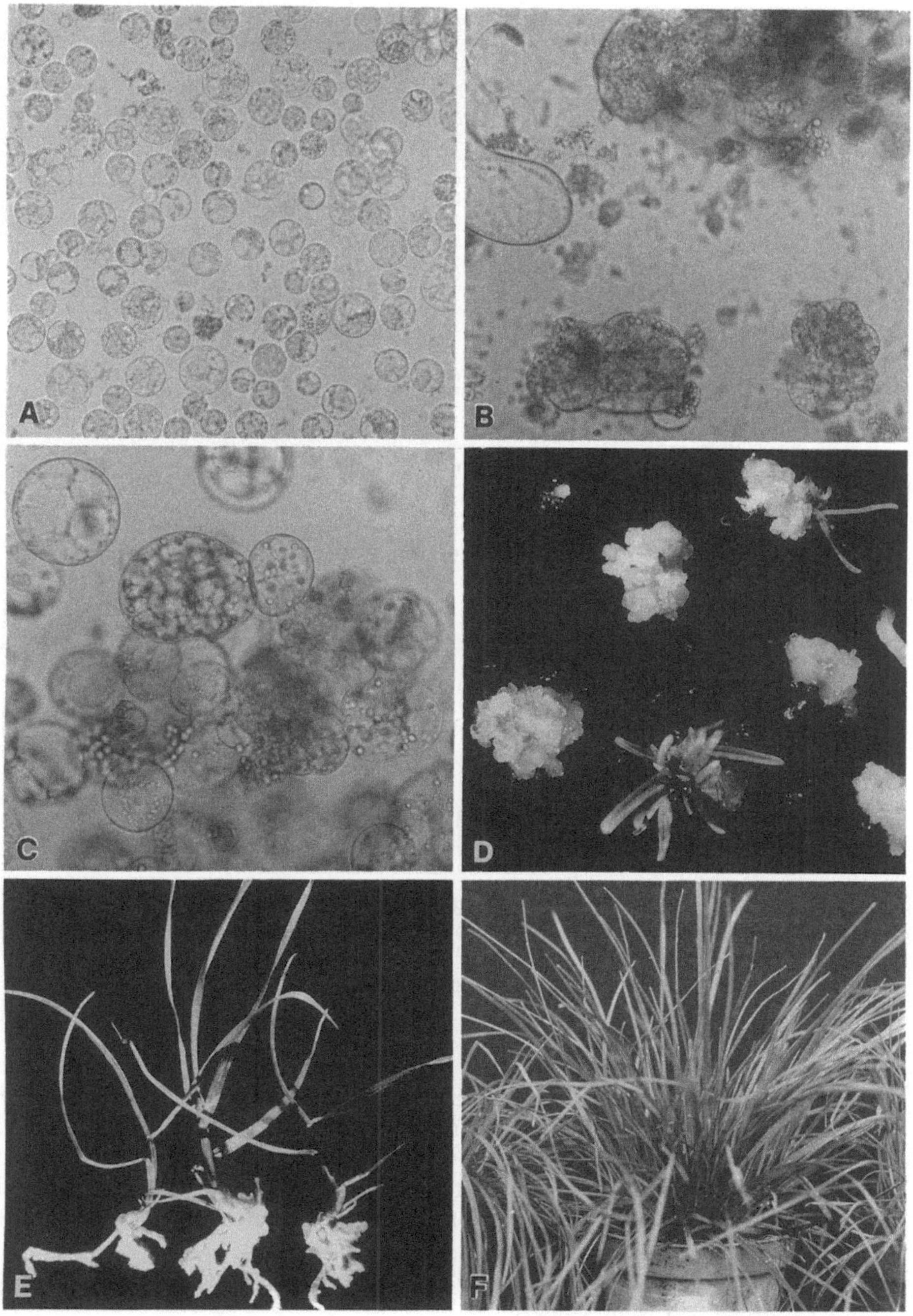

with shoot-tip callus (× 400). **D** Regenerating cell colonies of *L. perenne* cv. S23 derived from protoplasts isolated from a 7-week-old cell suspension initiated with shoot-tip callus (× 2). **E** Plantlets of *L. multiflorum* cv. Trident derived from protoplasts isolated from a 27-week-old cell suspension initiated with chopped embryos (× 1). **F** Established *L. perenne* plant cv. Aurora derived from protoplasts isolated from a 22-week-old cell suspension initiated with shoot-tip callus (× 0.5)

Table 4. Effect of different methods of cell suspension initiation on regeneration of plants from protoplasts isolated from 1 ml volumes of cell suspension colonies of a single selected *Festuca* genotype

Tissue used to start suspension	Shoot tips	Shoot tips in med 133[a]	Shoot tips in med 192[b]	Callus from CS colonies	Shoot tip callus
No. of lines	9	4	5	4 best lines	5
No. of isolations	25	13	12	11	18
Range in weeks after cell suspension started	12–18	12–18	12–18	5–9 14–18 in total	9–13 13–17 in total
Isolations from which green shoots were regenerated (%)	52%	54%	50%	91%	89%
Mean number of regenerating colonies per isolate	9.8	8.0	12.0	7.1	18.1
Isolations from which white shoots were regenerated	40%	38%	42%	100%	94%
Mean number of regenerating colonies per isolate	2.9	3.4	2.4	6.6	5.9
Isolations from which any shoots were regenerated (%)	60%	62%	58%	100%	94%
Mean number of regenerating colonies per isolate	10.5	9.1	12.0	13.1	22.9
Ratio of colonies regenerating green: white shoots	4.4:1	3.3:1	6:1	1:1	2.9:1

[a] MEDIUM 133 = MS medium 3% sucrose.
[b] MEDIUM 192 = MS medium 3% sucrose, 1.6% mannitol.

be isolated a few weeks sooner using the direct method, only 50% of lines were responsive compared to 90% where embryogenic cultures were selected for suspension initiation. Restarting cell suspensions from plated suspension colonies of good lines prolonged the regenerative life, although the ratio of albino colonies was increased. Increasing the osmotic potential of the culture medium with mannitol also improved subsequent protoplast regeneration in this experiment.

2.3 Cell Suspension Maintenance

Various media containing 2–3 mg/l 2,4-D: MS (Murashige and Skoog 1962), AA medium (Muller and Grafe 1978), and N6 medium (Chu et al. 1975) have been successfully used to maintain *Lolium* and *Festuca* cell suspension cultures. However, higher levels of 2,4-D (5–10 mg/l) are usually necessary to maintain growth and non-differentiation in *L. perenne* cultures, although 3 mg/l is sufficient for certain responsive genotypes. There is little advantage in using sugars other than sucrose, but increasing the osmotic potential with mannitol is sometimes helpful for regeneration, while Zaghmout and Torello (1992a) found temporarily increasing the sucrose level to 180 mM (6.2%) improved green shoot regeneration in year-old *F. rubra* suspension cultures.

2.4 Enzymes

The published methods for *Festuca* and *Lolium* protoplast culture differ widely in the mixtures of enzymes used, but a common feature is the use of Cellulase RS (Onozuka) which enables protoplasts to be isolated from material which would be relatively untouched by Cellulase R10. Mixtures containing the pectinase Pectolyase Y-23 are also generally more effective than those containing Macerozyme R10 alone, although Takamizo et al. (1990) used Macerozyme R10 to isolate *Festuca* protoplasts. The cells were also incubated statically rather than with gentle agitation and consequently the protoplasts took longer to be released. In the solution used by Dalton (1988a), most protoplasts are released within 6 h and those released between 6 and 24 h did not form cell colonies.

2.5 Other Factors Influencing Protoplast Yield

Apart from enzyme mixture, the major factors influencing yield are suspension type and age, speed of growth, and time of isolation. Protoplasts are usually isolated at the time of maximum growth rate, which in a suspension subcultured weekly, would normally be between days 3 and 4. Yield falls rapidly after 5 days.

Suspension type is influenced by species and embryogenic *Festuca* suspensions are more "friable" and easier to isolate protoplasts from than the more nodular *Lolium* suspension colonies. *Lolium temulentum* suspensions are very nodular and yields are consistently less than 5×10^6/g fr. wt. Mean suspension colony size decreases with age and the colonies become less compact, thereby allowing more enzyme activity and higher yields, but this is concurrent with loss of totipotency (Dalton 1988c). Fast-growing cell suspensions also yield more protoplasts, but inducing cell suspensions to grow faster, by increasing the dilution ratio or frequency of subculture causes colony types which are losing totipotency to be selected. Subjective assessments of "good" lines can also be inaccurate (Table 5). Variation in these factors is probably responsible for the differences in protoplast yield and plating efficiency observed between suspension lines initiated at the same time from the same genotype.

2.6 Purification Procedures

After incubation in enzyme, protoplasts must be passed through mesh to separate undigested cells. Protoplasts of the hexaploid species *Festuca arundinacea* are

Table 5. Results of subjective assessment of *Festuca arundinacea* cell suspension lines on yield, colony formation, and regeneration from protoplasts

Subjective assessment of *F. arundinacea* cell suspensions	Mean yield of protoplasts per g fr. wt.	Planting efficiency (%)	No. of colonies with green shoots per million protoplasts	No. of colonies with albino shoots per million protoplasts
Poor n = 5, 1 line	9.0×10^6	0.07 ± 0.05	0.8 ± 0.5	0.3 ± 0.7
Reasonable n = 23, 5 lines	9.7×10^6	0.08 ± 0.04	1.2 ± 1.0	0.3 ± 0.04
Good n = 5, 1 line	10.2×10^6	0.10 ± 0.02	1.8 ± 1.6	0.0 ± 0
Very good n = 7, 1 line	14.6×10^6	0.21 ± 0.04	0.6 ± 0.6	1.4 ± 2.0
	$p = 0.238$	$p = 0.000$	$p = 0.235$	$p = 0.025$

considerably larger (15–50 µm) than those of diploid *Lolium* species (10–25 µm) (Dalton 1988a). Takamizo et al. (1990) therefore used only a 50 µm mesh to purify *Festuca* protoplasts, but *Lolium* protoplasts definitely require a smaller mesh size (22.5 µm) to remove unwanted cells. Experiments to culture *Festuca* and *Lolium* cell debris in protoplast medium have never yielded any colonies, however, because the cells rapidly plasmolyze in such a high osmotic potential medium. Contaminating cells may therefore be less of a problem than was originally considered by tissue culturists.

Protoplasts are then washed several times in either culture medium or special washing solutions such as CPW salts (Frearson et al. 1973) or even a simple mixture of mannitol and $CaCl_2$ (Takamizo et al. 1990). The advantage in using special washing solutions is that they are simple to make and are autoclavable, while the density of culture media can cause the protoplasts to be difficult to pellet by centrifugation, especially media containing di-saccharides such as maltose. High density solutions (e.g., 21% sucrose in water) can be used to purify protoplasts by flotation. However, experiments indicate that there is little difference in the plating efficiency of the fractions which float or pellet although the floated fraction is freer of cell debris. Heat shocks have been used with advantage in rice (Abdullah et al. 1986) but the results of heat shock experiments have been erratic in *Festuca*, although they may be of benefit to *Lolium* protoplasts.

3 Culture of Protoplasts

3.1 Protoplast Density

Protoplasts have been cultured at initial densities of 10^5 to 10^6 per ml, but the density immediately declines more or less rapidly as the majority of protoplasts die

Table 6. Effect of protoplast density on colony formation and albino shoot regeneration in *Lolium perenne* protoplasts

n = 2, 1 line	Plating efficiency %	No. of colonies with shoots per million protoplasts	No. of shoots per million protoplasts	Ratio of shoots per colony
Cultured at 1×10^5 per ml	0.45 ± 0.20	104.4 ± 21.2	330 ± 81	3.2 ± 0.3
Cultured at 2×10^5 per ml	0.37 ± 0.08	31.0 ± 3.3	61 ± 12	2.0 ± 0.4
	$p = 0.498$	$p = 0.000$	$p = 0.001$	$p = 0.004$

or do not develop. Typically, a high number of regularly small colonies or a small number of variable colonies are formed with similar total cell dry-weights.

Within a line, comparisons between plating densities show that in cultures capable of high plating efficiencies, denser platings may not have a lower efficiency, but that colonies do not develop due to nutrient shortage which can reduce shoot regeneration (Table 6). Conversely, in marginal cultures, increasing the plating density can frequently mean the difference between a few colonies developing or none at all, indicating that protoplasts have a conditioning effect.

3.2 Nurse Cultures

Cells in cell suspension culture are in close proximity and the volume ratio of cell to medium may be 1:15. However, cultured protoplasts are physically separated and the cell : medium ratio may be only 1:40. Nurse cultures of cell suspension cells in close proximity, but physically separate from protoplasts, have been successfully used to improve plating efficiency in *L. perenne* (Creemers-Molenaar et al. 1991, Zaghmout and Torello 1992b) and in *F. rubra* (Zaghmout and Torello 1990) and *F. arundinacea* (Takamizo et al. 1990). Interestingly, in the latter work cell suspension cells of a different and nonmorphogenic line were more successful as nurse cells than the morphogenic parental protoplast line.

Attempts to culture single cells and small cell groups of 25–90 μm indicate that the high osmotic potential of protoplast media quickly plasmolyzes and prevents proliferation of intact cell suspension cells (Dalton 1988a). Nurse cells are therefore unlikely to be growing actively and are under stress themselves. However, experiments to grow cell suspensions and protoplasts in media of different osmotic potentials, indicate that in terms of growth, nonmorphogenic suspensions adapt more quickly and are more tolerant of extreme conditions than morphogenic suspensions (Dalton unpubl.).

3.3 Culture Vessel and Volume

Protoplasts have been cultured in various vessels, but with similar depths of agarose medium (0.9–1.2 mm). Warm agarose medium spreads easily and smaller volumes can be dispensed than with liquid medium. The smaller the vessel, however, the greater the loss of water through evaporation, although Creemers-Molenaar et al. (1989) were successful in culturing protoplasts in 35-mm Petri dishes. However, protoplasts cultured in the outer 20 mm square compartments of 25 compartment square Petri dishes (Sterilin) have sometimes failed through excess evaporation.

4 Culture Media and Conditions

4.1 Minerals and Vitamins

The published work shows that grass protoplasts will survive in standard media ranging from the relatively simple B5 medium (Gamborg et al. 1968) to the rich and complex Kao's medium (Kao 1977). For selected cultures the more simple media are adequate and may even discourage the formation of too many friable colonies (Dalton 1988a). However, Kao's medium can mean the difference between a few colonies forming and complete failure in situations where protoplasts have been difficult to isolate, such as those from very young cultures or from very nodular suspensions like those formed by *Lolium temulentum*. This medium therefore acts in a way similar to increasing plating density for difficult cultures.

4.2 Carbohydrates, Osmotica and Growth Regulators

Carbohydrates are usually the largest single ingredient to provide energy and osmotic stability. High osmotic potential is extremely important for the culture of grass protoplasts. Young suspension protoplasts of *Festuca* and *Lolium* do not survive unless cultured in a medium of at least 890 mOsm, although older cell suspension protoplasts form more colonies in media of 835 and even 700 mOsm.

Glucose or glucose with mannitol was used in all the published work, but replacement of glucose by maltose has improved both colony formation and regeneration in *Festuca* and *L. multiflorum*, while addition of 11% glucose as the sole osmoticum without mannitol improved colony formation in *Lolium perenne* although not significantly. Kao's or MS medium containing 13% maltose and 5% mannitol (890 mOsm) has also been used very successfully for regenerating *L. perenne* protoplasts. In fact, *Lolium perenne* protoplasts may require slightly higher osmotic potentials than either *F. arundinacea* or *L. multiflorum* and Creemers-Molenaar et al. (1989) used RY-2 medium (Yamada et al. 1986) adjusted to 920 mOsm with glucose while PC4 medium (Dalton 1988a) containing 7% glucose (900 mOsm) also improved shoot formation in *L. perenne*, although again not significantly. *Festuca* and *L. multiflorum* protoplasts, however, have been more responsive in MS medium containing 12% maltose and 5% mannitol (840 mOsm).

During protoplast culture, the osmotic potential of the medium is increased by evaporation, hydrolysis of di-saccharides, and the release of cell contents from disintegrating protoplasts, but is decreased by cell growth and division, starch accumulation, and the precipitation of media constituents.

Experiments to determine the changes in osmotic potential over the culture period indicate that the media starting at initial osmotic potentials of 880–920 mOsm, glucose based medium declines in osmotic potential by about 200 mOsm over 8 weeks while that of maltose-based medium rises by about 180 mOsm. By contrast, sucrose is progressively hydrolyzed, thus doubling its osmotic contribution, so that the osmotic potential of sucrose based medium rises by 660 mOsm.

Calculations to partition the effects of the various factors indicate that protoplasts grown in sucrose-based media accumulate more starch than maltose-based cultures of the same initial osmolarity. Through hydrolysis, sucrose-based cultures can increase in osmotic potential so much that the cells are killed (anything over 1200 mOsm).

2,4-D is usually used to promote cell division at levels of 0.1 to 1 mg/l. The higher level definitely promotes cell colony formation, but does not necessarily improve rates of regeneration.

4.3 Agarose

Protoplast colonies developing in liquid media are easily transferred to other media or harvested, e.g., for transient gene expression studies. However, colony counting is more accurate in agarose media because individual colonies are kept intact and separate and the risk of chimeral colony formation is reduced. Protoplast colonies forming on the surface of agarose are usually more morphogenic than those embedded within the agarose, indicating that physical constraints and aeration can affect morphogenesis.

Sea plaque agarose has been shown to be the best agarose for protoplast culture (Lorz et al. 1983). However, plating efficiency of *Festuca* protoplasts was seriously reduced by the addition of 0.5% agarose, although *Lolium* protoplasts were less affected (Table 7a). The addition of 1 to 1.2% agarose does not usually seriously reduce protoplast colony formation in *F. arundinacea* and can improve it in young cultures (Table 7b).

4.4 Dilution

Protoplasts are normally cultured for 4 to 6 weeks before colonies are transferred to a culture medium of lower osmotic potential and cultured in the light. Experiments to dilute protoplasts already in liquid medium with medium of lower osmotic potential failed to show any benefit, except in cases where numerous colony formation was depleting the medium. However, in agarose culture, transferring colonies embedded in agarose pieces to a larger Petri dish containing liquid medium (540 mOsm) at 4 weeks, improved both colony growth and embryogenesis compared with adding medium on top of the agarose or not diluting at all. The subsequent softening of the agarose in liquid medium also enabled colonies to continue growing and to break out of the agarose after transfer to solid medium and light at 6 weeks, rather than to remain embedded.

Table 7. Effect of agarose on (a) plating efficiency (%) in protoplast cultures of *Lolium* and *Festuca* and (b) plating efficiency (%) and regeneration in *Festuca*

a)

	F. arundinacea	*L. perenne*	*L. multiflorum*
30-week-old suspensions	4 lines	1 line	2 lines
Liquid medium	0.12 ± 0.13	0.17 ± 0.05	0.19 ± 0.17
0.5% Agarose	0.02 ± 0.02	0.18 ± 0.04	0.09 ± 0.10
	n = 12 p = 0.009	n = 6 p = 0.747	n = 6 p = 0.202

b)

F. arundinacea 12 weeks old 3 lines	Plating efficiency %	No. of colonies per million protoplasts with green shoots	No. of colonies per million protoplasts with white shoots
Liquid medium n = 18	0.02 ± 0.03	1.7 ± 2.2	0 ± 0
1.2% Agarose n = 10	0.12 ± 0.08	16.5 ± 23.0	1.0 ± 2.1
	p = 0.010	p = 0.090	p = 0.201

4.5 Culture Conditions

All the published methods cultured the protoplasts at 25 °C in the dark for 4 to 6 weeks. Transferring colonies to the light too soon causes browning or anthocyanin pigmentation and a loss of morphogenesis. However, maturation of embryoids after transfer to a lower osmotic potential medium can be helped by bright light, although Creemers-Molenaar et al. (1989) continued to subculture the colonies in the dark for several passages.

5 Regeneration from Protoplast Colonies

Embryogenesis and regeneration cannot be induced unless the protoplast colonies are genetically capable. Analysis of colonies which did not regenerate (Humphreys and Dalton 1991) indicate that somaclonal variation is responsible for many colonies developing as friable or nodular callus or forming degenerate "structures" and root hairs. The formation of albino shoots also indicates that a less extreme form of variation has occurred.

Protoplast colonies are substantially similar to cell suspension colonies and no longer require high osmotic potentials to keep cell membranes stable. Growth may be arrested, unless the osmotic potential is lowered by transfer to another medium. However, unsuitable media or poor conditions, such as excess condensation or overcrowding of colonies, can cause otherwise competent protoplast colonies to brown and die rather than form embryoids and subsequently shoots.

In experiments to look at various factors affecting embryogenic maturation and regeneration of shoots (Dalton and Thomas 1992), the use of Gelrite and a medium of a relatively high osmotic potential were the most important factors for the maturation of embryoids to a stage where germination was possible after transfer to regeneration medium. In this context it is important to assess protoplast treatments on the proportion of protoplasts which formed colonies capable of forming shoots, rather than comparing simply the numbers of shoots (Table 6). The number of separate shoots emerging from single protoplast colonies can vary widely (from 1 to 50) depending on subsequent protoplast colony treatment, time of scoring, and somaclonal variation. Many published results on plant regeneration are confusing in this respect and quote only total numbers of shoots.

6 Somaclonal Variation in Regenerated Plants

It is important that regenerated plants are normal and fertile. However, somaclonal variation amongst regenerated plants has been detected in iso-enzyme pattern, chromosome number, and general morphology.

Variation in iso-enzymes was common between *F. arundinacea* plants regenerated from suspension cultures (11%) and increased with time in culture (Humphreys and Dalton 1991), but variation occurred rarely in either cell suspension or protoplast regenerents of *L. perenne* or *L. multiflorum* (unpubl.). Many of the plants also contained fewer than the normal chromosome number, but *F. arundinacea*, being hexaploid, can apparently tolerate more somaclonal variation than diploid *Lolium* without losing the ability to regenerate plants. Despite this, a high percentage of *F. arundinacea* protoplast calli which were unable to regenerate plants were abnormal (29%), although protoplast-derived plants and albino shoots had the normal iso-enzyme phenotype. There was also no variation between plants derived from different shoots from the same protoplast colony (unpubl.). However, abnormalities could have been masked by the hyperploidy found in many of the regenerents studied.

Chromosome doubling was common in protoplast-derived regenerents of all three species, although there was evidence that in experiments in which plants were regenerated from relatively large numbers of colonies, the chromosome numbers tended to be normal. Only plants with normal chromosome numbers set seed, although several otherwise normal *F. arundinacea* regenerents produced viviparous tillers. Chromosome doubling has also been detected in plants regenerated directly from *L. perenne* cell suspensions, indicating that there are mechanisms other than protoplast fusion causing endo-polyploidy.

The most common types of morphology change were variegated or very dark leaves, stunted growth, premature death, increased disease susceptibility, and abnormal inflorescence development. In a comparison of 420 plants regenerated from 44 protoplast colonies of *L. multiflorum*, abnormal inflorescence development was the most common morphological change and the percentage of abnormal plants varied from 10–33% depending on the number of shoots regenerated per colony. There was a trend for a higher percentage of abnormal plants to be regenerated from colonies producing only one to three shoots, indicating that regeneration had been difficult, perhaps due to somaclonal variation. However, in colonies producing 4–50 shoots, at least one plant was always abnormal, indicating that somoclonal variation also occurred at the colony maturation stage.

There was also phenotypic variation, particularly in leaf color and disease susceptibility, amongst 24 *L. perenne* protoplast-derived plants examined. The amount of variation amongst protoplast regenerents was considerably higher than amongst regenerents from cell suspension colonies plated at the same time protoplasts were isolated (25% normal phenotype protoplast regenerents, 97% normal phenotype cell suspension regenerents in *L. perenne*; 21 and 100% respectively in *L. multiflorum*).

However, while 89–100% of protoplast regenerents of *L. multiflorum* and *L. perenne* produced abundant inflorescences and 83% and 88% respectively were fertile, even though sometimes abnormal in appearance, only 22% of 175 *F. arundinacea* regenerents produced inflorescences, although the plants were otherwise normal in appearance. By comparison, inflorescences were produced by 54% of 405 plants regenerated from cell suspension colonies plated at the same time that protoplasts were isolated. The mean number of heads per plant was also reduced from 4.4 in suspension-derived plants to 2.5 in protoplast-derived plants. Mean inflorescence length was significantly shorter in the protoplast-derived plants (129 mm compared to 143 mm, $p = 0.018$) and the proportion of plants with inflorescences which set viable seed was halved (25 vs. 13%).

7 Summary and Conclusions

Grass-breeders continue to breed for increased yield over a wide variety of managements, improved nutritional quality, wider distribution through improved tolerance to climatic factors and for lower losses in the field or during conservation through increased disease and pest resistance. There are therefore many potential applications for somatic hybridization and genetic transformation in *Lolium* and *Festuca* – somatic hybridization to combine or introgress the stress tolerances of the various *Festuca* species with the productivity, nutritional quality, and palatability of *Lolium* species and genetic transformation to insert genes conferring greater disease and pest resistance. To advance these applications, more fundamental research is continuing into the nature of cell wall synthesis, cell division, protein synthesis, virus infection, and stress responses using *Lolium* and *Festuca* protoplasts.

Regeneration is routinely possible from genotypes of the three most economically important *Lolium* and *Festuca* species, although regeneration efficiency is still less than ideal for the efficient regeneration of somatic hybrids or transformants. However, improvements in protoplast culture protocols, particularly in the quality of cell suspensions, in the understanding of the important role of osmotic potential, and in the regeneration of plants from protoplast colonies, have made it worthwhile to attempt genetic manipulation methods.

Many *Lolium* and *Festuca* species are capable of sexual hybridization and protoplasts of potential parent species seem able to thrive in the same media. Successful somatic hybridization between two species which readily sexually hybridize has been reported by Takamizo et al. (1991) and the applications can therefore be expected to widen as the technology becomes more routine and automated. Particular applications may be the introgression of stress-related genes (such as drought resistance) into a single parental background rather than complete somatic hybridization and cybridization to transfer cytoplasmic traits such as mitochondrial male sterility.

Transient genetic transformation with the marker gene GUS in *Lolium* and *Festuca* protoplasts have been reported by Last et al. (1991) and Hale (1988), and stable transformation of *Lolium* protoplasts with kanamycin resistance by Potrykus et al. (1985). However, none of these early results was obtained with protoplasts from cell suspensions capable of plant regeneration and, in the meantime, other transformation methods have been developing.

Surrogate transformation using genetically modified strains of the *Lolium* endophyte *Acremonium* to infect plants and subsequently the seed has been reported by Murray et al. (1992) and endophytes of the same genus also occur in *Festuca* (Kearney et al. 1991). Attempts have also been made to repeat work in maize (Gould et al. 1991) and to transform grass tissues using *Agrobacterium*-mediated methods. However, it is the bombardment of embryos and embryogenic cultures with DNA-coated particles which has been most successful in terms of transient gene expression and has worked in numerous *Lolium* and *Festuca* species (Dalton in prep). Regeneration of plants from tissues transformed by this method has yet to be reported, but in any case, direct transformation of regenerable protoplasts may still prove necessary for transformation with genes which are difficult to detect or select at the whole cell level and the number of transformation events may also be higher than with other methods. Protoplast-derived plants are also less likely to be chimeral than plants derived from tissues.

The recent work of Wang et al. (1992) shows that transformation of *Festuca* protoplasts and subsequent regeneration of plants is possible using young embryogenic cultures and appropriate selection methods. Similar results in *Lolium* should soon follow. However, the consequences of releasing genetically modified *Lolium* and *Festuca* cultivars requires careful consideration, because of the vast quantities of pollen produced by these species and the ubiquity of wild relatives.

The main technical problem in these genetic manipulation procedures is now the high incidence of somaclonal variation found in regenerents both from cell suspensions and from protoplasts of *Lolium* and particularly of *Festuca.* The rapid build-up of variation also reduces the useful lifespan of cultures, while the low fertility of many regenerated plants does not augur well for presenting breeders with range of fertile somatic *Lolium-Festuca* hybrids or transformants.

Various treatments, such as the use of o-acetylsalicylic acid and anti-oxidant enzymes, are reported to improve colony formation from *Lolium* protoplasts (Creemers-Molenaar et al. submitted, b) perhaps by preventing further DNA damage. However, this may be tackling the wrong end of the problem, because protoplasts are of necessity isolated from fast-growing cell suspensions, which are likely to be genetically variant, even if the plants regenerated are morphologically normal as they usually are in *Lolium.* Means of isolating protoplasts direct from meristematic tissue such as shoot-tips (so far impractical), or of prolonging the genetic stability of cell suspensions are therefore still required, and in this the use of anti-oxidants may have potential (Dalton 1992), particularly oxalic acid, vitamin C and β-carotene.

Studies on keeping *Lolium* suspensions in the cold for several weeks have been tried with some success (Creemers-Molenaar et al. 1992), although the short life of *Lolium* and *Festuca* cell suspensions and the relatively short time that they can be left in cold conditions means that the lifespan cannot be stretched very much by these means. Cryo-preservation of suspensions identified as responsive would be a better option. However, since instability can occur within days of initial meristem culture (e.g., cultured *Lolium* shoot tips can be unable to regenerate or to regenerate only albino shoots), the only way at present to be reasonably sure of obtaining a high frequency of fertile and apparently normal plants is to continually initiate new cultures from responsive genotypes.

8 Protocol for the Isolation, Culture, and Regeneration of *Festuca* or *Lolium* Protoplasts

Without a suitable suspension, protoplasts cannot be isolated or regenerated, but beyond using suitable isolation and culture protocols for the protoplasts, such as high osmolarity media, there is little that can be done to manipulate that stage of the process. However, the regenerative potential of protoplast colonies can be destroyed by bad handling.

The most important factors for plant regeneration from grass protoplasts therefore are (a) to find a responsive genotype, (b) to initiate a fast-growing, but regenerative cell suspension, (c) to isolate and culture the protoplasts before regenerative ability is lost, and (d) to regenerate protoplast colonies as quickly and carefully as possible.

There is wide genotypic variation in the ability to form embryogenic cultures, cell suspensions and protoplasts so selection is necessary at each stage.

8.1 Induction of Embryogenic Cultures and Cell Suspensions

1. Surface sterilize tillers or embryos if necessary (Table 8)
2. Culture dissected shoot-tips, immature inflorescences (cut up), sectioned mature embryos (particularly the coleoptile area) or whole immature embryos onto callus induction medium (Table 9). Culture at 25 °C in dim light (30 $\mu E/m^2/s$).

Table 8. Surface sterilization methods for grass tillers and embryos

Tissue	Pre-treatment	Sterilant	Treatment
In vivo tillers	Wash, trim roots and leaves	Sodium hypochlorite solution (7% available chlorine)	15 min followed by several rinses in sterile water
Mature embryos	–	Sodium hypochlorite solution (14% available chlorine)	60 min followed by several rinses in sterile water
Immature embryos	Remove caryopses from inflorescence 16 days after pollination	0.01% Mercuric chloride solution	15 min followed by several rinses in sterile water

Table 9. Media for embryogenic callus, cell suspension and protoplast culture in *Lolium* and *Festuca*

	Medium	Carbohydrate	Growth regulators	Agar
Callus induction medium	Murashige and Skoog	3% Sucrose	3 mg/l 2,4-D, 0.2 mg/l BAP, 100 mg/l casein hydrolysate (4 mg/l 2,4-D for *L. perenne*)	0.8% Sigma agar
Cell suspension medium	Murashige and Skoog	3% Sucrose	2 mg/l 2,4-D (3 mg/l 2,4-D for *L. perenne*)	–
Protoplast washing medium	10 mM $CaCl_2 \cdot 2H_2O$	11% Mannitol	–	–
Protoplast culture medium[a]	Kao or Murashige and Skoog + 10 mM $CaCl_2 \cdot 2H_2O$	12 or 13% Maltose + 5% Mannitol or 11% Glucose	1 mg/l 2,4-D 0.1 mg/l zeatin	1.2% Sea plaque agarose
Protoplast colony medium[a]	Murashige and Skoog + 10mM $CaCl_2 \cdot 2H_2O$	6.84% Glucose	0.1 mg/l 2,4-D 0.1 mg/l zeatin	0.3% Gelrite
Germination medium	Murashige and Skoog	3% Sucrose	0.2 mg/l kinetin	0.8% Sigma agar
Enzyme solution[b]	Murashige and Skoog + 10mM $CaCl_2 \cdot 2H_2O$	3% Sucrose 8% Mannitol	Enzymes: 1% Cellulase RS 1% Meicelase 0.3% Macerozyme R10 0.1% Pectolyase Y23 (all Onozuka)	–

Media adjusted to pH 6.0 before autoclaving at 121° C 15 min.
[a] Media made up double strength and filter sterilized, before adding to double strength and autoclaved Sea plaque agarose or Gelrite solution.
[b] pH not adjusted and enzyme solution centrifuged at 2000 rpm for 10 min before filter sterilization to remove solids.

3. Select embryogenic tissues after 3–5 weeks, transfer to 2–5 ml cell suspension medium in small flasks or bottles (50–60 ml) and shake at 100 rpm at 25 °C in dim light (30 $\mu E/m^2/s$) or darkness.
4. Slowly increase the volume to 10 ml and/or replace medium until there is sufficient tissue to transfer to a 250-ml flask with 50 ml medium (2–4 weeks).
5. Begin to select small dense colonies weekly and replace medium at 1:4 ratio.

8.2 Direct Cell Suspension Induction

1. To initiate single genotype cultures, place single dissected shoot tip, immature inflorescence, immature embryo, or coleoptile from a mature embryo into 0.8 ml cell suspension medium in a flat-bottomed universal tube, shake at 100 rpm, and thereafter increase the volume etc. as above.
2. To select responsive genotypes from amongst many genotypes, dissect 20–30 embryos together. Chop groups of mature embryos with a scalpel blade sufficiently to cut each embryo into two or three pieces, but culture immature embryos without chopping. Place the embryos in 2 ml cell suspension medium in 50–60-ml flasks or bottles and thereafter treat as above.

8.3 Protoplast Isolation

1. After 5 to 8 weeks or when suspension growth is sufficiently fast (at least × 5 per week), select 1 ml volume of small cell colonies 3 to 5 days after subculture and then remove medium with a pipet, before adding 10 ml enzyme mixture.
2. Roll colony/protoplast/enzyme mixture slowly (6 rpm) for 6 h at 25 °C, then wash through 90 and 20–25 µm meshes with washing medium.
3. Pellet the protoplasts at 800 rpm for 5 min and resuspend in washing medium. At this stage the protoplasts can be stored overnight at 2–4 °C. Otherwise, pellet the protoplasts again and resuspend in a known volume to be counted with a haemocytometer.

8.4 Protoplast Culture

1. Pellet the protoplasts once more and resuspend at 2×10^5 protoplasts per ml in liquified protoplast culture medium (37 °C).
2. Dispense the protoplasts in 2 ml aliquots into 55-mm Petri dishes, seal with Nescofilm and culture in darkness at 25 °C.
3. After 4 weeks, transfer the protoplasts in agarose (in one piece if possible) to 10 ml liquid protoplast colony medium in 90-mm Petri dishes.
4. Culture the colonies for a further 2 weeks in darkness.

8.5 Embryoid Maturation and Regeneration

1. Pick out compact colonies and transfer to solidified protoplast colony plating medium.
2. Culture the colonies at 20 °C in bright light (110 $\mu E/m^2/s$) to encourage embryoid maturation.
3. Transfer colonies with mature embryoids to regeneration medium to germinate shoots.
4. Culture shoots at 20 °C in less bright light (75 $\mu E/m^2/s$) before transferring to soil.
5. Wash roots free of medium and plant in moist Levington's compost. Cover with plastic cups until plants are established.

Acknowledgments. The author thanks Ian Thomas, Mike Humphreys, Alan Lovatt, and members of the Cell Manipulation Group past and present for interesting discussion and information, particularly Dr. P. Morris for Fig. 2. Thanks also to Alan Lovatt for his technical help with Fig. 1.

References

Abdullah R, Cocking EC, Thompson JA (1986) Efficient plant regeneration from rice protoplasts through somatic embryogenesis. Bio/technology 4:1087–1090

Chu CC, Wang CC, Sun CS, Hsu C, Yin KC, Chu CY, Bi FY (1975) Establishment of an efficient medium for anther culture of rice through comparative experiments on the nitrogen sources. Sci Sin 18:659–668

Creemers-Molenaar J, Loeffen JPM, van der Valk P (1988) The effect of 2,4-Dichlorophenoxyacetic acid and donor plant environment on plant regeneration from immature inflorescence-derived callus of *Lolium perenne* L. and *Lolium multiflorum* L. Plant Sci. 57:165–172

Creemers-Molenaar J, van der Valk P, Loeffen JPM, Zaal MACM (1989) Plant regeneration from suspension cultures and protoplasts of *Lolium perenne* L. Plant Sci. 63:167–176

Creemers-Molenaar J, Loeffen JPM, van Rossum M, Colijn-Hooymans CM (1992) The effect of genotype, cold storage and ploidy level on the morphogenic response of perennial ryegrass (*Lolium perenne* L.) suspension cultures. Plant Science 83:87–94

Creemers-Molenaar J, Hall RD, Krens FA (submitted a) Assymetric protoplast fusion aimed at intraspecific transfer of cytoplasmic male sterility (cms) in *Lolium perenne* L. Mol Gen Genet

Creemers-Molenaar J, van Eeuwijk FA, Krens FA (submitted b) Optimizing the culture of perennial ryegrass protoplasts by conditioned medium, o-acetylsalicylic acid and anti-oxidants. J. Plant Physiol

Dalton SJ (1988a) Plant regeneration from cell suspension protoplasts of *Festuca arundinacea* Schreb. (tall fescue) and *Lolium perenne* L. J. Plant Physiol 132:170–175

Dalton SJ (1988b) Plant regeneration from cell suspension protoplasts of *Festuca arundinacea* Schreb., *Lolium perenne* L. and *L. multiflorum* Lam. In: Puite KJ, Dons JJM, Huizing HJ, Kool AJ, Koornneef M, Krens FA (eds) Progress in plant protoplast research. Kluwer Academic Publ Dordrecht pp 49–52

Dalton SJ (1988c) Improved regeneration from grass protoplasts. Welsh Plant Breed Stn Annu Rep 1988:9

Dalton SJ (1992) The use of anti-oxidants to prolong the regenerative ability of *Lolium* cell suspension cultures. J. Exp Bot Suppl 43:250, p 69 (1992 SEB Meeting Abstracts)

Dalton SJ, Thomas ID (1992) A statistical comparison of various factors on embryogenic proliferation, morphogenesis and regeneration in *Lolium temulentum* cell suspension colonies. Plant Cell Tissue Organ Culture 30:15–29

Dennis ES, Gerlach WL, Pryor WL, Bernetzen JL, Inglis A, Llewelyn D, Sachs MM, Ferl RJ, Peacock WJ (1984) Molecular analysis of the alcohol dehydrogenase (Adh1) gene of maize. Nucl Acids Res 12:3983–4000

den Nijs APM, van der Maas HM, Hensgens LAM (1991) Towards artificial fertilisation and transformation in perennial ryegrass. In: den Nijs APM, Elgersma A (eds) Fodder crops breeding: achievements, novel strategies and biotechnology. Pudoc Wageningen 1991:135–140

Frearson EM, Power JB, Cocking EC (1973) The isolation, culture and regeneration of *Petunia* leaf protoplasts. Dev Biol 33:130–137

Gamborg OL, Miller RA, Ojima K (1968) Nutrient requirements of suspension cultures of soybean root cells. Exp Cell Res 50:151–158

Gould J, Devey M, Hasegawa O, Ulian EC, Peterson G, Smith RH (1991) Transformation of *Zea mays* L. using *Agrobacterium tumefaciens* and the shoot apex. Plant Physiol 95:426–434

Hale AD (1988) Transient expression of a bacterial gene in grass protoplasts. Welsh Plant Breed Stn Annu Rep 1988:8–9

Humphreys MW, Dalton SJ (1991) Stability at the phosphoglucoisomerase (PGI/2) locus in *Festuca arundinacea* plants regenerated from cell suspension and protoplast culture. Genome 34:59–65

Humphreys MW, Dalton SJ (1992) Stability at the phosphoglucoisomerase (PGI/2) locus in *Lolium multiflorum* ($2n = 4x = 28$) × *Festuca arundinacea* ($2n = 6x = 42$) plants regenerated from cell suspension. Genome 35:461–467

Jefferson RA, Kavanagh A, Bevan MW (1987) GUS fusions: β-glucuronidase as a sensitive and versatile fusion marker in higher plants. EMBO Journal 6:3901–3907

Kao KN (1977) Chromosomal behaviour in somatic hybrids of soybean and *Nicotiana glauca*. Mol Gen Genet 150:225–230

Karp A (1991) On the current understanding of somaclonal variation. Oxford Surv Plant Mol Cell Biol 7:1–58

Kearney JF, Parrott WA, Hill NS (1991) Infection of somatic embryos of tall fescue with *Acremonium coenophialum*. Crop Sci 31:979–984

Last DI, Brettell RIS, Chamberlain DA, Chaudhury AM, Larkin PJ, Marsh EL, Peacock WJ, Dennis ES (1991) pEmu: an improved promotor for gene expression in cereal cells. Theor Appl Genet 81:581–588

Lee B, Murdoch K, Kreis M, Jones MGK (1989) A method of large-scale progeny screening of putative transformed cereals. Plant Mol Biol Rep 7(2):129–134

Lewis EJ (1982) Cytogenetic aspects of interspecific hybridization in relation to breeding in the ryegrass/fescue complex. Welsh Plant Breed Stn Annu Rep 1982:218–235

Lörz H, Larkin PJ, Thomson J, Scowcroft WR (1983) Improved protoplast culture and agarose media. Plant Cell Tissue Organ Culture 2:217–226

McElroy D, Zhang W, Cao J, Wu R (1990) Isolation of an efficient actin promotor for use in rice transformation. Plant Cell 2:163–171

Muller AJ, Grafe R (1978) Isolation and characterisation of cell lines of *Nicotiana tabacum* lacking nitrate reductase. Mol Gen Genet 161:67–76

Murashige T, Skoog F (1962) A revised medium for rapid growth and bioassay with tobacco tissue cultures. Physiol Plant 15:473–497

Murray FR, Latch GCM, Scott DB (1992) Surrogate transformation of perennial ryegrass, *Lolium perenne*, using genetically modified *Acremonium* endophyte. Mol Gen Genet 233:1–9

Ougham HJ, Thomas AM, Thomas BJ, Roberts PC, Mutinda C, Hayward MDH, Dalton SJ (1992) Leaf development in *Lolium temulentum* L.: characterisation of a slow-to-green mutant. New Phytol 122:261–272

Potrykus I, Saul MW, Petruska J, Paszkowski J, Shillito RD (1985) Direct gene transfer to cells of a graminaceous monocot. Mol Gen Genet 199:183–188

Radilofe-Rajoelina S, Alibert G, Planchon C (1990) Continuous plant regeneration from established embryogenic cell suspension cultures of Italian ryegrass and tall fescue. Plant Breed 104:265–271

Stafford A (1991) Genetics of cultured plant cells. In: Stafford A, Warren G (eds) Plant cell and tissue culture. Open University Press, Milton Keynes, pp 25–47

Takamizo T, Suginobu K-I, Ohsugi R (1990) Plant regeneration from suspension culture derived protoplasts of tall fescue (*Festuca arundinacea* Schreb.) of a single genotype. Plant Sci 72:125–131

Takamizo T, Spangenberg G, Suginobu K, Potrykus I (1991) Intergeneric somatic hybridization in Gramineae: Somatic hybrid plants between tall fescue (*Festuca arundinacea* Schreb.) and Italian ryegrass (*Lolium multiflorum* Lam.). Mol Gen Genet 231:1–6

Wang ZY, Takamizo T, Iglesias VA, Osusky M, Nagel J, Potrykus I, Spangenberg G (1992) Transgenic plants of tall fescue (*Festuca arundinacea* Shreb.) obtained by direct gene transfer to protoplasts. Bio/technology 10:691–696

Webb KJ, Chamberlain DA, Woodcock S, Armstead I, Dale PJ (1985) Fusion of protoplasts. Welsh Plant Breed Stn Annu Rep 1985:98

Yamada Y, Zhi-Qi Y, Ding-Tai T (1986) Plant regeneration from protoplast-derived callus of rice (*Oryza sativa* L.) Plant Cell Rep 5:85–88

Zaghmout OMF, Torello WA (1990) Isolation and culture of protoplasts from embryogenic suspension cultures of red fescue (*Festuca rubra* L.). Plant Cell Rep 9:340–343

Zaghmout OMF, Torello WA (1992a) Restoration of regeneration potential of long-term cultures of red fescue (*Festuca rubra* L.) by elevated sucrose levels. Plant Cell Rep 11:142–145

Zaghmout OMF, Torello WA (1992b) Plant regeneration from callus and protoplasts of perennial ryegrass (*Lolium perenne* L.). J. Plant Physiol 140:101–105

I.6 Regeneration of Plants from Protoplasts of *Lotus* spp. (Birdsfoot trefoil)

M. Niizeki[1]

1 Introduction

There are approximately 200 species of *Lotus* distributed throughout the world (Grant 1965). The greatest diversity occurs in the Mediterranean basin, considered the regional center of origin. *Lotus corniculatus* L. (birdsfoot trefoil) is native to Europe and parts of Asia. Nineteenth-century reports show that it grew naturally in many pastures and was a good feed for cattle and horses. However, it was not until after 1900 that it was cultivated in Europe (MacDonald 1946).

Birdsfoot trefoil is a perennial cross-pollinating tetraploid legume ($2n = 4x = 24$) which is at present used for pasture, hay, and silage in the north central and northeastern United States and eastern Canada. The feeding value of this plant is almost equal to that of *Medicago sativa* L. (alfalfa) (Seaney and Henson 1970). In addition to its use as a major pasture legume, either alone or with various grasses, birdsfoot trefoil is used extensively on new highway slopes, for soil improvement, controling erosion, and for beautification (Grant and Marten 1985). In Britain it is widely distributed and grows on soils too poor to support legumes such as *Trifolium pratensis* L. (red clover) and alfalfa (Ahuja et al. 1983). In addition to North America, Canada, and Europe, *L. corniculatus* and to some extent *L. tenuis* Waldst. & Kit. and *L. uliginosus* Schkuhr are also grown in Australia, New Zealand, and South America (Grant 1965).

Bridsfoot trefoil grows on many different types of soil, from clay to sandy loams. It is quite adaptable to prevailing soil conditions and is able to grow on dry, infertile, and acid or mildly alkaline and saline soils. It is also more resistant to water-logged soil conditions than alfalfa (Grant and Marten 1985).

Plant regeneration from birdsfoot trefoil calli was established in an anther culture study by Niizeki and Grant (1971). In regenerated plants, variants, such as resistance to the herbicide 2,4-dichlorophenoxyacetic acid (2,4-D), were found to be useful under field conditions (Swanson and Tomes 1980). However, this variation arising from calli and cell suspensions is restricted to changes of the plant genotype. Thus, suitable protoplast systems are essential for exploiting interspecific and intergeneric gene flows. Birdsfoot trefoil has problems which need to be solved. These include lack of seedling vigor, and the dehiscence of the seed pods as soon as they mature, causing seed loss. The latter problem is so serious in seed production

[1] Plant Breeding Laboratory, Faculty of Agriculture, Hirosaki University, Hirosaki, Aomori-ken 036, Japan

Biotechnology in Agriculture and Forestry, Vol. 22
Plant Protoplasts and Genetic Engineering III (ed. by Y.P.S. Bajaj)

that somatic cell hybridization was attempted to introduce the seed pod indehiscence trait from a wild *Lotus* species, *L. conimbricensis* Brot. (syn. *L. coimbrensis* Willd.) into birdsfoot trefoil (Wright et al. 1987). Furthermore, since *Lotus* contains tannins in its leaves, it should be a useful fusion partner in somatic hybridization attempts to transfer this bloat-safe character into other bloat-inducing forage crops such as alfalfa and *Trifolium repens* L. (white clover) which are currently of more economic value (Ahuja et al. 1983). In this chapter the conditions of protoplast isolation, culture, and plant regeneration for *Lotus* spp., together with the establishment of various protoplast systems suitable for the genetic manipulation of somatic cells are discussed.

2 Protoplast Isolation

Aseptic seeds of birdsfoot trefoil (cv. Viking) were germinated on Nitsch and Nitsch (1969) medium without growth regulators (designated N-0 medium). About 1 cm segments of hypocotyls from 10-day-old seedlings were transferred onto the basic medium of MS (Murashige and Skoog 1962) containing 4 mg/l NAA and 2.5 mg/l kinetin (designated M-1N medium) at 25 °C (Niizeki and Saito 1986). The induced calli were usually pale green and subcultured at intervals of 4 weeks. Subcultured fresh callus (1–2 g) was used for the isolation of protoplasts. At 10–15 days after transplantation onto the fresh medium, the calli were suitable for obtaining viable protoplasts. A solution containing 4% Cellulase Onozuka RS (Kinki Yakult, Nishinomiya, Japan), 1% Macerozyme R10 (Kinki Yakult, Nishinomiya, Japan), 0.2% Pectolyase Y23 (Seishin Pharmaceutical, Tokyo, Japan) and 0.7 M mannitol (pH 5.8) was used for the protoplast isolation. The enzyme-callus mixture was incubated at 28 °C in a shaker bath (60 shakes/min). The optimum time for protoplast isolation was 3–4 h in the enzyme solution. The protoplasts were separated from undigested cell clumps by eight layers of cotton gauze. The enzyme was removed by four successive washings with 0.7 M mannitol solution (pH 5.8) by centrifugation at 80 *g* for 4 min each. The protoplast yield was $1.5 - 2 \times 10^6$/g of callus tissue over several isolations. When the protoplasts were isolated intact, most of them appeared to be spherical, and a few showed transparent vacuoles (Fig. 1A).

Ahuja et al. (1983) obtained protoplasts from leaf mesophyll, and roots, hypocotyls, and cotyledons of in vitro birdsfoot trefoil seedling, cv. Leo and Maitland. Enzyme solutions consisting of 2% Cellulysin (Calbiochem, California, USA) and 0.1% Pectolyase Y23, or consisting of 2% Cellulysin, 0.25% Driselase (Kyowa Hakko Kogyo, Tokyo, Japan), 0.5% Rhozyme HP150 (Röhm and Haas, Philadelphia, USA) and 0.15% Pectolyase Y23 were used for leaf mesophyll. They also used an enzyme solution consisting of 4% Cellulysin, 0.8% Driselase, 2% Rhozyme HP150 and 1% Macerozyme R10 for the isolation of mesophyll protoplasts. Enzyme solutions consisting of 0.8% Cellulase Onozuka R10 (Kinki, Yakult, Nishinomiya, Japan), 0.4% Driselase, 0.8% Rhozyme HP150 and 0.4% Macerozyme R10, or consisting of 4% Meicelase (Meiji Seika, Tokyo, Japan), 2% Rhozyme HP150, and 0.3% Macerozyme R10 were used for the seedling roots,

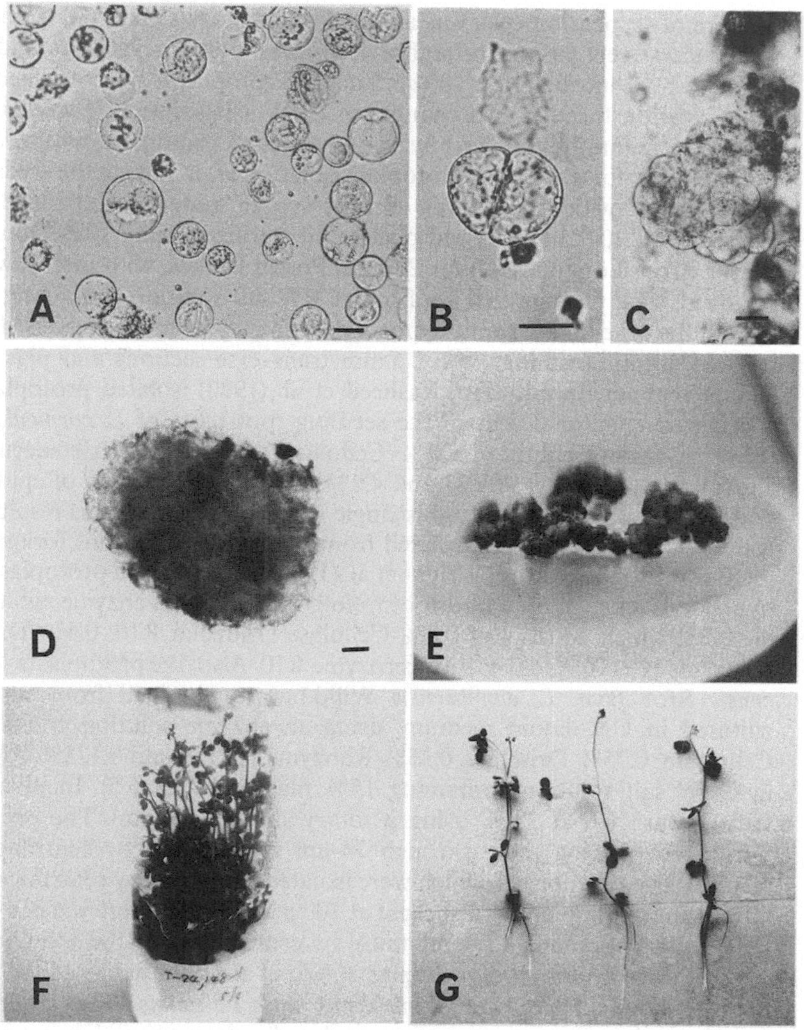

Fig. 1A–G. Isolation and culture of protoplasts of birdsfoot trefoil. **A** Isolated protoplasts. **B** Initiated cell division which was observed within 7 days of culture. **C** A formed cell cluster which was observed 10 days after addition of fresh medium. **D** A colony composed of numerous cells after 1 month of culture. **E** Calli formed for a period of 1 month after transplantation of colonies to the callus culture medium. **F** Shoot formation at 1.5 months after transplantation of calli to the regeneration medium. **G** Complete plantlets which were obtained about 6 months after the initiation of protoplast culture. *Bars* 20 μm. (Niizeki and Saito 1986)

hypocotyls, and cotyledons. Protoplast yields were higher with the former solution. Cell suspensions, which were initiated by transfer of hypocotyl-derived callus to UM liquid medium (Uchimiya and Murashige 1974), were a convenient source material for protoplast isolation using an enzyme solution containing 4% Meicelase, 2% Rhozyme HP150 and 0.3% Macerozyme R10. Enzyme incubation was carried out at 28 °C for 6–16 h in the dark without agitation, while an enzyme mixture of suspension cells was agitated on a rotary shaker (80 rpm). Protoplasts were successively passed through nylon sieves (pore sizes, 64 µm and 20 µm) and pelleted by centrifugation (80 *g*, 5 min). Protoplasts were washed by further centrifugation through two changes of CPW salt solution (Frearson et al. 1973) with 13% mannitol, followed by two changes of protoplast culture medium. The optimum age for maximum protoplast isolation after sowing the seeds of birdsfoot trefoil varied with the explant used as source material. Expanded leaves were excised from 30- to 50-day-old plants and sterilized using 10% Domestos bleach (Lever Bros., London, UK) for 20 min. Peeled leaflets, with their lower epidermis removed, were plasmolyzed for 1 h by CPW salt solution containing 13% mannitol. On the other hand, roots, hypocotyls, and cotyledons were excised from 2- to 7-day-old aseptic seedlings into 0.5 mm transverse sections and plasmolyzed in a similar manner. In addition, Rasheed et al. (1990) isolated protoplasts from the specialized epidermal cells of the seedling root hairs of *L. corniculatus*, cv. Leo, using an enzyme solution of 1% Cellulase YC (Seishin Pharmaceutical, Tokyo, Japan), 0.1% Pectolyase Y23 and 4.2% mannitol. The study of epidermal protoplasts in culture is limited to this single report to date, but this result confirms the totipotency of protoplasts isolated from the root hairs of this forage legume.

For protoplast fusion, Wright et al. (1987) also obtained protoplasts from sliced hypocotyls (ca. 2 mm) of birdsfoot trefoil, cv. Leo, in an enzyme solution (Ahuja et al. 1983) which consists of 0.8% Cellulase Onozuka R10, 0.4% Driselase, 0.8% Rhozyme HP150, and 0.4% Macerozyme R10. Also, the protoplasts of *L. conimbricensis* Brot. (syn. *L. coimbrensis* Willd.) were obtained from suspension cells cultured in UM liquid medium, using an enzyme solution consisting of 0.5% Cellulysin, 0.25% Driselase, 0.25% Rhozyme HP150 and 0.125% PASE dissolved in CPW salt solution containing 13% mannitol (pH 5.5). Incubation for both species was 16 h at 28 °C using a rotary shaker (50 rpm). The protoplasts were purified by sieving them through 74-µm screens and by centrifugation (100 *g*, 10 min). The protoplasts, which were isolated from cells by centrifugal flotation on CPW salts containing 21% sucrose at 100 *g* for 10 min, were washed three times in CPW salts containing 13% mannitol by centrifugation (100 *g*, 5 min).

By the electrofusion of protoplasts, Aziz et al. (1990) achieved the production of hybrid plants between *L. corniculatus* and *L. tenuis*. The latter species was transformed to kanamycin resistance by *Agrobacterium rhizogenes* carrying the neomycin phosphotransferase II gene. The seedlings (6 to 7 days old) of *L. corniculatus*, cv. Leo, were placed in the dark at 4 °C for 48 h and then cotyledons were cut transversely into 0.5 mm pieces and plasmolyzed for 1 h in a CPW salt solution with 13% mannitol. The composition of the enzyme solution was 0.8% Cellulase Onozuka R10, 0.4% Driselase, 0.4% Macerozyme R10 and 0.8% Rhozyme HP150 (pH 5.8). Kanamycin-resistant cells of *L. tenuis* were maintained by subculture at intervals of 7 days in MS liquid basal medium containing 0.54 mg/l BA, 2 mg/l NAA and 2 mg/l 2,4-D with 100 µg/ml of kanamycin sulfate. Five grams fresh weight of harvested cells were incubated in 20 ml of an enzyme solution consisting of 2% Rhozyme HP150, 2% Meicelase, and 0.03% Macerozyme R10 in CPW salt solution with 13% mannitol, pH 5.8. Incubation for protoplast isolation of both species was 16 h on a horizontal shaker (30 rpm) at 25 °C. Following incubation, the enzyme-protoplast mixtures were filtered through 64-µm (*L. corniculatus*) and 45-µm (*L. tenuis*) nylon sieves and were then subjected to centrifugation (80 *g*, 5 min). Protoplasts were resuspended in CPW salts solution supplemented with 21% sucrose and were then centrifuged (80 *g*, 10 min). Following this, the collected bands of floating protoplasts remaining intact were washed using CPW salts solution containing 13% mannitol.

3 Protoplast Culture and Plant Regeneration

The isolated protoplasts of birdsfoot trefoil at a density of 1.2×10^5/ml were cultured in thin layers of a modified KM8P liquid medium (Kao and Michayluk 1975) containing 0.5 mg/l BA instead of zeatin without coconut milk (Niizeki and Saito 1986). The culture dishes were kept at 25 °C under dark conditions. Within 2–4 days, most of the protoplasts aggregated into a number of clumps. The protoplasts initiated their division within 7 days (Fig. 1B), but the percentage of dividing cells could not be determined due to the difficulty in counting protoplasts in the aggregated portions. The effect of Sea Plaque agarose (FMC Corp., Rockland, USA) was evaluated and an aggregation-preventing effect was obtained by the addition of 1.25% agarose to the modified KM8P medium (Niizeki unpubl.). At the time of the first cell division, fresh modified KM8P medium was added and the cultures were continued under light conditions (18.9 $\mu E/m^2/s^1$). After about 10 days, many of the cells divided two or three times and subsequently formed cell clusters (Fig. 1C). Colonies composed of numerous cells were then formed actively. In the case of culture using agarose, plating efficiency was 11.2%. Nurse cells of birdsfoot trefoil calli were very effective for cell division, and plating efficiency increased to 20.8%. After 1 month of culture, the colonies were visible to the naked eye and were able to be transferred to the M-1N medium (Fig. 1D). The colonies formed calli for a period of 1 month (Fig. 1E). The calli derived from the protoplasts were transferred to Miller's basic medium (1963) containing IAA and BA (Table 1). Many adventitious buds were formed on the calli, especially on those cultured on medium I containing 1.5 mg/l IAA and 1.5 mg/l BA. When transplanted to the N-0 medium, these adventitious buds produced shoots which further developed into complete young plants (Fig. 1F). Root formation was significantly promoted by using Gelrite (Scott Laboratories, California, USA) instead of agar (Fig. 1G).

Table 1. Adventitious bud formation on Miller's basic medium with different concentrations of IAA and BA. (Niizeki and Saito 1986)

Medium type	IAA (mg/l)	BA (mg/l)	No. of adventitious buds formed
I	1.5	1.5	25 (58.1)[a]
II	1.5	4.0	12 (27.9)
III	4.0	1.5	6 (14.0)
			Total 43 (100)

[a] Figures in parentheses are in percentages.
Ten calli which were about 500 mg each in fresh weight were transplanted onto each medium.

Ahuja et al. (1983) and Wright et al. (1987) obtained colonies in almost the same way. Initially, Ahuja et al. (1983) cultured protoplasts at 1×10^5/ml in liquid KM8P medium. In some cases, seedlings prior to protoplast isolation or isolated protoplasts in culture medium were maintained at 4 °C for 48 h in the dark. Placing the seedlings under a cold treatment significantly increased the plating efficiency of cotyledon protoplasts from 32% to 60–70%. Likewise, cold treatment of isolated protoplasts before culture had a similar effect. In protoplast culture, the media were diluted at intervals of 5 days with KM8P or KM8P medium previously mixed with cell culture medium A (Kao and Michayluk 1980). Finally, protoplasts were maintained in cell culture medium A alone. Cultures were incubated in the dark for 72 h at 27 °C, and were subsequently incubated under a light intensity of 800 lx. Mesophyll protoplasts showed a low plating efficiency, and did not develop into cell colonies. On the other hand, Webb et al. (1987) found protoplasts which derived from young, fully expanded leaves in an enzyme consisting of 2% Cellulase Onozuka R10, 0.5% Macerozyme R10, and Pectolyase Y23 were cultured in an agarose medium based on B5M (Swanson and Tomes 1980) with 3% sucrose, 1% Sea Plaque agarose, 11% mannitol, 0.1 mg/l 2,4-D, and 0.1 mg/l BA. The solidified agarose drops were immersed by Kao's liquid protoplast medium (1977) supplemented with 1.25% mannitol. Using this method, highly totipotent colonies were produced from the leaf protoplasts. However, in the experiment of Ahuja et al. (1983), protoplasts from seedling explants and cell suspensions, in both liquid KM8P and protoplast culture medium A, divided to form cell colonies without much difficulty. Plant regeneration occurred from protoplast-derived calli within 14 days of transferring 30- to 35-day-old colonies to MS basal agar medium supplemented with 0.2, 0.5 or 1 mg/l BA or zeatin. Almost the entire surface of calli proliferated into the shoots. Rapid shoot development resulted in low levels of cytokinins (0.2 mg/l BA or zeatin). Birdsfoot trefoil plantlets rooted readily on being transferred to a hormone-free medium.

In the case of Wright et al. (1987), fused protoplasts between *L. corniculatus* and *L. conimbricensis* were also cultured in KM8P medium at 1×10^5 protoplasts/ml. Protoplast cultures were initially incubated in the dark at 28 °C for 5 days after plating. After 5 days, protoplast cultures were diluted with KM8P medium previously mixed with UM liquid medium and moved to a 16-h day-length at 28 °C. Colonies from protoplasts with fusion treatments were transferred after 28 days of culture to agar solidified plant regeneration medium containing MS basic medium supplemented with 0.2 mg/l BA. Several shoots from each regenerable callus were transferred to MS medium containing sucrose, but without growth regulators, and allowed to take root. These plantlets seemed to be somatic hybrids, because hypocotyl protoplasts of *L. corniculatus* were inactivated with iodoacetate to inhibit cell division prior to fusion with *L. conimbricensis*, callus derived from which does not have regeneration ability. Regenerated plants were transplanted into peat pellets and grown in a growth chamber at 22 °C with a 16-h day-length under light conditions until they were large enough to be transplanted into pots and grown in a green house.

Aziz et al. (1990) cultured the electrofused protoplasts of *L. corniculatus* and *L. tenuis* at 5×10^4/ml in a modified medium of Kao (1977) with glucose increased to 100 g/l, sucrose increased to 250 mg/l and thiamine HCl reduced to 1 mg/l. The medium was solidified with 1.2% Sea Plaque agarose. The agarose was cut into four blocks per dish, and the blocks floated in the same liquid medium. The osmoticum of the culture medium was reduced every 5 days. After 3 weeks in the dark, the dishes were placed in continuous illumination at 25 °C. Three weeks later, the colonies that developed were placed on agar solidified B5 basal medium (Gamborg et al. 1968) with 0.05 mg/l BA and 100 μg/l kanamycin. Following 5 to 6 weeks of culture on selection medium, surviving microcalli were transferred to B5 basal medium with 0.05 mg/l BA, but without kanamycin, for plant regeneration. Shoots with a length of approximately 6 cm after 7 weeks on regeneration medium, were transferred for rooting to MS basal medium without growth regulators. The plantlets produced were putative somatic hybrids, because the protoplasts of the kanamycin-resistant callus of *L. tenuis* were treated with iodoacetate prior to their fusion with *L. corniculatus* protoplasts which are sensitive to kanamycin.

4 Conclusion and Prospects

Birdsfoot trefoil is easy to regenerate from callus culture both through organogenesis (Niizeki and Grant 1971; Swanson and Tomes 1980) and somatic embryogenesis (Mariotti et al. 1984; Damiani et al. 1985, 1990). Plant regeneration from the protoplast-derived calli of birdsfoot trefoil was also reported by Ahuja et al. (1983), and Niizeki and Saito (1986). Without doubt, *Lotus* protoplasts show the most prolific regeneration among any forage legume protoplast system reported to date. Thus, the prolific regeneration capability of *Lotus* should facilitate plant recovery in genetic manipulation programs. Niizeki et al. (1990b) regenerated a protoclonal population from calli derived from a single protoplast (Fig. 2). Variations among their cytogenetical, morphological, and agronomic traits of this population, in which preexisting variation in explants was excluded by using a single protoplast initially, were evaluated and compared with those of the seed produced population. In this study, new phenotypic variants with higher values than those of the plant initially used for protoplast isolation were observed. Somatic hybrid plants were produced by Wright et al. (1987), using protoplast fusion of *L. corniculatus* cv. Leo and *L. conimbricensis* Brot. (syn. *L. coimbrensis* Willd), which is a species with a pod indehiscence trait (McGraw and Beuselinck 1983). However, these hybrid plants were unable to introduce the indehiscence trait because of their male and female sterility. Aziz et al. (1990) produced somatic hybrid plants from kanamycin-resistant cell colonies, recovered following electrofusion between *L. corniculatus* and *L. tenuis*. The hybrid plants exhibited heterosis and were more proficient in terms of nitrogenase activity as compared to the parental plants, but were infertile in seed setting. The stems and leaves of the somatic hybrids contained tannins at a lower concentration than in *L. corniculatus*. However, they still retained the important bloat-safe characteristic. This observation supports the proposal that somatic hybridization may be an approach for transferring genes for tannin synthesis into bloat-inducing forage legumes. Niizeki et al. (1990a) also obtained somatic hybrids between *L. corniculatus*, cv. Viking, and *Glycine max*, cv. Harosoy. In this case the soybean chromosomes were eliminated by X-irradiation. The morphology of the hybrid plants resembled that of the birdsfoot trefoil derived

Fig. 2. A protoclonal population derived from a single protoplast. (Niizeki unpubl.)

from parent calli. However, the hybrid plants were usually an erect type in contrast to the creeping habit of normal birdsfoot trefoil and, in addition, the plants did not flower. Although these experiments were not completely successful, genetic manipulation programs involving *Lotus* may still have the capability to improve forage crops by biotechnological methods using protoplasts.

5 Protocol (Niizeki and Saito 1986)

5.1 Protoplast Isolation

1. Seeds are sterilized by 70% ethanol for several seconds and followed by sodium hypochlorite (Cl concentration; 0.5%) for 10 min. The seeds are then washed five times with sterilized distilled water and germinated on Nitsch and Nitsch medium (1969) without growth regulators (N-0 medium).
2. About 1 cm of hypocotyl segments from 10-day-old seedlings are transferred onto the MS basic medium (Murashige and Skoog 1962) + 4 mg/l NAA + 2.5 mg/l kinetin (M-1N medium).
3. The induced calli are subcultured at intervals of 4 weeks.
4. Fresh callus (1–2 g) subcultured for 10–15 days is incubated in a solution of 4% Cellulase Onozuka RS + 1% Macerozyme R10 + 0.2% Pectolyase Y23 + 0.7M mannitol (pH 5.8) for 3–4 h at 28 °C in a shaker bath (60 shakes/min), in diffused light.
5. Protoplasts are separated by eight layers of cotton gauze and the enzyme is removed by four successive washings with a solution of 0.7 M mannitol (pH 5.8) by centrifugation at 80 *g* for 4 min each.

5.2 Protoplast Culture

1. Protoplasts (1.2×10^5/ml) are cultured in thin layers of a modified KM8P medium containing 0.5 mg/l BA instead of zeatin and without coconut milk and culture dishes are kept at 25 °C in the dark.
2. Droplets of fresh culture medium (0.2–0.5 ml) are added to each dish after the beginning of the first division (within 7 days) and the cultures are continued under light conditions (18.9 $\mu E/m^2/s^1$).
3. After 1 month of culture, colonies which are visible to the naked eye are transferred to M-1N medium.
4. Green calli, 4–8 mm in diameter (approx. 1 month in age), are transferred to regeneration medium.

5.3 Plant Regeneration

1. Many adventitious buds are formed on the regeneration medium, Miller's basic medium + 1.5 mg/l IAA + 1.5 mg/l BA.
2. About 2 weeks after the appearance of adventitious buds, the shoot elongated sufficiently to be dissected.
3. Excised shoots are planted on the N-0 medium solidified by Gelrite.
4. Roots develop in about 2 to 4 weeks and rooted plants are transferred to vermiculite + half-strength MS basal salts and acclimatized at 25 °C and 70% humidity in the light for 1–2 months.
5. The acclimatized plants are transferred to a green house.

Acknowledgment. The author expresses his thanks to Dr. W. F. Grant and Suyanee Vessabutr, Department of Plant Science, MacDonald Campus of McGill University, for providing information and reviewing this manuscript.

References

Ahuja PS, Hadiuzzaman S, Davey MR, Cocking EC (1983) Prolific plant regeneration from protoplast-derived tissues of *Lotus corniculatus* L. (birdsfoot trefoil). Plant Cell Rep 2:101–104

Aziz MA, Chand PK, Power JB, Davey MR (1990) Somatic hybrids between the forage legumes *Lotus corniculatus* L. and *L. tenuis* Waldst. et Kit. J Exp Bot 41:471–479

Damiani F, Mariotti D, Pezzotti M, Arcioni S (1985) Variability among plants regenerated from tissue culture of *Lotus corniculatus* L. Z. Pflanzenzuchtg 94:332–339

Damiàni F, Pezzotti M, Arcioni S (1990) Somaclonal variation in *Lotus corniculatus* L. in relation to plant breeding purposes. Euphytica 46:35–41

Frearson EM, Power JB, Cocking EC (1973) The isolation, culture and regeneration of *Petunia* leaf protoplasts. Dev Biol 33:130–137

Gamborg OL, Miller RA, Ojima K (1968) Nutrient requirements of suspension cultures of soybean root cells. Exp Cell Res 50:151–158

Grant WF (1965) A chromosome atlas and interspecific hybridization index for the genus *Lotus* (Leguminosae). Can J. Genet Cytol 7:457–471

Grant WF, Marten GC (1985) Birdsfoot trefoil. In: Heath ME, Barnes RF, Metcalfe DS (eds) Forages. Iowa State University Press, Ames, pp 98–108

Kao KN (1977) Chromosomal behaviour in somatic hybrids of soybean-*Nicotiana glauca* L. Mol Gen Genet 150:225–230

Kao KN, Michayluk MR (1975) Nutritional requirements for growth of *Vicia hajastana* cells and protoplasts at a very low population density in liquid medium. Planta 126:105–110

Kao KN, Michayluk MR (1980) Plant regeneration from mesophyll protoplasts of alfalfa. Z Pflanzenphysiol 96:135–141

MacDonald HA (1946) Birdsfoot trefoil (*Lotus corniculatus* L.)-its characteristics and potentialities as a forage legume. Cornell Agric Exp Stn Mem 261:1–182

Mariotti D, Pezzotti M, Falistocco E, Arcioni S (1984) Plant regeneration from leaf-derived callus of *Lotus corniculatus* L. cv. Franco. Genet Agrar 38:219–223

McGraw RL, Beuselinck PR (1983) Growth and seed yield characteristics of birdsfoot trefoil. Agron J 75:443–446

Millers CO (1963) Kinetin and kinetin-like compounds. In: Linskens HF, Tracey MV (eds) Moderne Methoden der Pflanzen-Analyse, vol 6. Springer, Berlin Heidelberg New York, pp 192–202

Murashige T, Skoog F (1962) A revised medium for rapid growth and bioassays with tobacco tissue culture. Physiol Plant 15:473–497

Niizeki M, Grant WF (1971) Callus, plantlet formation from cultured anthers of *Lotus* and *Nicotiana*. Can J Bot 49:2041–2051

Niizeki M, Saito K (1986) Plant regeneration from protoplasts of birdsfoot trefoil, *Lotus corniculatus* L. Jpn J Breed 36:177–180

Niizeki M, Cai K, Kihara M, Nakajo S, Harada T (1990a) Somatic cell hybrids between birdsfoot trefoil and soybean. Lotus Newslett 21:14–17

Niizeki M, Ishikawa R, Saito K (1990b) Variation in a single protoplast- and seed-derived population of *Lotus corniculatus* L. Theor Appl Genet 80:732–736

Nitsch JP, Nitsch C (1969) Haploid plants from pollen grains. Science 163:85–87

Rasheed JH, Al-Mallah MK, Cocking EC, Davey MR (1990) Root hair protoplasts of *Lotus corniculatus* L. (birdsfoot trefoil) express their totipotency. Plant Cell Rep 8:565–569

Seaney RR, Henson PR (1970) Birdsfoot trefoil. Adv Agron 22:119–157

Swanson EB, Tomes DT (1980) In vitro responses of tolerant and susceptible lines of *Lotus corniculatus* L. to 2,4-D. Crop Sci 20:792–795

Uchimiya H, Murashige T (1974) Evaluation of parameters in the isolation of viable protoplasts from cultured tobacco cells. Plant Physiol 54:936–944

Webb KJ, Woodcock S, Chamberlain DA (1987) Plant regeneration from protoplasts of *Trifolium repens* and *Lotus corniculatus*. Plant Breed 98:111–118

Wright RL, Somers DA, McGraw RL (1987) Somatic hybridization between birdsfoot trefoil (*Lotus corniculatus* L.) and *L. conimbricensis* Willd. Theor Appl Genet 75:151–156

I.7 Regeneration of Plants from Protoplasts of *Lycium barbarum* L. (Wolfberry)

Y.I. Ratushnyak, V.A. Rudas, and N.M. Piven[1]

1 Introduction

Lycium barbarum L. (2n = 24), is a woody bush spread all over Eurasia and North Africa. It is highly branched, 1–2.5 m high with single, reverse lanceolate smooth-edged leaves. The flowers are bell-shaped and violet-pink in color; the red, egg-shaped berries are edible. Wild *L. barbarum* forms a thicket, but it is often cultivated as an ornamental plant. It is also used as a medicinal (Niu et al. 1983) and melliferous plant, highly resistant to different *Phytophthora* races. The berries contain ascorbic acid (27–90 mg%) (Grossgaim 1946) and a high percentage of dry substance.

A number of investigations were aimed at studying *Lycium* in vitro. Regenerated plants of different ploidy were obtained from *L. chinense* endosperm callus (Gu et al. 1985) and *L. barbarum* unfertilized ovules (Qin et al. 1985). Haploid plants were induced from anther culture of *L. barbarum* and *L. chinense*, but they diploidized (2n = 24) (Fan et al. 1982). Callus capable of plant regeneration was induced from hypocotyls, shoots, and ovules of *L. barbarum* (Niu et al. 1983). Only callus was induced from *L. chinense* protoplasts (Sun et al. 1982).

With modern technologies for plant regeneration from protoplasts, it also became possible to obtain somatic hybrids and genetic transformants for woody species (Grosser et al. 1988a, b; Kobayashi et al. 1988; James et al. 1989). We consider *L. barbarum* to be a potential source for obtaining somatic hybrids with *Lycopersicon*, as well as for experiments on genetic transformation.

2 Plant Material and Callus Culture

Seeds of *L. barbarum* were sterilized successively by treating them in 70% ethanol (30–40 s) followed by 3% sodium hypochlorite (6–8 min), washed three times in sterile distilled water and germinated in Petri dishes on a wet filter paper. Plantlets 8–10 days old were transferred to the half-strength hormone-free MS-3 medium (Table 1) and grown at 25 °C, illumination 2000–3000 lx with 16-h photoperiod.

Several media based on MS mineral elements were tried for callus induction from leaf segments and young internodes. By experimenting with the growth

[1]Institute of Cell Biology and Genetic Engineering, Ukrainian Academy of Sciences, Acad. Lebedev Str., 1, 252650, Kiev-GSP-22, Ukraine

Biotechnology in Agriculture and Forestry, Vol. 22
Plant Protoplasts and Genetic Engineering III (ed. by Y.P.S. Bajaj)

Table 1. Components of the media used (mg/l). (After Ratushnyak et al. 1989, 1990)

	Medium[a]					
	W-S-S	S-S-S	PRM	MS-1	MS-2	MS-3
Macronutrients						
NH_4NO_3	1280	–	MS[b]	MS	MS	1/2MS
KNO_3	–	950				
$CaCl_2$ $2H_2O$	600	600				
KH_2PO_4	170	85				
KCl	300	–				
Micronutrients	B5[c]	B5	MS	MS	MS	MS
Vitamins						
Nicotinic acid	1	W-S-S	1	1	1	–
Thiamine-HCl	10		5	2	2	1
Pyridoxine-HCl	1		1	1	1	–
Adenine sulfate	–		15	2	2	–
L-glutamine	100		–	20	20	–
Riboflavine	–		–	0.2	0.2	–
Yeast extract	100		–	–	–	–
Casein hydr.	500		–	500	500	–
Myo-inositol	100		100	100	100	100
Others						
Xylose	100		–	–	–	–
Glucose, g/1	72		–	–	–	–
Mannitol, g/1	–		36, 4	–	–	–
Sucrose, g/1	–		20	20	20	20
Serva agar, g/1	–		8	7	7	7
Growth regulators[d]						
NAA	2		–	–	0.2	0.1
BA	0.5		2.5	–	–	–
2,4-D	0.2		–	0.4	1	–
IAA	–		0.2	–	–	–
pH = 5.7						

[a] W-S-S and S-S-S media were filter-sterilized by using Millipore filter. PRM, MS-1, MS-2 and MS-3 were autoclaved for 15 min at 120 °C and 1.05 kg/m^2.
[b] Medium MS = Murashige and Skoog (1962).
[c] Medium B5 = Gamborg et al. (1968).
[d] Abbreviations: NAA = α-naphthaleneacetic acid; BA = 6-benzylaminopurine; 2,4-D = 2,4-dichlorophenoxyacetic acid; IAA = indole-3-acetic acid.

substances, a suitable variant of MS-1 medium (Table 1) was chosen. Callus induced on MS-1 medium was notable for its intensive growth, was soft and had a friable consistency and white color (Fig. 1). The efficiency of callus formation was 96%. A suitable medium was chosen also for callus subculture. Callus induced on MS-1 medium grew on different variants of MS medium. MS-2 medium (Table 1) proved to be the most suitable for long-term callus culture (Fig. 2). Callus was induced within 1 month on MS-1 and then subcultured every 20 days on MS-2 in the dark at 25 °C. This callus was used as a source for isolating protoplasts.

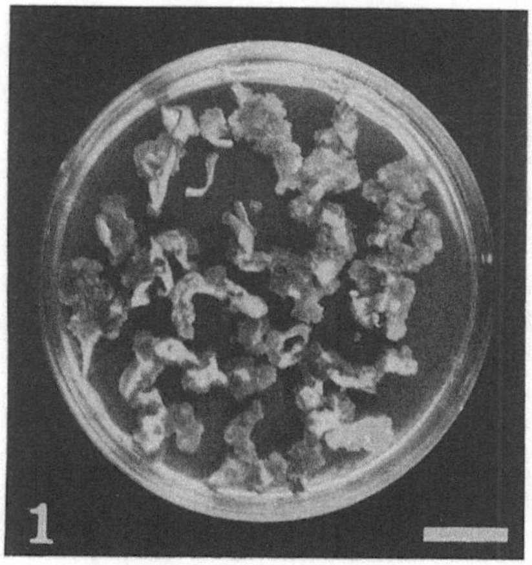

Fig. 1. Callus induction from leaf segments on MS-1 medium after month cultivation. *Bar* 15 mm. (Ratushnyak et al. 1990)

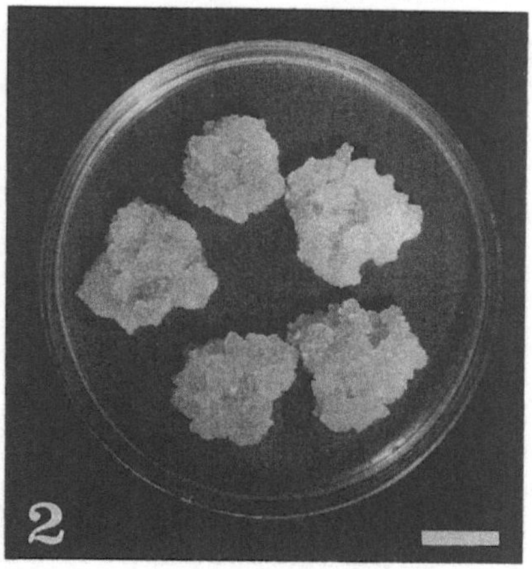

Fig. 2. One-month callus culture on MS-2 medium. *Bar* 15 mm. (Ratushnyak et al. 1990)

3 Mesophyll and Callus Protoplast Isolation

Mesophyll protoplasts were isolated from aseptically growing plants in the enzyme mixture (Ratushnyak et al. 1989), containing 0.2% Cellulase Onozuka R10 (Serva), 0.1% Driselase (Sigma), 0.5 M sucrose and 5 mM $CaCl_2$, and the callus protoplasts from callus culture in the enzyme mixture (Ratushnyak et al. 1990), containing 2.2% Cellulase Onozuka R10, 1.2% Pectinase, 0.8% Macerozyme R10 (Serva), 1% Driselase (Sigma), 0.8% Cellulysine (Calbiochem), 0.5 M sucrose, and 5 mM $CaCl_2$. Enzyme treatment (pH = 5.6) was carried out in the dark at 25 °C for 12–14 h for mesophyll and for 18 h for callus protoplasts.

The protoplasts were filtered through a 100-μm nylon sieve, and centrifuged for 3 min at 100 *g*. The floating protoplasts were collected, washed twice with W5 salt solution (Medgyesy et al. 1980), and pelleted for 2 min at 60 *g*. Plating mesophyll and callus protoplast density in liquid nutrient medium was 10^3–10^5 cell/ml.

4 Mesophyll Protoplast Culture and Plant Regeneration

The nutrient media W-S-S, S-S-S (Sidorov et al. 1984), PRM (Piven and Makhorina 1985) (Table 2), TM-2, TM-3, TM-4 (Shahin 1984), and K_3 (Kao et al. 1974) were tried with respect to the mesophyll protoplast cultivation and plant regeneration ability. The washed protoplasts (Fig. 3A), transferred into Petri dishes containing one of the liquid W-S-S, S-S-S, K_3, or TM-2 media, were incubated in the dark for 5 days. Then corresponding fresh medium was added to each dish and

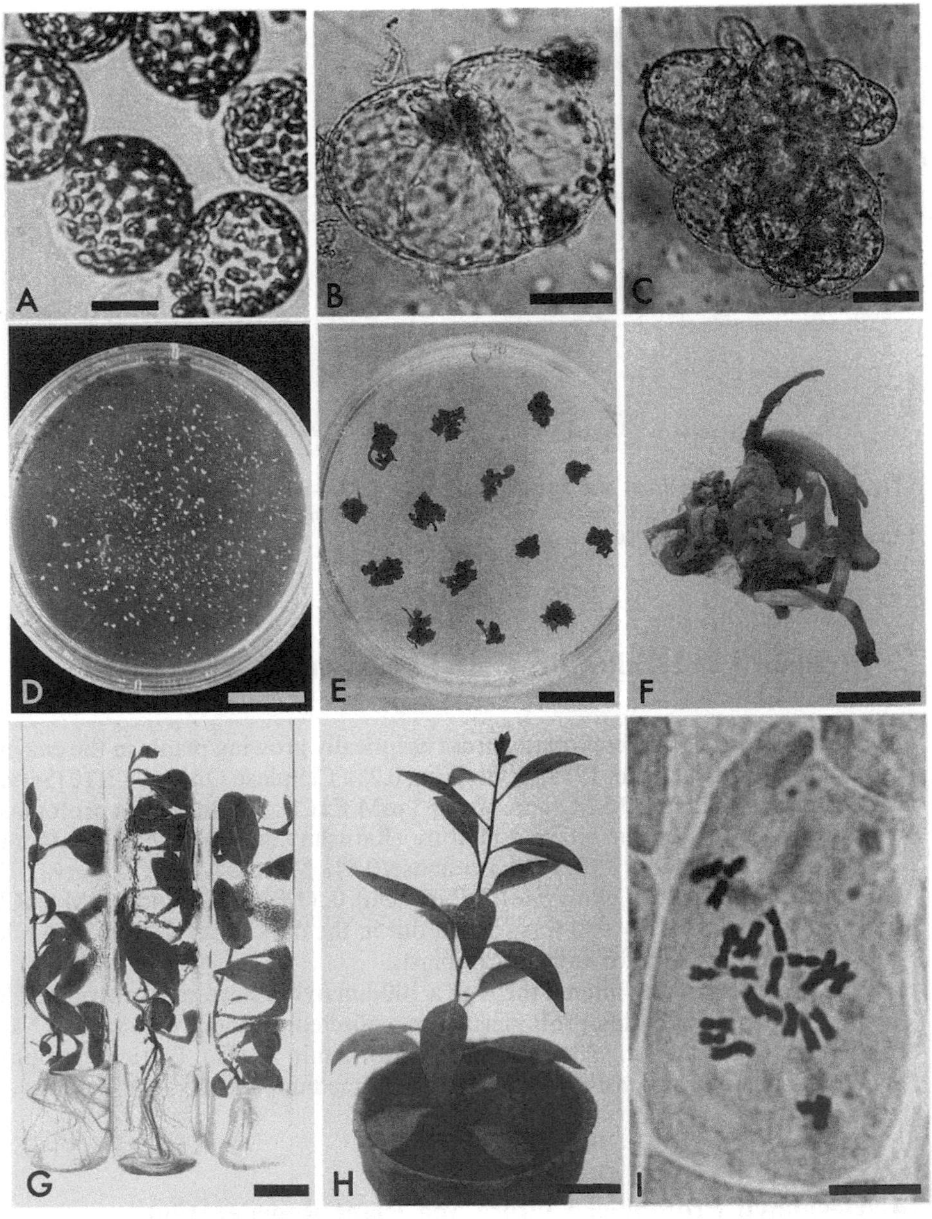

Fig. 3A–I. Plant regeneration from leaf mesophyll protoplasts of *Lycium barbarum*. **A** Protoplasts isolated from leaf mesophyll (medium TM-2). **B** First protoplast division on third day (medium TM-2). **C** Ten-day cell colony from single protoplast (medium TM-2). **D** Formation of minicalli (13th cultivation day, medium TM-3). **E** Induction of morphogenesis after 5 weeks (medium TM-4). **F** Shoot regeneration from 6-week-old protoclones (medium TM-4). **G** Rooted regenerated plants in 9 weeks after protoplasts isolation (hormone-free MS-3 medium). **H** Plant 14L 6 months old in the soil. **I** Diploid number of chromosomes (2n = 24) in the root tip cell of the regenerant 9L. *Bars* 45 μm (**A, B**), 2 μm (**C**), 18 mm (**D, E**), 5 mm (**G**), 50 mm (**H**) and 80 μm (**I**). (Ratushnyak et al. 1989)

Table 2. Stages of regeneration of *L. barbarum* plant from mesophyll and callus protoplasts. (After Ratushnyak et al. 1990)

Stages	Media used	Duration of cultivation, days		%	
		Mesophyll protoplast	Callus protoplast	Mesophyll protoplast	Callus protoplast
Cell colony formation	TM-2	10	15	9.8	5.5
Minicallus formation	TM-3	5	8	98	97
Regeneration of shoots	TM-4	30	35	50	80
Rooting of shoots	MS-3	30	20	50	82.6

cultivation continued with 16-h photoperiod (700–900 lx) at 25 ± 2 °C. Only a few colonies were obtained on S-S-S and K_3 media but none on W-S-S. The best results regarding the protoplast division ability was achieved using TM-2 medium. Two to 3 days later, resynthesis of the cell wall took place and the cells increased in size 1.5–2 times. Cell divisions occurred on the third day (Fig. 3B). The highest plating efficiency (10%) was achieved using TM-2 medium. Plating efficiency was determined as the number of transplanted colonies divided by initial protoplast number and multiplied by 100. Cell colonies differed substantially in morphology, density, and size (Fig. 3C).

After 10–12 days of protoplast cultivation, the regenerating colonies were placed on the surface of solidified TM-3 medium and cultured under bright illumination (3000–4000 lx) for 16-h photoperiod at 25 ± 2 °C. Five to 6 days later, white minicalli reached 1–2 mm in diameter (Fig. 3D).

To induce organogenesis, PRM and TM-4 media were used. During the first series of experiments, plant regeneration was achieved on PRM medium. After transfer of visible cell colonies to the medium, they enlarged, turned green, and thickened. A month later, after bud formation, the protoclones were transferred to the fresh nutrient medium. The small single shoots bearing two or three leaves were regenerated. The regeneration index was 80–90%; but after transplanting the shoots to the hormone-free MS-3 medium, most of them were lost. As a result, only 21 rooted plants were rescued from 48 protoclones, some of which are shown in Fig. 3G.

Minicalli transferred from TM-3 onto TM-4 medium during the second series of experiments differed in the character of regeneration. The surviving colonies turned green and initiated shoots. Transplantation to the fresh medium activated the process (Fig. 3E, F). Approximately 50% of the regenerated plants were rooted on the hormone-free MS-3 medium. The regeneration rate varied for different protoclones, but as far as the TM-4 medium was concerned, the regeneration rate was double that of the PRM medium (1–1.5 months instead of 2–3 months). The abnormal shoots did not produce roots and were excluded from the experiment. In general, it took 2.5–4 months to regenerate plants from isolated protoplasts (Table 2).

For cytologic analysis, more than 50 metaphase plates were studied for each plant. Root tips for chromosome analysis were treated with 0.03% colchicine, fixed with acetoalcohol (1:3), and stained with 1% acetoorcein. All the plants investigated were diploid (2n = 24; Fig. 3I). The regenerated plants did not differ from the initial plants phenotypically. They retained the specific morphology and color of leaves, flowers and fruits, growth rate, and rooting ability, and rooted easily in soil

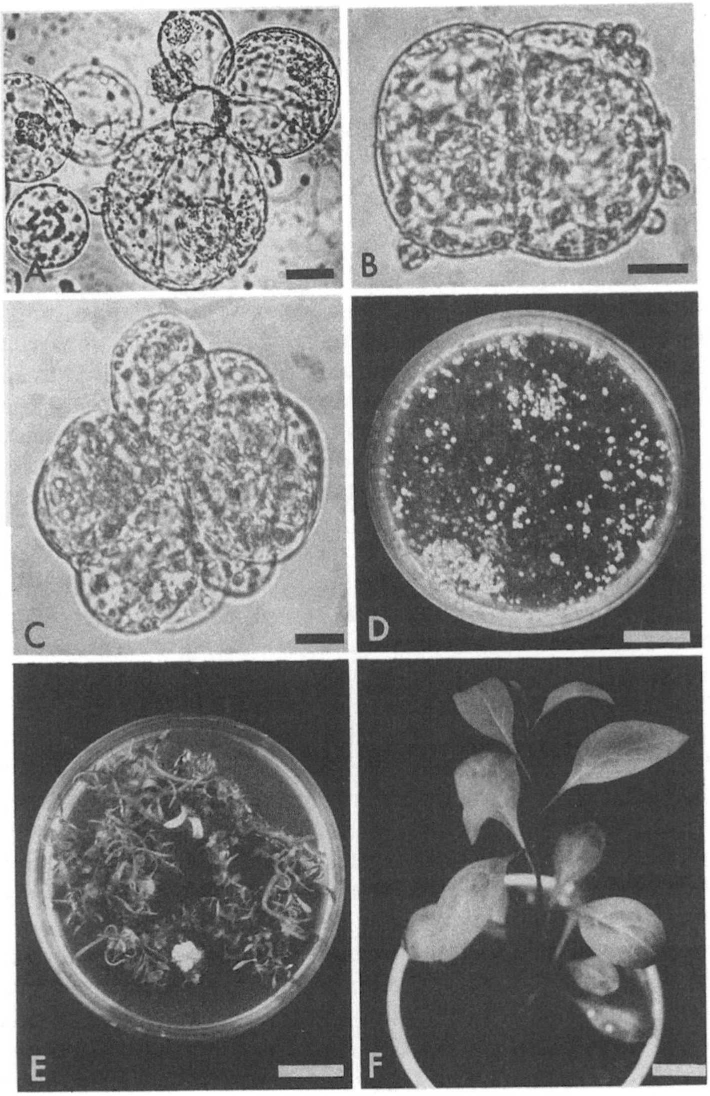

Fig. 4A–F. Plant regeneration from callus protoplasts. **A** Protoplasts isolated from callus (medium TM-2). **B** First protoplast division on 5th day (medium TM-2). **C** Nine-day cell colony (medium TM-2). **D** Three-week-old cell colonies on TM-3 medium. **E** Organogenesis on TM-4 medium in 2 months after protoplasts isolation. **F** Eight-month-old regenerated plant in soil. *Bars* 26 μm (**A**), 30 μm (**B**), 7 μm (**C**), 19 mm (**D, E**) and 14 mm (**F**). (Ratushnyak et al. 1990)

(Fig. 3H). The karyotypic stability of the regenerated plants reveals the fact that the procedures of plant regeneration from the protoplasts applied do not result in an alteration of the chromosome number.

5 Culture of Callus-Derived Protoplasts

Plant regeneration from callus protoplasts (Fig. 4A) was achieved using TM-2 (protoplast culture), TM-3 (minicallus formation), and TM-4 (morphogenesis) media (Shahin 1984). After 3 days of culture, more than 80% of the living protoplasts resynthesized a cell wall and increased in size up to 150 μm on TM-2 medium. First divisions (Fig. 4B) of protoplasts were observed on the 5th day. The plating efficiency of the callus protoplast reached 5 to 6%, which is approximately half that of mesophyll protoplasts. After 15 days of culture, the compact cell colonies (Fig. 4C) were transferred to TM-3 intermediate nutrient medium. White minicalli of 2 to 3 mm (Fig. 4D) were transferred to TM-4 regeneration medium. During the first 2 weeks, the minicalli grew intensively and turned green. Bud formation and regeneration of the shoots (Fig. 4E) were very intensive, and the morphogenesis index reached 80%. The shoots were separated from the callus and transferred to MS-3 hormone-containing medium (Table 1). In total, 19 out of the 23 shoots obtained for each protoclone were rooted. The plants were transferred to soil (Fig. 4F). All the plants possess morphological traits characteristic of *L. barbarum*. On the whole, the "callus culture-protoplast-plant" cycle took about 3 months (Table 2).

6 Summary and Conclusions

Our results demonstrate the possibility of *L. barbarum* plant regeneration from mesophyll and callus protoplasts with high efficiency using Shahin (1984) medium. We have obtained good results in rooting of plant regenerants. The high rooting efficiency for protoclones was obtained on both hormone-free (for shoots derived from mesophyll protoplasts) and hormone-containing 0.1 mg/l NAA (for shoots derived from callus protoplasts) medium. Completely formed and rooted plants were transferred to the soil. Cytological and morphological analysis of the regenerated plants revealed the relative genetic stability of this species in the process "plant-protoplast-plant". The results obtained allow us to conclude that *L. barbarum* protoplasts have high regeneration potential and can be used for genetic engineering experiments.

Acknowledgement. The authors thank Dr. N.N. Cherep for correcting the English version of this manuscript.

References

Fan YH, Zang SY, Zhao JF (1982) Induction of haploid plants in *Lycium chinense* Mill. and *L. barbarum* L. from anther culture. Hereditas (China) 4:25–26

Gamborg OL, Miller RA, Ojima K (1968) Nutrient requirements of suspension cultures of soybean root cells. Exp Cell Res 50:151–158

Grosser JW, Gmitter FG, Chandler JL (1988a) Intergeneric somatic hybrid plants from sexually incompatible woode species: *Citrus sinensis* and *Citrus distieha*. Theor Appl Genet 75:397–401

Grosser JW, Gmitter FG, Chandler JL (1988b) Intergeneric somatic hybrid plants of *Citrus sinensis* cv. Hamlin and *Poncirus trifoliata* cv. Flying Dragon. Plant Cell Rep 7:5–8

Grossgaim AA (1946) Plant resources of the Caucasus. Azerbaijan SSR Publishers, Baku, 671 pp

Gu SR, Gui YL, Xu TY (1985) Induction of endosperm plantlets in *Lycium*. Acta Bot Sin 27:106–109

James DJ, Passey AJ, Barbara DJ, Bevan M (1989) Genetic transformation of apple (*Malus pumila* Mill.) using a disarmed Ti-binary vector. Plant Cell Rep 7:658–661

Kao KN, Constabel F, Michayluk MR, Gamborg OL (1974) Plant protoplast fusion and growth of intergeneric hybrids cells. Planta 120:215–227

Kobayashi S, Ohgawara T, Ohgawara E, Oiyama I, Ishii S (1988) A somatic hybrid plant obtained by protoplast fusion between novel orange (*Citrus sinensis*) and satsuma mandarin (*Citrus unshiu*). Plant Cell Tissue Organ Cult 14:63–69

Medgyesy P, Menczel L, Maliga P (1980) The use of cytoplasmic streptomycin resistance: chloroplast transfer from *Nicotiana tabacum* into *Nicotiana sylvestris*, and isolation of their somatic hybrids. Mol Gen Genet 179:693–698

Murashige T, Skoog F (1962) A revised medium for rapid growth and bioassays with tobacco tissue cultures. Physiol Plant 15:473–497

Niu DS, Shao QQ, Wang L, Qin JS, Chen SP, Cao SF, Wang DZ (1983) Induction of calli of hypocotyl, shoot and young ovary of *Lycium barbarum* and plant regeneration. Hereditas (China) 5:24–26

Piven NM, Makhorina OK (1985) Plant regeneration from callus protoplasts of *Lycopersicon peruvianum* var. *dentatum*. Sov Cytol Genet 19:271–277

Qin JS, Wang L, Chen SP, Wang DZ, Nui DS, Jiang XC, Shao QQ (1985) Establishment of autotetraploid of *Lycium barbarum*. Acta Genet Sin 12:200–203

Ratushnyak YaI, Piven NM, Rudas VA (1989) Protoplast culture and plant regeneration in *Lycium barbarum* L. Plant Cell Tissue Organ Cult 17:183–190

Ratushnyak YaI, Rudas VA, Piven NM (1990) Regeneration of *Lycium barbarum* L. plants from leaf tissue, callus culture and callus protoplasts. Plant Cell Rep 9:84–87

Shahin EA (1984) Isolation and culture of protoplasts: tomato. In: Vasil IK (ed) Cell culture and somatic cell genetics of plants, vol 1, Academic Press, New York, pp 370–380

Sidorov VA, Kuchko AA, Gleba YuYu (1984) Genetic modification of *Solanum tuberosum* through protoplast culture and fusion. In: Röhlich P and Bócsy E (eds) Tissue culture and RES. Academiai Kiado, Budapest, pp 529–534

Sun YR, Li WB, Huang MJ, Li XH (1982) Callus formation from protoplasts of chinese wolfberry (*Lycium chinense*) Acta Bot Sin 24:477–479

I.8 Regeneration of Plants from Protoplasts of *Pelargonium* spp. (Geranium)

K.B. DUNBAR[1]

1 Introduction

1.1 Importance and Distribution of *Pelargonium* Species

Pelargonium species comprise one of the most popular groups of flowering potted plants in the world (Wood 1966). Most cultivated *Pelargonium* plants are grown for their colorful showy flowers, others for their scented foliage and exotic leaf shapes. The genus *Pelargonium* contains approximately 300 species, most of which are native to South Africa (Moore 1971). *Pelargonium zonale* was first introduced into Europe in 1609 (Ewart 1980). Currently, *P.* × *hortorum* L.H. Bailey (garden geranium), *P.* × *domesticum* L.H. Bailey (regal geranium), and *P. peltatum* (ivy geranium) are the most economically important *Pelargonium* species. Records of the crosses that produced the hybrid species were not kept. *Pelargonium* × *hortorum* is thought to have originated from four wild species: *P. zonale*, *P. inquinans*, *P. frutetorum*, and *P. scandens*. *Pelargonium* × *domesticum* is thought to have originated from seven wild species: *P. angulosum*, *P. betulinum*, *P. capitatum*, *P. cordifolium*, *P. cucullatum*, *P. fulgidum*, and *P. grandiflorum* (Ewart 1980).

1.2 Significance and Use of Protoplasts from *Pelargonium* Species

Of the *Pelargonium* species most important in the US floriculture industry, *P.* × *domesticum* is considered the most beautiful; however, problems with cultivation and flowering have prevented it from becoming as popular as *P.* × *hortorum* in most parts of the USA (Hanniford and Holcomb 1982). *Pelargonium* × *domesticum* requires a long treatment of cool night temperatures to flower, but is resistant to bacterial blight of geranium (Ewart 1980; Hanniford and Holcomb 1982). *Pelargonium* × *hortorum* does not require a cold treatment to flower, but is susceptible to bacterial blight of geranium (Ewart 1980). Bacterial blight of geranium is considered the most serious disease of *P.* × *hortorum* (Nelson and Nichols 1982). A hybrid between *P.* × *hortorum* and *P.* × *domesticum* would be very desirable, but attempts to sexually cross these species have been unsuccessful (Ewart 1980).

[1] Division of Natural Science and Mathematics, Dalton College, 213 North College Drive, Sequoya 127A, Dalton, GA 30720, USA

Biotechnology in Agriculture and Forestry, Vol. 22
Plant Protoplasts and Genetic Engineering III (ed. by Y.P.S. Bajaj)

1.3 Review of Protoplast Isolation and Regeneration of *Pelargonium*

There are few reports of the culture and plant regeneration from protoplasts of *Pelargonium* species (Kameya 1975; Abo El-Nil and Hilderbrandt 1976; Yarrow et al. 1987; Dunbar and Stephens 1989a; Dunbar and Stephens 1991). Protoplasts of *Pelargonium* spp. have been isolated and cultured from leaf mesophyll (Kameya 1975), anther- and leaf-derived callus (Abo El-Nil and Hildebrandt 1976; Dunbar and Stephens 1991), and suspension cultures (Yarrow et al. 1987; Dunbar and Stephens 1989a). Plant regeneration from protoplasts has been reported from a single cultivar of *P.* × *domesticum* (Melissa) (Dunbar and Stephens 1991) and *P. peltatum* (Ivy CB) (Yarrow et al. 1987). Plants of *P.* × *hortorum* have been regenerated from protoplasts of the variegated cultivar Mrs. Parker (Yarrow et al. 1987; Dunbar and Stephens 1989a) (Fig. 1) and an unnamed variegated cultivar (Kameya 1975). Although there are reports of regeneration of plants from protoplasts of the three major cultivated *Pelargonium* species, a great deal of work needs to be done to develop and improve protoplast isolation and culture techniques for *Pelargonium*.

Fig. 1. Variegated plantlet regenerated from green sector protoplasts of *Pelargonium* × *hortorum* cv. Mrs. Parker

2 Isolation of Protoplasts

2.1 Mesophyll Protoplasts of *P.* × *hortorum*

Kameya (1975) regenerated plants from leaf mesophyll protoplasts of a variegated cultivar of *P.* × *hortorum*. This procedure was initiated with surface disinfected leaf tissue. Tissue was dried for 2 h at 30 °C prior to removing the epidermis and placement in an enzyme solution (Kameya 1975). The tissue was incubated and protoplasts isolated as described by Kameya and Uchimiya (1972) and Kameya (1973).

2.2 Suspension Culture-Derived Protoplasts of *Pelargonium* Species

Yarrow et al. (1987) also reported the regeneration of plants from a variegated cultivar of *P.* × *hortorum* (Mrs. Parker). With a single procedure, Yarrow et al. (1987) isolated protoplasts from suspension cultures of *P.* × *hortorum*, *P. peltatum* (Ivy CB), and *P. aridum*. For this procedure, callus was initiated from leaf explants on a Murashige and Skoog (MS) medium (1962) supplemented with 2 mg/l naphthaleneacetic acid (NAA), 0.5 mg/l 6-benzylaminopurine (BAP), 3% sucrose, and 0.8% agar. Suspension cultures were initiated by transferring callus pieces to identical MS medium minus agar and then culturing on a rotary shaker (80 rpm) at 23 °C. Suspension cultures were harvested and isolated in an enzyme mixture of 2% Rhozyme HP150 (Rhome and Haas Co., USA), 4% Meicelase (Meija Seika Ltd., Japan), 13% mannitol, and CPW salts (Frearson et al. 1973) (Yarrow et al. 1987). After a 16-h incubation at 23 °C on a rotary shaker (15 rpm), the protoplasts were passed through a nylon sieve (64 μm pore size), centrifuged at 100 *g* for 8 min, and then resuspended in KMP8/KP (Yarrow et al. 1987) liquid medium. This protocol yielded $0.5–2.0 \times 10^5$ protoplasts per g of tissue (Table 1) (Yarrow et al. 1987).

Table 1. Protoplast isolation, culture, and regeneration from *Pelargonium* species

Species	Source tissue	Yield ($\times 10^5$)	Plating efficiency	Plant regeneration	References
P. × *hortorum*,	Leaf mesophyll	N.R.	40–50	(+)	Kameya (1975)
P. × *hortorum*,	Callus tissue	N.R.	6	(−)	Abo El-Nil and Hildebrandt (1976)
P. × *hortorum*,	Suspension culture	0.5–2	30	(+)	Yarrow et al. (1987)
P. peltatum,	Suspension culture	0.5–2	55	(+)	Yarrow et al. (1987)
P. aridum	Suspension culture	0.5–2	20	(+)	Yarrow et al. (1987)
P. × *domesticum*	Callus tissue	2–3	3.5	(+)	Dunbar and Stephens (1991)

Yield = # of protoplasts/gram of tissue. N.R. = not reported. Plating Efficiency = % of dividing cells per total protoplasts plated. (−) = plants not regenerated. (+) = plants regenerated.

2.3 Callus-Derived Protoplasts of *P.* × *domesticum*

A procedure for plant regeneration from protoplasts of *P.* × *domesticum* has recently been described (Dunbar and Stephens 1991). Leaf explants from cv. Melissa were used to initiate green organogenic callus tissue as described by Dunbar and Stephens (1989b). To induce white friable callus from which protoplasts could be isolated, it was necessary to culture the organogenic callus in the dark on MS medium supplemented with 3 mg/l NAA, 2 mg/l BAP, 3% sucrose, and 0.8% agar. The white friable callus was broken into 1 mm^3 pieces and incubated in an enzyme mixture of 2% Cellulysin (Behring Diagnostics, USA), 0.5% Macerase (Behring Diagnostics, USA), 0.5 M sucrose, in CPW salts at 23 °C on a rotary shaker (46 rpm). After a 6-h incubation, it was necessary to pass the callus, suspended in enzyme solution, in and out of a Pasteur pipet to release the protoplasts. Protoplasts were passed through a 62-μm nylon sieve, and then centrifuged at 100 *g* for 10 min. Protoplasts were collected from the surface of the enzyme solution in W5 solution (Menczel et al. 1981). The protoplasts were centrifuged at 36 *g* for 10 min, the W5 solution was discarded, and the pellet was resuspended in KMP8/KP liquid medium (Yarrow et al. 1987). This protocol yielded an average of 2.7×10^5 protoplasts per g of callus (Table 1) (Dunbar and Stephens 1991).

3 Culture of Protoplasts

3.1 Mesophyll Protoplasts of *P.* × *hortorum*

The inorganic salt concentration in the protoplast culture medium affects cell division of protoplasts isolated from *P.* × *hortorum* leaf mesophyll (Kameya 1975). Medium containing 1/10 × dilution of MS inorganic salts, 1 mg/l indole-3-acetic acid (IAA), 1 mg/l BAP, 10% mannitol, and 0.2% sucrose was identified as the optimal medium for initiation of cell division. Cell wall formation and first division occurred within 20 days of plating protoplasts at 2×10^4 protoplasts/ml at 25 °C and 20 μmol/m^2/s. To initiate second division and colony formation, the cells were then plated onto MS medium supplemented with 1 mg/l IAA, 1 mg/l BAP, 5% mannitol, 2% sucrose, and 0.8% agar. After 40–50 days, protoplast-derived colonies were transferred to the same medium minus the mannitol. Callus developed within 60 days (Kameya 1975).

3.2 Suspension Culture-Derived Protoplasts of *Pelargonium* Species

Suspension culture-derived protoplasts of *P.* × *hortorum*, *P. peltatum*, and *P. aridum*, were cultured at a density of 2.5×10^4 protoplasts/ml in a complex liquid medium (KMP8/KP) at 25 °C and 9 μmol/m^2/s (Yarrow et al. 1987). First divisions were observed 5–10 days after plating. After 6 days, additional KMP8/KP medium was added to the cultures, and then the cultures were solidified with an equal volume of molten KMP8/KP medium containing 0.6% agar. Colonies formed within 15–20 days and the colonies, in agar blocks, were transferred to an MS medium supplemented with 0.1 mg/l NAA, 0.5 mg/l BAP, 3% sucrose, 4.5% mannitol, and 0.8% agar (Yarrow et al. 1987).

3.3 Callus-Derived Protoplasts of *P.* × *domesticum*

Callus-derived protoplasts of *P.* × *domesticum* (Fig. 2a) were cultured at a density of 1×10^5 protoplasts/ml in liquid KMP8/KP (Yarrow et al. 1987) medium layered onto the same medium with 0.6% agarose (Sea Plaque, FMC, Rockland, MD, USA). These cultures were fed every 7 to 10 days with fresh KMP8/KP liquid medium. After 8 weeks, the glucose in the feeding medium was reduced to 30 g/l, and after the 12th week, the glucose was reduced to 15 g/l and the sucrose was raised to 15 g/l. Under these culture conditions, first division (Fig. 2b) occurred within 48 h, 50 cell colonies were observed after 8 weeks in culture, and p-calli 12 mm^3 were observed within 15 weeks (Dunbar and Stephens 1991).

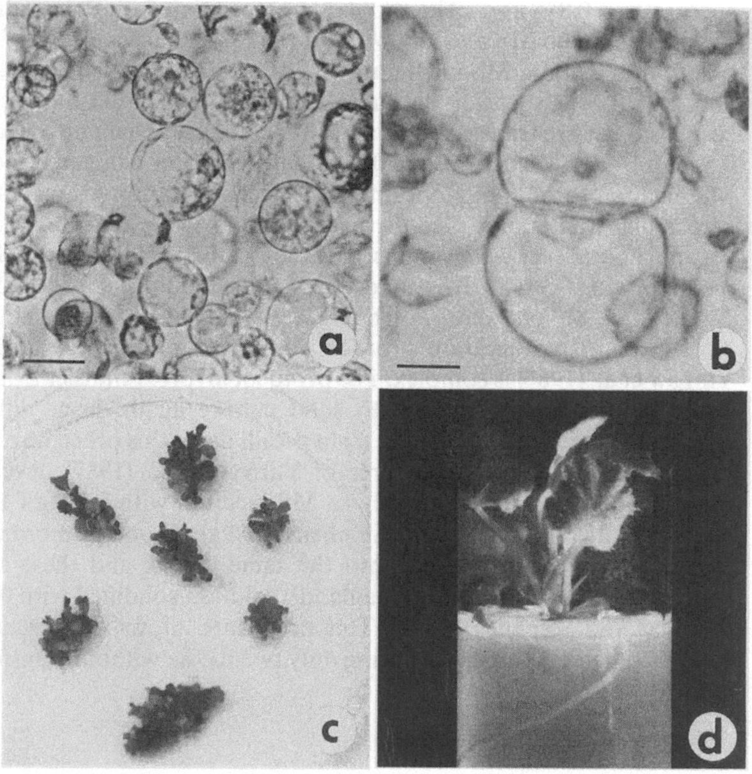

Fig. 2a–d. Development of *Pelargonium* × *domesticum* cv. Melissa from callus-derived protoplasts to plants. **a** Callus-derived protoplasts 15 min after isolation. *Bar* 10 µm. **b** First division 3 days after isolation. *Bar* 5 µm. **c** Shoot development on callus tissue 30 days after transfer to MS medium with 0.2 mg/l BAP. **d** Plantlet 3–4 weeks after transfer to half-strength MS medium supplemented with 0.1 mg/l NAA

4 Regeneration of Plants

4.1 Regeneration from Mesophyll-Derived-P-Calli of *P.* × *hortorum*

Kameya (1975) reported plantlet regeneration from leaf mesophyll-derived callus 20–30 days after transferring the callus to an MS medium without phytohormones. Green plants were regenerated from green sectors on leaves, while albino plants were regenerated from the white sectors. Variegated plants were not recovered from either green or white leaf sectors (Kameya 1975).

4.2 Regeneration from Suspension Culture-Derived Protoplasts of *Pelargonium* Species

The plant regeneration protocol used for suspension culture-derived p-calli was dependent on *Pelargonium* species (Yarrow et al. 1987). P-calli from *P. aridum* were transferred to an MS medium containing 0.05 mg/l NAA, 0.5 mg/l BAP, and 3% sucrose, and 0.8% agar. The shoots that regenerated on this medium were excised and rooted on an MS medium without phytohormones. P-calli of *P. peltatum* were placed into a liquid MS medium with 2 mg/l NAA, 0.5 mg/l BAP, and 3% sucrose and grown as suspension cultures. The shoot primordia that developed in these suspensions were transferred to an MS medium supplemented with 0.1 mg/l NAA, 0.5 mg/l BAP, 3% sucrose, and 0.8% agar. As shoots elongated they were rooted in MS medium without phytohormones. The regeneration protocol for *P.* × *hortorum* was similar to that of *P. peltatum* except that after the subculture in liquid medium, organogenic tissues were trimmed of excess callus and transferred first to liquid MS medium with 0.5 mg/l BAP and 3% sucrose and then to liquid MS containing 2 mg/l GA_3. Shoots were rooted as for *P. aridum* (Yarrow et al. 1987).

The plant regeneration protocol for p-calli of *P.* × *hortorum* described by Yarrow et al. (1987) is complex. The protocol involves transfer from solid culture to liquid culture and then back to solid culture again. Four different media are required to regenerate a plantlet from p-calli using this procedure. However, p-calli regenerated using the procedures of Yarrow et al. (1987) developed shoot primordia when transferred directly to MS medium with 0.5 mg/l BAP, 3% sucrose and 0.8% agar (Dunbar and Stephens 1989a). Shoot primordia developed and elongated when subcultured onto the same media, and these shoots developed roots when transferred onto Hoagland's solution solidified with 0.7% agar (Fig. 1) (Dunbar and Stephens 1989a). This procedure allows the regeneration of plants from p-calli of *P.* × *hortorum* using only two media without transfer to liquid shake culture.

4.3 Regeneration from P-Calli of *P.* × *domesticum*

Protoplast-derived calli of *P.* × *domesticum* were transferred to an MS medium supplemented with 3 mg/l NAA, 2 mg/l BAP, 1 g/l casamino acids, and 0.8% agar. Shoot primordia developed 4 weeks after callus was transferred to an MS medium with 0.2 mg/l BAP, 2% sucrose, and 0.8% agar (Fig. 2c). Shoot primordia were subcultured onto the same medium and shoots elongated. Shoots developed roots within 21 days after transfer to half-strength MS medium supplemented with 0.1 mg/l NAA, 1% sucrose, and 0.9% agar (Fig. 2d) (Dunbar and Stephens 1991).

5 Plastid Inheritance

Studies of plant regeneration from protoplasts of *P.* × *hortorum* have utilized plants with green/white variegated foliage (Kameya 1975; Yarrow et al. 1987; Dunbar and Stephens 1989a). Kameya (1975) reported that only green plants were regenerated from protoplasts isolated from green sectors on leaves and only albino plants were regenerated from white sectors. Yarrow et al. (1987) also reported that only albino plants were regenerated from white sectors of leaves, while both green and albino plants were regenerated from the green sectors. Variegated shoots were not regenerated from protoplasts in either of these studies (Kameya 1975; Yarrow et al. 1987). However, using modifications (Dunbar and Stephens 1989a) of the protocols described by Yarrow et al. (1987), variegated shoots were regenerated from protoplasts of green sectors of the cultivar Mrs. Parker (Dunbar and Stephens, unpublished) (Fig. 1).

The green sectors of leaves of cv. Mrs. Parker contain both green and albino cells. These sectors are a mottle of light and dark green shades. Plants regenerated from these sectors can be either green or albino, if they arise from single cells. The white sectors contain only albino cells and so give rise to only white plants. The variegated shoots regenerated from the green sectors may have been the result of somaclonal variation or segregation of green and mutant plastids that came together in a spontaneous protoplast fusion. Using a liquid shake culture approach to regenerate plants from p-calli reduced the proliferation of disorganized growth (Yarrow et al. 1987) which should reduce somaclonal variants at this stage in the protocol. The regeneration of shoots from p-calli on a solid MS medium with 0.5 mg/l BAP (Dunbar and Stephens 1989a) produced additional callus while shoots developed, and this may result in additional somaclonal variation.

6 Conclusion

Although relatively little work has been reported of protoplast techniques for *Pelargonium*, protocols for plant regeneration from the three species most economically important to the floriculture industry have been developed. These protoplast isolation and regeneration techniques can be improved and need to be extended to additional cultivars and species. Only a single known cultivar of the species *P.* × *hortorum*, *P.* × *domesticum* and *P.* × *peltatum* have been regenerated from protoplasts.

With the protoplast isolation and regeneration procedures presently developed, protoplast fusion protocols can be developed to regenerate somatic hybrids between different *Pelargonium* species. A somatic hybrid between *P.* × *domesticum* and *P.* × *hortorum* would be very desirable (Dunbar and Stephens 1991). Attempts to sexually cross these species have been unsuccessful. A somatic hybrid between these two hybrid species could be used to transfer bacterial blight resistance from *P.* × *domesticum* to *P.* × *hortorum* and the ability to flower without a cold treatment from *P.* × *hortorum* to *P.* × *domesticum*. *Pelargonium* × *domesticum* also has large throated flowers and many flower colors not found in *P.* × *hortorum* (Ewart 1980).

7 Protocols

7.1 Plant Regeneration from Protoplasts of *P.* × *hortorum* Using the Procedures of Yarrow et al. (1987) with Modifications

1. Fast-growing suspension cultures of cv. Mrs. Parker, maintained as described by Yarrow et al. (1987), were collected in a 250-µm nylon sieve 5–7 days after a subculture, and 8–10 g of tissue were placed into 100 × 15-mm Petri dish with 20 ml of an enzyme mixture of 2% Rhozyme HP150 (Rhome and Haas Co., USA), 4% Meicelase (Meija Seika Ltd., Japan), 13% mannitol (Yarrow et al. 1987), and CPW salts (Frearson et al. 1973). The suspension was incubated for 16 h at 23 °C on a rotary shaker at 15 rpm in the dark.
2. After the incubation protoplasts were passed through a 62-µm nylon sieve and centrifuged at 100 g for 8 min. Protoplasts were collected from the top of the enzyme solution by layering a W5 salts solution on top of the enzyme solution and gently removing the protoplasts in the W5 solution from the interface between the two solutions with a Pasteur pipet. The protoplasts were then spun down at 100 g for 8 min and the pellet was resuspended in a 21% sucrose and CPW salts solution. The protoplasts were floated in the 21% sucrose solution by centrifugation as before. The protoplasts were collected by layering KMP8/KP liquid medium (Yarrow et al. 1987) on top of the sucrose solution and removing the protoplasts in the medium with a Pasteur pipet.
3. The protoplasts were cultured at a density of 2.5×10^4 protoplasts/ml in 5 ml of KMP8/KP liquid medium in 60 × 15-mm Petri dishes, at 25 °C, and 9 µmol/m^3/s. After 6 days, the protoplasts were transferred to 100 × 15-mm Petri dishes containing an additional 5 ml of KMP8/KP medium. The cultures were then solidified with an equal volume of molten KMP8/KP medium (Yarrow et al. 1987) containing 0.6% agar (Difco, Detroit, MI, USA).
4. After 15–20 days, colonies in agar blocks were transferred to an MS medium supplemented with 0.1 mg/l NAA, 0.5 mg/l BAP, 3% sucrose, 4.5% mannitol, and 0.8% agar. P-calli that developed on this medium were transferred directly to MS medium with 0.5 mg/l BAP, 3% sucrose, and 0.8% agar. Shoot primordia developed and elongated when subcultured onto the same medium, and these shoots developed roots when transferred onto Hoagland's solution solidified with 0.7% agar (Dunbar and Stephens 1989b).

7.2 Plant Regeneration from Protoplasts of *P.* × *domesticum* (Dunbar and Stephens 1991)

1. Leaf explants from regal geranium cv. Melissa were used to initiate green organogenic callus tissue as described (Dunbar and Stephens 1989b). This callus was transferred to an MS medium with 3 mg/l NAA, 2 mg/l BAP, 3% sucrose, with 0.8% agar (Sigma Chemical Co., St. Louis, MO), adjusted to pH 5.8, and dispensed into 100 × 15-mm polystyrene Petri dishes (20 ml each). Callus was incubated at 25 °C in the dark, and white friable callus was selected for subculture every 3 weeks (Dunbar and Stephens 1991).

2. Seven to 11 g of callus was harvested 10 to 12 days after subculture and placed into a 100 × 15-mm polystyrene Petri dish. The tissue was broken into 1 mm^3 pieces with forceps and 15 ml of an enzyme mixture of 2% Cellulysin (Behring Diagnostics, La Jolla, CA), 0.5% Macerase (Behring Diagnostics, La Jolla, CA), 0.5 M sucrose, in CPW salts adjusted to pH 6.0 was poured over the tissue. Dishes were incubated for 6 h at 23 °C in the dark on a rotary shaker at 46 rpm. After incubation, protoplasts were released by slowly passing the callus suspended in enzyme solution in and out of a Pasteur pipet two times. Protoplasts were separated from undigested callus tissue by filtration through a 62-μm nylon sieve. The protoplasts were centrifuged at 100 g for 10 min in 15 × 125-mm test tubes. Protoplasts were collected in 6 ml of W5 solution by layering the W5 solution on top of the enzyme solution and removing the protoplasts from the interface between the two solutions in the W5 solution. An equal volume of W5 solution was added to the protoplast suspension and the protoplasts were centrifuged at 36 g for 10 min (Dunbar and Stephens 1991). The pellet was suspended into 3 ml of filter sterilized KMP8/KP liquid medium (Yarrow et al. 1987).
3. Protoplasts were diluted to 1×10^5 protoplasts/ml, and 4 ml aliquots of the protoplast suspension were poured into 60 × 15-mm polystyrene Petri dishes that contained 3 ml of KMP8/KP medium with 0.6% agarose (Sea Plaque, FMC, Rockland, MD). Protoplast cultures were fed every 7 to 10 days by the addition of 0.5 ml of fresh KMP8/KP liquid medium. After 8 weeks, the glucose in the feeding medium was reduced to 30 g/l, and after the 12th week, the glucose was reduced to 15 g/l and the sucrose was raised to 15 g/l (Dunbar and Stephens 1991).
4. After 14 weeks, 1–2 mm^3 protoplast-derived calli were transferred to 100 × 15-mm Petri dishes containing an MS medium with 3 mg/l NAA, 2 mg/l 6-BAP, 1 g/l casamino acids, 3% sucrose, 0.8% agar (Sigma Chemical Co., St. Louis, MO), and adjusted to pH 5.8. After 1 month, callus tissue was transferred to MS medium with 0.2 mg/l BAP, 2% sucrose, 0.8% agar, adjusted to pH 5.8. Callus with shoot primordia was subcultured onto the MS medium for shoot elongation. Shoots were transferred to culture tubes (25 × 150 mm) containing 15 ml of half-strength MS medium supplemented with 0.1 mg/l NAA, 1% sucrose, 0.9% agar, and adjusted to pH 5.8. After 60 days, rooted plantlets were transferred to Bacto Professional Planting Mix (Michigan Peat Co., Houston, Texas) in 8-cm plastic pots and covered with a plastic bag for 1 week (Dunbar and Stephens 1991).

References

Abo El-Nil MM, Hildebrandt AC (1976) Cell wall regeneration and colony formation from isolated single geranium protoplasts in microculture. Can J Bot 54:1530–1534

Dunbar KB, Stephens CT (1989a) Regeneration of *Pelargonium* × *hortorum* plants from protoplasts and screening protoclones for resistance to bacterial blight. In Vitro 25:31A

Dunbar KB, Stephens CT (1989b) Shoot regeneration of hybrid seed geranium (*Pelargonium* × *hortorum*) and regal geranium (*Pelargonium* × *domesticum*) from primary callus cultures. Plant Cell Tissue Organ Cult 19:13–21

Dunbar KB, Stephens CT (1991) Plant regeneration from callus-derived protoplasts of *Pelargonium* × *domesticum*. Plant Cell Rep 10:417–420

Ewart LC (1980) Utilization of flower germ plasm. HortScience 16:135–137

Frearson E, Power JB, Cocking EC (1973) The isolation, culture and regeneration of *Petunia* leaf protoplasts. Dev Biol 33:130–137

Hanniford GC, Holcomb EJ (1982) Regal geraniums. In: Mastalerz JW, Holcomb EJ (eds) Geranium III. Pennsylvania Flower Growers. University Park, PA, pp 161–169

Kameya T (1973) The effects of gelatin on aggregation of protoplasts from higher plants. Planta 115:77–82

Kameya T (1975) Culture of protoplasts from chimeral plant tissue of nature. Jpn J Genet 50:417–420

Kameya T, Uchimiya H (1972) Embryoids derived from isolated protoplasts of carrot. Planta 103:356–360

Menczel L, Nagy F, Kiss ZsR, Maliga P (1981) Streptomycin resistant and sensitive somatic hybrids of *Nicotiana tabacum* and *Nicotiana kightiana*: correlation of resistance to *N. tabacum* plastids. Theor Appl Genet 59:191–195

Moore HE (1971) Plant classification 3: Taxonomy of *Pelargoniums* in cultivation. In: Mastalerz JM (ed)

Geraniums, a Penn State Manual. Pennsylvania Flower Growers, University Park, PA, pp 14–15
Murashige T, Skoog F (1962) A revised medium for rapid growth and bioassays with tobacco cultures. Physiol Plant 15:473–597
Nelson PE, Nichols LP (1982) Bacterial blight. In: Mastalerz JW, Holcomb EJ (eds) Geraniums III. Pennsylvania Flower Growers. University Park, PA, pp 221
Wood HJ (1966) In: *Pelargoniums*: a complete guide to their cultivation. Wheaton & Co., Exeter, Great Britain, pp 17
Yarrow SA, Cocking EC, Power JB (1987) Plant regeneration from cultured cell-derived protoplasts of *Pelargonium aridum*, *P.* × *hortorum* and *P. peltatum*. Plant Cell Rep 6:102–104

I.9 Regeneration of Plants from Protoplasts of Pea (*Pisum sativum* L.)

R. LEHMINGER-MERTENS[1] and H.-J. JACOBSEN[2]

1 Introduction

Pea, with its numerous varieties and cultivars, is an important protein and vegetable crop. It is used for human and animal nutrition and might be of additional future value as an industrial crop due to some unique features of its starch. Pea is a classical model for physiological and genetic studies; it represents a highly autogamous species, and, since Mendel, it has a well worked-out genetics. Thus pea breeding is not too problematic if compared to partially outcrossing species like the faba bean, and other important grain legumes of the temperate zones. However, as in all other crops, the breeder wants to achieve some goals in his work faster than is possible with conventional means. This, for instance, can be obtained by transferring specific genes of interest directly into the relevant varieties. The application of transformation technology, however, requires, among other needs, the availability of species-adapted plant cell technologies like protoplast regeneration and plant transformation (see also de Kathen and Jacobsen 1993). In addition, some specific objectives, like, for instance, the application of in vitro selection systems, can only be achieved by means of plant cell and tissue culture, where the induction of somaclonal variation frequently is an inherent feature.

For aspects of fundamental research, where the pea plant has served intensively as a model, it should not be overlooked that the availability of an efficient in vitro regeneration system leading from a single cell to a mature and fertile plant provides a fascinating tool to study the control of developmental processes, i.e., the role of hormones and other physical and chemical factors. This is of particular importance since especially in pea many morphogenetic mutants have been characterized (Marx 1987).

Thus for both applied and fundamental aspects, it appears to be necessary to establish a regeneration system from protoplasts up to fertile plants.

In recent years, divisions in pea protoplast isolated from different tissues have been reported by various authors (Constabel et al. 1973; Gamborg et al. 1975; von Arnold and Eriksson 1976, 1977; Landgren 1976; Jia 1982). Since the success of regenerating plants from protoplasts via organogenesis was rather limited (Puonti-Kaerlas and Eriksson 1988; Lehminger-Mertens and Jacobsen 1989a), the development of a regeneration scheme via embryogenesis was expected to hold more

[1] Institut f. Genetik, Universität Bonn, Kirschallee 1, 53115 Bonn, Germany
[2] Lehrgebiet Molekulargenetik, Universität Hannover, Herrenhäuserstr. 2, 30419 Hannover, Germany

Biotechnology in Agriculture and Forestry, Vol. 22
Plant Protoplasts and Genetic Engineering III (ed. by Y.P.S. Bajaj)

promise (Lehminger-Mertens and Jacobsen 1989b). To achieve this, we established a protoplast system which allows culture and regeneration of protoplasts from nearly all genotypes. The system is reliable, efficient, and easy to handle. Moreover, the source of protoplasts is available all year around under standardized conditions.

2 Isolation of Protoplasts

2.1 Preculture

Dry seeds of a given pea genotype (*Pisum sativum* ssp. *sativum*) were surface sterilized with 70% ethanol for 30 s followed by a treatment with 4 or 5% NaOCl for about 5 min. Sterilized seeds were rinsed at least four times in sterile water and imbibed for more than 12 h, generally overnight. After imbibition, testae of the swollen seeds could easily be peeled off. From the remaining pea embryo, the two storage cotyledons were carefully removed, avoiding any injury to the axillary shoot buds of the cotyledons (see Fig. 1). The isolated embryo axes were cultured on agar (0.7%) solidified hormone-free B5 standard medium (Serva) or hormone-free B5 standard medium enriched with 10 mM NH_4Cl. Both media were adjusted at pH 5.7 and the embryo axes were grown in a growth chamber at 26 ± 2 °C in the dark or low light not more than 500 lx.

When the epicotyls and shoot apex reached about 1 cm in length, the apex was removed (see Fig. 1) and subjected to protoplast isolation (see Table 1). The remaining embryo axes (radicle, hypcotyl, and axillary shoot buds of cotyledons) were kept on the same medium for a further 4–5 days. After this growth period, two axillary shoots of about 1–2 mm from a single embryo axis could be used as best tissue sources for protoplast isolation.

Since every 4–5 days new adventitious shoot buds were formed on the embryo axes, protoplast could be isolated more than ten times from one single preculture before the embryonic axes were exhausted. Storing the preculture in Petri dishes at 4 °C in the dark for weeks or even several months was possible without losing protoplast quality.

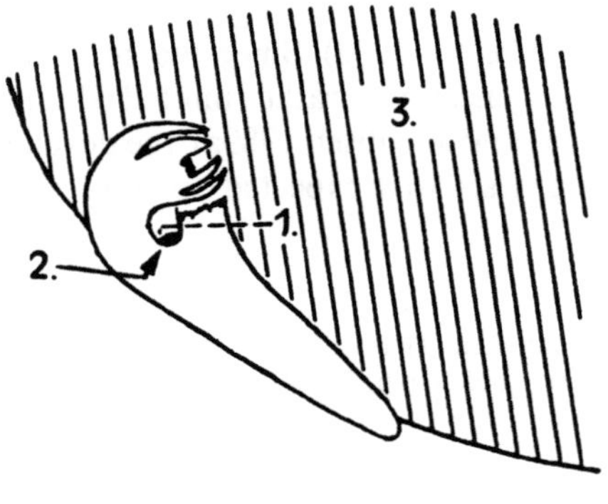

Fig. 1. Schematic cross section of a pea seed. *1* cut area for removal of embryo axis; *2* cotyledonary node meristem; *3* cotyledon

Table 1. Modified LP medium (LP*)

Inorganics	mg/l	Organics	mg/l
$CaCl_2$	660	Casein hydrolysate	1000
KH_2PO_4	170	Folic acid	0.5
KNO_3	950	Nicotinic acid	2
$MgSO_4 \times 7H_2O$	185	Pyridoxine	1
NH_4NO_3	600	Thiamine	5
$MnSO_4 \times 4H_2O$	2.23	Glycine	2
H_3BO_3	0.63	Pantothenate	0.87
Na_2EDTA	18.7		
$FeSO_4 \times 7H_2O$	13.9	**Sugars**	**g/l**
KI	0.75	Cellobiose	0.25
$Na_2MoO_4 \times 2H_2O$	0.025	Fructose	0.25
$CuSO_4 \times 5H_2O$	0.0025	Mannose	0.25
$CoCl_2 \times 6H_2O$	0.0025	Rhamnose	0.25
MES	586	Ribose	0.25
meso-Inositol	250	Xylose	0.25
		Sucrose	15
		Agarose	3
Hormones	**μM**		
Picloram	see text		
Kinetin	see text	pH 5.7	

2.2 Protoplast Isolation

Lateral buds and small shoots from the cotyledon-free embryos were harvested in hormone-free LP* medium and cut into fine slices. After removal of the LP* solution, the tissue was incubated in an enzyme mixture comprising 3% Macerozyme R-10, 4% Cellulase R-10 (both Serva) or 4% Cellulase Y-C, 0.1% Pectolyase Y-23 (both Seishin Pharmaceutical), 3 mM 2-(N-morpholino)-ethanesulfonic acid (MES) in LP*-medium at 500 mOsmol/kg and pH 5.5.

Incubation normally lasted for 12–15 h on a discontinuous rotary shaker (30 s at 60 rpm and 10 min intervals) at 27 °C in the dark. Incubation time could be reduced to 5–7 h when the enzyme mixture was desalted with a Sephadex G-25M PD 10 column (Pharmacia) at 4 °C. The yield of protoplasts was routinely high (1.7×10^7 protoplasts/g freshweight) and the viability was determined at $\geq 75\%$ (Gaff and O'Kong Ogola 1971).

After incubation, the enzyme mixture was diluted with buffered seawater (3 mM MES, Winnex synthetic sea salt from Wiegandt GmbH; 500 mOsmol/kg commonly used for seawater aquariums) and filtered through a series of sieves (500, 100 and 40 μm pore size) to remove nondigested tissues. The protoplast suspension was freed of cell debris and enzyme residues by centrifugation (35 *g* for 5 min) followed by resuspension in osmotically buffered synthetic seawater three times.

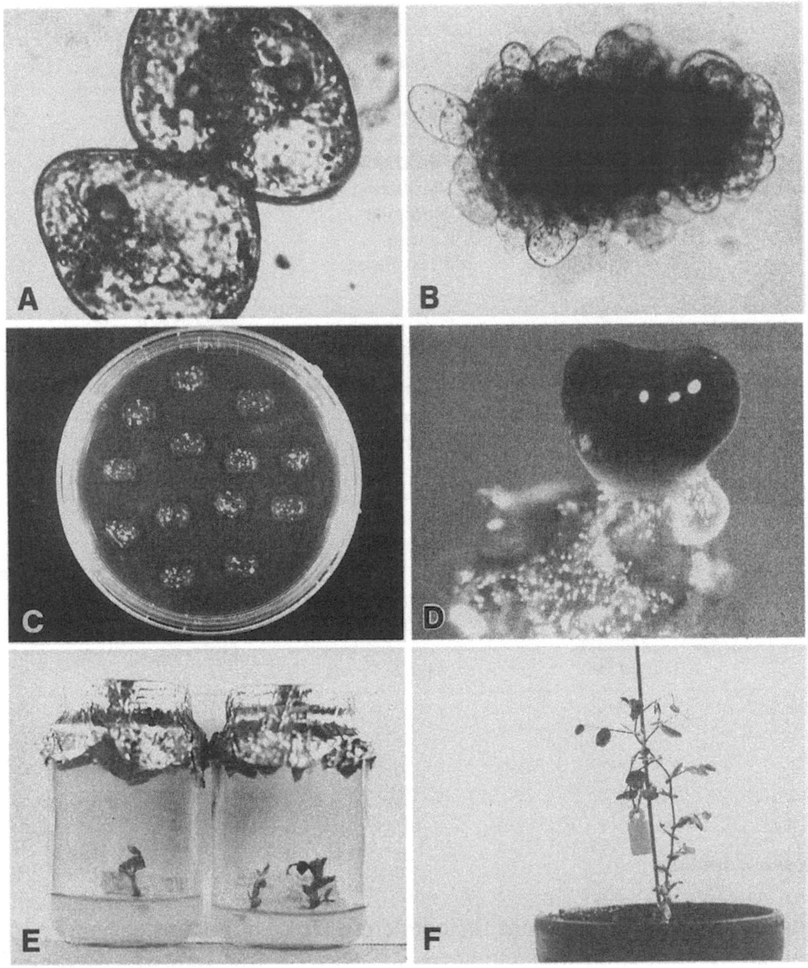

Fig. 2. **A** Two-cell stage of regenerating protoplasts. **B** Microcallus after 3 weeks. **C** Visible microcalli in agarose drops. **D** Somatic embryo after 3 weeks on induction medium. **E** Shoot development in somatic embryos. **F** Protoplast-derived pea regenerant

3 Protoplast Culture (Fig. 2)

The final pellet was resuspended in buffered LP* medium at a plating density of 1×10^5 protoplasts/ml in liquid medium or fixed in 50-l agarose droplets. Approximately 40 droplets were placed in a Petri dish (94/16 mm). After the droplets had solidified, 10 ml of liquid LP* medium were added (Gupta and Durzan 1987; Tan et al. 1987). The Petri dishes were maintained in the dark for 1 week, followed by a slow increase in the intensity of diffuse light for the next 2 weeks.

At day 3 after protoplast isolation, the surrounding liquid medium of the droplets was replaced completely by fresh LP* medium; in liquid cultures, medium should be changed as carefully as possible. At 5–6-day intervals, changing the medium again is necessary to prevent acidification and accumulation of possible growth-inhibiting compounds. With every change of medium, the inositol concentration and thus the osmolarity is reduced stepwise (60 g l, 50 g l, 45 g l). Fixing the protoplasts in agarose droplets offers the unique possibility of removing deleterious compounds, easily stabilizing the sugar concentration and pH, as well as providing a simple transfer to solid medium.

Whenever unbuffered media were used, the pH dropped from 5.8 to 3.9 within 3 days. As a result, cell division dramatically decreased in number and the high proton concentration supported the budding of nearly all protoplasts. The same observation was made with other standard media tested (B5, MS, or K8p). The protoplast medium designed for pea mesophyll protoplasts (LP medium, von Arnold and Eriksson 1977) was sufficient to induce some cell divisions, but only a few viable microcalli were formed.

About 3 weeks after the initiation of the protoplast culture in LP* medium, the free floating agarose drops were transferred onto the surface of B5 medium solidified with 0.6% highly purified agar (Sigma), mannitol 3%, 0.4 μM Picloram and 4.7 μM kinetin at pH 5.7. At this time, the plating efficiency (PE) was calculated as the percentage of macroscopically visible colonies obtained from all viable protoplasts at day 1 (Table 1).

The small microcalli could be picked individually 3 weeks later and transferred to fresh B5 medium. When the colonies had reached a size of about 2 mm (3 weeks after), they could be used for regeneration experiments.

The protoplast cultivation system up to the microcallus stage, as presented here, was found to work without any changes or problems on all cultivars of both *Pisum sativum* (Birte, Belman, Bodil, Bohatyr, Consort, Dippe's Gelbe Victoria, Finale, Madria, Maxi, Solara, Sum), representing a great genetic variability, as well as *P. sativum* ssp. *arvense*.

4 Regeneration of Plants

4.1 Embryogenesis

For embryo induction, the protoplast-derived calli were transferred onto an "induction medium". The basal media used were either MS-, LP*- or B5 salts with 2% sucrose and 0.6% highly purified agar (Sigma) enriched with $CaCl_2$, NH_4Cl, casein hydrolysate and/or mannitol in some experiments. To induce somatic embryo formation, we added strong chlorinated auxins to the different basal media like Picloram or 2,4-D as the sole phytohormone in the range of 4–15 μM (Table 2; Kysely et al. 1987, 1988; Kysely and Jacobsen 1990).

Modified MS medium containing the MS salts, B5 vitamins, 0.1% casein hydrolysate, 4–5% mannitol (MSC^{4M} and MSC^{5M}), 7–8 M Picloram and in some genotypes 3 M ABA was the most effective induction medium for all cultivars tested so far (Table 3). From all batches of protoplast-derived calli, relatively high amounts were found to be embryogenic, normally bearing more than one embryo on their surface. In general, all cultivars tested were capable of forming somatic

Table 2. Cell division and plating efficiencies (PE) of protoplasts isolated from different tissues and genotypes of pea

Genotype	Division rate after 14 days (%)		PE(%)	
	Apex	Axillary buds	Apex	Axillary buds
Birte	41	70–80	2.0	4.0
Belman	70–76	93–98	1.1	6.2
Bodil		83[a]		1.1
Dippes Gelbe Victoria		81–88		3.7
Finale		41[a]		2.7
P.s. arvense	n.d.	23–31	n.d.	n.d.

n.d.: not determined
[a] One experiment, all other values are based on at least five independent experiments.

Table 3. Embryo induction in protoplast-derived calli of *P. sativum.* ssp. *arvense* on different basal media (Lehminger-Mertens and Jacobsen 1989b)

Medium	Auxin	(M)	Somatic embryos
MS′	2,4-D	4	–
MSC	2,4-D	1–5	–
	Picloram	6	–
MSC + 3% mannitol	2,4-D	4	–
	Picloram	6	–
MSC + 4% mannitol	2,4-D	4	+
B5	2,4-D	5–8	–
	Picloram	5	–
B5 + 3% mannitol	2,4-D	5–8	–
	Picloram	4–8	–
B5 + 4% mannitol	Picloram	5	–

embryos, with only one exception: in the variety Madria, we never observed the formation of somatic embryos. In some recent experiments, embryogenesis rates of 80–90% from the variety Maxi could be obtained, while Belman and Birte exhibited regeneration frequencies between 17 and 30%. The calli of the semi-leafless cultivars Bodil, Finale, and Solara responded with frequencies of only approximately 0.3–4% under the given conditions. The duration of the induction period should not greatly exceed 3 weeks, because this strong auxin regime promotes further callus formation and also inhibits embryos development on the calli at later stages (Griga et al. 1986; Kysely et al. 1987). In order to support initiation and differentiation of somatic embryos to late globular stages, calli were transferred from the induction medium to hormone-free MS medium enriched with B5 vitamins, 4–5% mannitol and 10 mM $CaCl_2$.

Kysely et al. (1987) and Kysely and Jacobsen (1990) matured somatic pea embryos on BA-, kinetin-, and zeatin-containing media, but whenever any cytokinin was used for protoplast-derived somatic embryos, the embryos recallused. Adding ABA in different concentrations at this phase of the regeneration procedure in all cases inhibited their further development. Furthermore, ABA-treated calli with embryos on their surface turned brown on top and died.

As soon as somatic embryos reach their globular stage they should be removed from the calli. Carefully separated and transferred onto MS or B5 medium containing 1.5–2.9 μM GA_3, they continue to develop. GA_3 also seems to promote embryo maturation, since embryos grown on GA_3 media were clearly elongated. Under the GA_3 regime, germination could be obtained very easily within a few days, but GA_3 treatment should exceed 5 days, since otherwise shoot elongation was inhibited and no normal somatic seedling was obtained.

Whenever the development of somatic embryos was normal, somatic seedlings with primary leaves could be transferred to sterile soil, covering the glass vessels with cotton wool. In order to harden up the plantlets slowly, the cotton wool was carefully removed in thin layers over a period of time. After this acclimatization period, plants could be potted under nonsterile conditions and be maintained under greenhouse conditions until flowering and seed setting.

After a period of 70–100 days on hormone-free medium, the formation of somatic embryos could successfully be induced for a second time. For this second induction cycle, only viable calli have to be selected. In this second run, the frequency of somatic embryo formation was comparable to that found in the first cycle.

4.2 Organogenesis

When the protoplast-derived calli had reached about 2 mm in diameter, they could also be used for the induction of organ formation. Calli were plated on MS media containing casein hydrolysate 0.1%, mannitol 3%, and 2-iP or zeatin at 0.3–0.46 μM. The addition of 3 μM ABA usually doubles the induction rate of shoot buds, and the formation of roots on the calli was also stimulated. After the third subculture (subculture carried out monthly) on the same shoot induction medium, elongating shoots became visible. These shoots could be kept and propagated on hormone-free B5 medium for years, but have failed to root. Similar results have been reported by Puonti-Kaerlas and Eriksson (1988). Thus, regeneration of pea plants from protoplasts via somatic embryogenesis seems to be a more effective and promising way than regeneration via organogenesis.

5 Protocol

1. Preculture: cotyledon-free and decapitated embryo axes on B5 with 10 mM NH_4Cl.
2. Protoplast isolation: cotyledonary shoots and shoot buds are harvested, plasmolyzed, and cut into slices in hormone-free LP* medium. Replace LP* medium with LP* medium containing Macerocyme R10 (3%), Cellulase R10 (4%) and Pectolyase Y23 (0.1%) and place on a discontinuous rotary shaker at 27 °C in the dark.
3. Filter protoplast solution through a series of filters (500, 100, and 40 μM), centrifuge the solution 5 min at 35 *g* and wash with buffered seawater three or four times.
4. Distribute protoplasts in LP* agarose medium and pitpet 50-l droplets in sterile Petri dishes. Float agarose drops in liquid LP*.
5. Change medium: first at day 3 with LP*, then every 5 days with reducing stepwise inositol concentration.
6. Plate the agarose drops on solid B5 with mannitol 3% and 0.4 μM Picloram and 4.7 μM kinetin. After 3 weeks transfer to the same medium.

7. Induction of somatic embryogenesis: calli (2 mm in diameter) are transferred to MS medium enriched with B5 vitamins, casein hydrolysate (0.1%), mannitol (4–5%), Picloram (7–8 μM), and ABA (3 μM) for 3 weeks.
8. Transfer the calli after induction for maturation to MS medium with B5 vitamins, mannitol (4–5%) and $CaCl_2$ (10 M).
9. Remove globular somatic embryos from calli and germinate them on GA_3-containing medium for about 5 days (2.9 M), then transfer to hormone-free medium.
10. After formation of primary leaves and development of the root system on hormone-free medium, transfer somatic seedlings to sterile soil and adapt them slowly to nonsterile conditions and normal humidity.

References

Arnold S von, Eriksson T (1976) Factors influencing the growth and division of pea mesophyll protoplasts. Physiol Plant 36:192–196

Arnold S von, Eriksson T (1977) A revised medium for pea mesophyll protoplasts. Physiol Plant 39:257–260

Constabel F, Kirkpatrick JW, Gamborg OL (1973) Callus formation from mesophyll protoplast of *Pisum sativum*. Can J Bot 51:2105–2106

Gaff F, O'Kong Ogola (1971) The use of non-permeation pigments for testing the survival of cells. J Exp Bot 22:756–758

Gamborg O, Miller RA, Ojima K (1968) Nutrient requirements of suspension cultures of soybean root cells. Exp Cell Res 50:151–158

Gamborg OL, Shyluk J, Kartha KK (1975) Factors affecting the isolation and callus formation in protoplasts from the shoot apices of *Pisum sativum* L. Plant Sci Lett 4:285–292

Griga M, Tejklove E, Novak FJ, Kubalakova M (1986) In vitro clonal propagation of *Pisum sativum* L. Plant Cell Tissue Organ Cult 6:95–104

Gupta PK, Durzan DJ (1987) Somatic embryos from protoplasts of loblolly pine proembryonal cells. Bio/Technology 5:710–712

Jia S (1982) Factors affecting the division frequency of pea mesoplhyll protoplasts. Can J Bot 60:2192–2196

Kao KN, Michayluk MR (1975) Nutritional requirements for growth of *Vicia hajastana* cells and protoplasts at a very low population density in liquid media. Planta 126:105–110

Kathen A de, Jacobsen H.-J (1993) Transformation of pea (*Pisum sativum* L.). In: Bajaj YPS (ed) Biotechnology in agriculture and forestry, vol 23. Plant protoplasts and genetic engineering IV. Springer, Berlin Heidelberg New York pp 331–347

Kysely W, Mysers JR, Lazzeri PA, Collins GB, Jacobsen HJ (1987) Plant regeneration via somatic embryogenesis in pea. Plant Cell Rep 6:305–308

Kysely W, Lehminger–Mertens R, Jacobsen H-J (1988) Gen- und biotechnologische Methoden bei der züchterischen Verbesserung von Körnererbsen. Bio Eng. 4:32–34

Kysely W, Jacobsen HJ (1990) Somatic embryogenesis from pea embryos and shoot apices. Plant Cell Tissue Organ Cult: 20:7–14

Landgren CR (1976) The influence of culture conditions on mitotic activity in protoplasts derived from *Pisum* root cortical explants. Protoplasma 87:49–69

Lehminger-Mertens R, Jacobsen HJ (1989a) Protoplast regeneration and organogenesis from pea protoplasts. In Vitro Cell Dev Biol 25:571–574

Lehminger-Mertens R, Jacobsen HJ (1989b) Plant regeneration from pea protoplasts via somatic embryogenesis. Plant Cell Rep. 8:379–382

Marx GA (1987) A suite of mutants that modify pattern formation in pea leaves. Plant Mol Biol Rep 5 (3):311–335

Murashige T, Skoog F (1962) A revised medium for rapid growth and bioassys with tobacco tissue cultures. Physiol Plant 15:473–497

Puonti-Kaerlas J, Eriksson T (1988) Improved protoplast culture and regeneration of shoots in pea (*Pisum sativum* L.). Plant Cell Rep 7:242–245

Tan MMC, Rietveld EM, van Marrewijk GAM, Kool AJ (1987) Regeneration of leaf mesophyll protoplasts of tomato cultivars (*Lycopersicon esculentum*): factors important for efficient protoplast culture and plant regeneration. Plant Cell Rep 6:172–175

I.10 Regeneration of Plants from Protoplasts of *Pyrus* spp. (Pear)

S. J. OCHATT[1]

1 Introduction

1.1 Importance and Distribution of *Pyrus* Genotypes

The earliest record of pear cultivation in Europe is that provided by Homer who, around 1000 B.C., wrote that pears were "one of the gifts of the gods". By circa 300 B.C., pear cultivation was well established in Greece with distinct cultivars propagated by grafting and cuttings. France and Belgium were the main centers for pear improvement during its main developmental phase (1750–1850), where the fruit characters (with the exception of texture) found in today's commercially grown cultivars were established. Several of the cultivars which are now among the most economically important globally [e.g., Williams' Bon Chrétien (syn. Bartlett), Conference, etc.] developed at this time, and were derived from selections among open-pollinated seedlings of common cultivars (Vavrà and Orel 1971).

Nowadays, pear is grown in all temperate areas of the world and, among deciduous fruit tree species, is second only to apple in terms of world production. Pears are consumed either fresh, cooked, dried or as preserves. Only 4 of the 22 existing primary *Pyrus* species (Rehder 1967) are cultivated. *Pyrus communis* L., the European or common pear, is grown in Europe, America, and Oceania, and the Asian pear or Nashi, *P serotina* (syn. *P. pyrifolia*), is traditionally grown in China and Japan and, in recent years, also in Europe and America. Although to a lesser extent and restricted to colder areas of China and Japan, *P. bretschneideri* and *P. ussuriensis* are also grown commerically. Other *Pyrus* species grown include *P. calleryana* (used as an ornamental and as a rootstock), *P. betulaefolia* (used as a rootstock) and *P. nivalis* (a perry pear). The world production of pears reached 8.9 million metric tons in 1984: 4.6 in Europe, 2.9 in Asia and 1.1 in America, with South Africa and Australia as the main producers in Africa and Oceania, respectively (FAO Production Yearbook for 1984).

1.2 Significance and Use of Protoplasts for *Pyrus*

As compared to the other subfamilies within the Rosaceae, the Pomoideae have a relatively high basic chromosome number (x = 17). All species in the genus *Pyrus*

[1] I.N.R.A., Station d'Amélioration des Espèces Fruitières et Ornementales, Centre d'Angers, B.P. 57, 49071 Beaucouzé Cédex, France

Biotechnology in Agriculture and Forestry, Vol. 22
Plant Protoplasts and Genetic Engineering III (ed. by Y.P.S. Bajaj)

are diploid ($2n = 2x = 34$), with most pear varieties being self-sterile, due to gametophytic incompatibility. This, in turn, precludes reproduction of pear varieties true from seed, due to their high level of heterozygocity. This fact, coupled with the long juvenile periods (of average 6 years) and the space and labor needed to study large progenies, places pears amongst the most suitable crop plants for genetic improvement by somatic methods, such as those based on biotechnology, including protoplast-based approaches. Thus, the production of novel genotypes will be enhanced by means of soma/protoclonal variation or by overcoming natural barriers to sexual crossing through somatic cell hybridization.

Breeders of pear scions share many objectives in common, since new cultivars must possess basic pomological characteristics in order to be commercially acceptable. New cultivars must have, for example, adequate climatic adaptation to thrive in the regions where they are to be cultivated. In some cases, emphasis may be towards cold hardiness, whereas for others it may be drought, salinity, or heat tolerance. In addition to climatic adaptation, disease resistance is an important consideration, particularly in view of the devastating effects of diseases such as the fire blight [*Erwinia amylovora* (Burr.) Winslow et al.], which was first detected in North America but has now spread also throughout Europe (van der Zwet and Keil 1979).

In pear rootstocks, modern objectives for breeding include the development of rootstocks that induce a similar degree of size control and precocity as found with the rootstock Quince A (*Cydonia oblonga*), the currently most widely used rootstock. Extended graft compatibility, winter hardiness, and disease resistance are all present objectives aswell (Layne and Quamme 1975).

1.3 General Background to Protoplast Technology with *Pyrus* Genotypes

The published work on *Pyrus* species protoplasts, involving five different genotypes, e.g., four cultivars of common pear (*P. communis* L.) and a form of wild pear rootstock (*P. communis* var *pyraster* L.) is summarized in Table 1.

The first example of whole plant regeneration from protoplasts of a top-fruit tree was for the form of wild pear above (Ochatt and Caso 1986), which was soon after followed by the first description of spontaneous protoclonal variation, for woody species, amongst those pear rootstock regenerants (Ochatt 1987). Thus, six different protoclones were identified from a large population of protoplast-derived tissues, and they were characterized by a differential growth rate, from the callus stage, which was later correlated with differences in the leaf phenotype of the regenerated shoots. In turn, rooting was found to be erratic and inconsistent, and it was shown that an abnormal phenotype was sometimes associated with changes in the ploidy level, these being ultimately correlated with a lack of rooting ability for such regenerated shoots (Ochatt 1987).

Some months later, Revilla et al. (1987) developed a general strategy for the successful isolation of large numbers of highly viable mesophyll protoplasts for numerous deciduous nut and top-fruit tree genotypes, including wild pear and the common pear scion Williams. Also in 1987, Rech et al. (1987) described the beneficial effects of the electroporation of protoplasts, isolated from embryo-

Table 1. Protoplast-based studies with *Pyrus* genotypes

Species	Cultivar/Clone	Result obtained	References
P. communis var. *pyraster* L.	Wild pear form	Plant regeneration	Ochatt and Caso (1986)
		Protoclonal variation	Ochatt (1987)
		Protoplast isolation	Revilla et al. (1987)
		Somatic hybrid plants (+ Colt cherry)	Ochatt et al. (1989)
		Tissue compatibility in vitro	Ochatt and Patat-Ochatt (1990)
P. communis L.	Conference	Enhanced growth after electroporation	Rech et al. (1987)
		Rhizogenic callus	Ochatt and Power (1988a)
		Shoot bud regeneration	Ochatt (1990)
	Williams'	Protoplast isolation	Revilla et al. (1987)
		Plant regeneration	Ochatt and Power (1988b)
		Co-culture with *Erwinia amylovora*	Brisset et al. (1990)
		Cold storage	Ochatt and Patat-Ochatt (1991)
	Passe Crassane	Co-culture with *Erwinia amylovora*	Brissel et al. (1990)
		Cold storage	Ochatt and Patat-Ochatt (1991)
		Plant regeneration	Ochatt et al. (1992)
	Old Home	Co-culture with *Erwinia amylovora*	Brisset et al. (1990)
		Cold storage	Ochatt and Patat-Ochatt (1991)
		Plant regeneration	Ochatt et al. (1992)
	Comice	Plant regeneration	Ochatt (unpubl.)
	OHF333	Plant regeneration	Ochatt (unpubl.)

derived callus tissues of the scion cultivar Conference, on the division of protoplast-derived cells (cf. Sect. 6).

It was not until the following year, however, that organogenesis was first induced from calli deriving from protoplasts of scion cultivars of the common pear. Thus, Ochatt and Power (1988a) reported the induction of rhizogenesis from callus tissues derived from protoplasts isolated from (immature embryo-derived) callus of Conference pear. Later that year, these same authors succeeded in regenerating complete, autotrophic plants from mesophyll protoplasts of the cultivar Williams'Bon Chrétien (Ochatt and Power 1988b), the most economically important common pear scion globally (Fig. 1). Plants were also regenerated, later on, from the protoplast-derived roots produced from Conference pear protoplasts (Ochatt 1990), by employing a strategy that had proven successful in this context for other rosaceous fruit tree genotypes (Ochatt 1993).

Most recently, Ochatt et al. (1992) have reported on the recovery of intact, leaf protoplast-derived plants for two other common pear genotypes, the scion cultivar Passe Crassane and the rootstock cultivar Old Home (Fig. 1). The latter, in turn, was the first pear rootstock belonging to the species *P. communis* L., where a protoplast-to-tree system is available. Also of relevance in this chapter, was the comparison of the protoplast regenerants with micropropagated axenic shoots of the respective genotypes, by means of assessments of the banding profiles for a

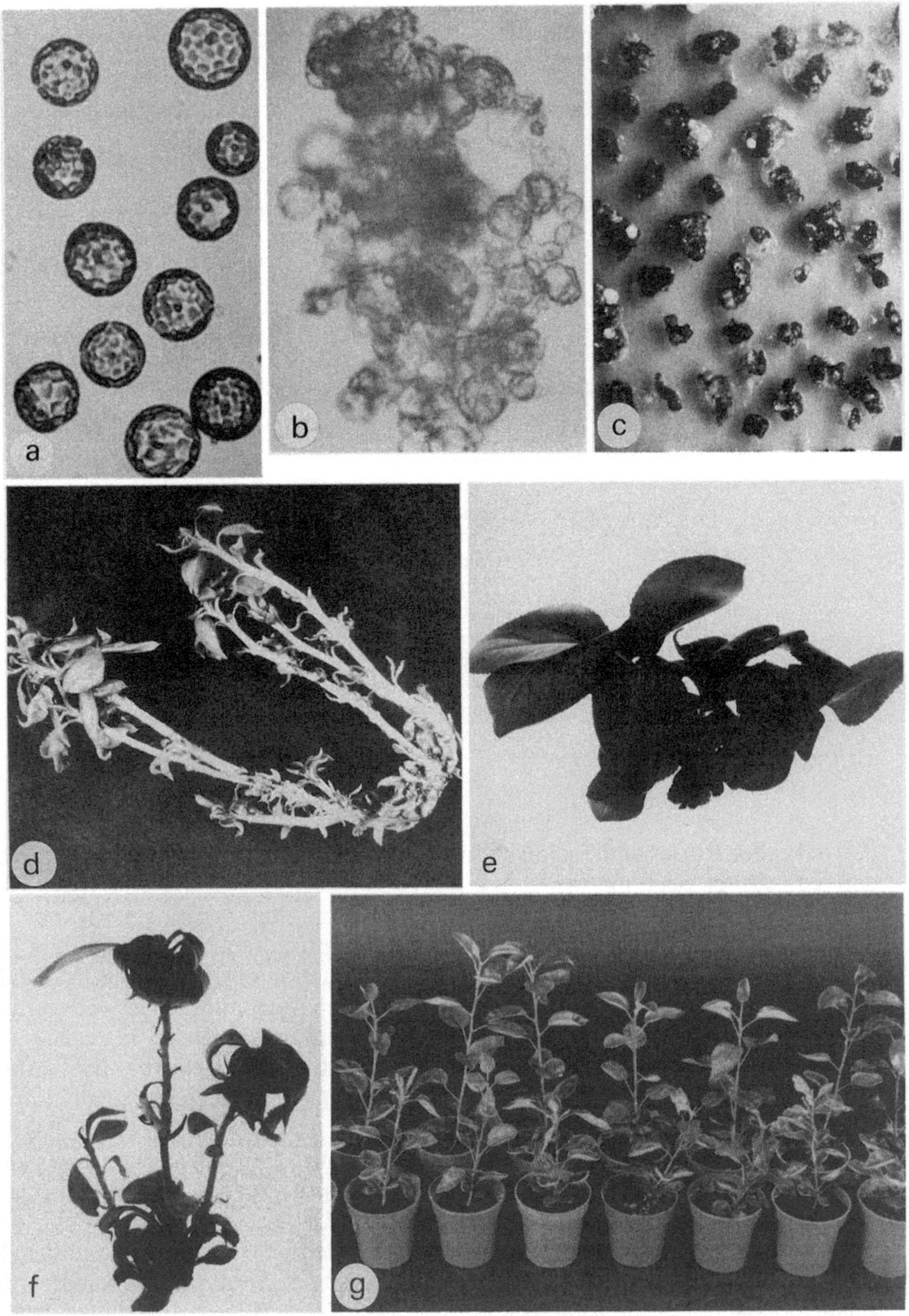

Fig. 1a–g. Plant regeneration from common pear mesophyll protoplasts (*Pyrus communis* L.) **a** Freshly isolated protoplasts of Old Home pear. **b** A protoplast-derived microcallus of Williams' pear (day 60). **c** Differentiated protoplast-derived calli of Passe Crassane pear (day 110). **d** Shoot regeneration from protoplast-derived callus of Williams' pear (day 150). **e** Regenerated protoplast-derived shoots of Old Home pear (day 160). **f** Protoplast-derived shoots of Passe Crassane pear (day 180), still attached to the regenerated callus. **g** A sample of protoplast-derived trees of Williams' pear 1 month after ex vitro transfer

large number of isoenzymatic systems. In this context, all the regenerated protoclones assessed were homogeneous and true to type, both phenotypically and with respect to the banding patterns for the 11 isozyme systems studied.

In addition, pear protoplasts were also used for studies of host-pathogen interaction with the bacterium responsible for fire blight, *Erwinia amylovora*, and a novel methodology for the precocious selection of plants according to their responses vis a vis the pathogen was developed (Brisset et al. 1990). Also, the effects of cold-storing protoplasts on their subsequent viability and competence for growth and regeneration was studied (Ochatt and Patat-Ochatt 1991).

Finally, the remaining published work with pear protoplasts was concerned with the fusion of leaf protoplasts of wild pear (*P. communis* var. *pyraster* L.) with cell suspension protoplasts of Colt cherry (*Prunus avium* x *pseudocerasus*), with the subsequent regeneration of novel somatic hybrid rootstocks for fruit trees (Ochatt et al. 1989), that were further examined for their graft-compatibility with several genotypes of the subfamilies Pomoideae and Prunoideae (Ochatt and Patat-Ochatt 1990).

2 Isolation of Protoplasts

In their work with wild pear, Ochatt and Caso (1986) compared field- and in vitro-grown leaves as protoplast source, and found that both yield and viability of protoplasts were adversely affected when using leaves of field-grown trees. Thus, leaves taken from in vitro (rooted or unrooted) shoot cultures (initiated from the same adult, 15-year-old trees) released twice as many viable protoplasts as compared to field-grown sources, with those taken from in vitro-rooted plants providing protoplast with the highest viability. Subsequently, all later experiments published on *Pyrus* protoplasts used similar, in vitro-produced source tissues. Interestingly, though, the difference in favor of rooted in vitro shoots compared to axenic shoot cultures, as verified with wild pear, were not found for any of the common pear genotypes studied to date. For common pear, the time elapsed from the last transfer of shoots appeared to play a larger role, in terms of viability of the isolated protoplasts, than the use of leaves from rooted or unrooted shoots as the protoplasts source. The use of leaves taken from shoots after at least 4 weeks from the last subculture has consistently provided the largest yields and coupled with the highest protoplast viability for all common pear genotypes so far.

For all the genotypes and source tissues of common pear, the handling of tissues prior to their enzymatic digestion was a key to success in the isolation of large yields of highly viable protoplasts. In this respect, a plasmolysis of such tissues for at least 1 h in the same solution as used for isolation but devoid of enzymes, has proven best. The effects of such a treatment were particularly noticeable in terms of protoplast viability, which was largely reduced for non-plasmolyzed tissues.

For enzymatic digestion, a ratio of 1 g fr. wt. of tissues per 10 ml of enzyme solution is routinely employed. For common pear, enzymes were dissolved in CPW salts medium (Power et al. 1989) supplemented with 0.7 M mannitol as the

Table 2. Optimum protoplast isolation strategies for various *Pyrus* genotypes

Species	Genotype	Protoplast Source[a]	Enzyme mixture[b] (%, w/v)	Yield (× 10^6/g fr. wt.)	Viability (%)
Wild pear	Wild type	FGL	ONO(0.5), DRI(0.1), MAC (0.1)	6.08	78
		VGL	As above	2.56	51
Common pear	Conference	EC	MEI(2), RHO(2), MAC(0.03)	5.81	82
	Williams'	VGL	ONO(1), PEC(0.1)	9.79	83
	Passe Crassane	VGL	ONO(1), HEMI(1), PEC(0.1), MAC(0.2)	4.80	92
	Old Home	VGL	ONO(1), HEMI(1), PEC(0.1)	7.55	94

[a] EC: embryo callus; FGL: field-grown leves; VGL: in vitro-grown leaves.
[b] DRI: Driselase; HEMI: Hemicellulase; MAC: Macerozyme R-10; MEI: Meicelase; ONO: Cellulase Onozuka R-10; PEC: Pectolyase Y-23; RHO: Rhozyme HP-150.

osmoticum, whereas for wild pear the enzyme solution typically contained MS salts medium (Murashige and Skoog 1962), at 10% strength, and included 0.35 M sucrose. Irrespective of the genotype or source tissue digested, MES-buffer (at 5 mM) and the antioxidant PVP-10 (at 1 %, w/v) were a requisite for the successful isolation of viable protoplasts. Under the optimum isolation conditions for each genotype, the typical protoplast yields ranged between 0.5 and 1 × 10^7 protoplasts/g fr. wt. of digested tissue, and were coupled with a protoplast viability of at least 80 %.

An enzyme mixture containing (w/v) 1 % Cellulase Onozuka R-10, 1 % Hemicellulase and 0.1 % Pectolyase Y23 (Revilla et al. 1987) will generally support the isolation of workable yields of viable leaf protoplasts for most pear genotypes studied so far. However, the optimum composition of enzyme mixtures varies with each genotype. Table 2 gives the best enzyme mixtures as employed to date for a range of pear genotypes and tissue sources, together with the typical yield and viability of protoplasts obtained in each case.

3 Culture of Protoplasts

3.1 Protoplast Size and Its Relationship with Culture Requirements

Assessments of the size of isolated protoplasts were deemed relevant, as characterization in this respect is a prerequisite for successful manipulation in the context of fusion, and there is also a clear relationship between protoplast size and their ability to withstand electric pulses in the context of electroporation, either for foreign gene uptake or for the enhancement of protoplast cultural responses (Rech et al. 1987; see also Sect. 6). As a spin-off of such assessments, a relationship was also found between protoplast size and the requirement for use of a semi-solid or liquid medium for sustained growth during the initial culture stages.

As compared with protoplasts of other rosaceous fruit tree species, pear protoplasts in general are relatively large in size (e.g., > 20 μm), irrespective of the source tissue. The one exception to this was for leaf protoplasts of Passe Crassane (Ochatt et al. 1992) which, on average, measured less than 15 μm in diameter. In this respect, it was found that small (< 20 μm in diameter) protoplasts grew best in agarose-solidified medium, while protoplasts beyond this size exhibited sustained growth in liquid medium. In this context, for the smaller Passe Crassane leaf protoplasts, a liquid-over-agar culture strategy had to be adopted for optimum responses (Ochatt et al. 1992).

In all published reports, pear protoplasts were plated at a relatively high initial density ($0.5–2.5 \times 10^5$/ml medium), in line with data for most other woody plant protoplast systems. Optimum responses, for all genotypes and tissue sources, were for protoplasts that were maintained either under a continuous or photoperiodic (16/8 h, light/dark) light regime.

3.2 Protoplast Culture Media

The most interesting feature during culture of protoplasts of all *Pyrus* genotypes assessed to date is their great sensitivity to the presence of ammonium ions in the medium. In this respect, only NH_4^+ -free media have supported sustained division of the protoplast-derived cells. For some particularly sensitive genotypes (Passe Crassane, Old Home; Ochatt et al. 1992), inclusion of NH_4^+ during the initial culture stages would even preclude the regeneration of cell walls by the cultured protoplasts.

All *Pyrus* protoplasts studied so far have responded best on NH_4^+-free MS-based media, with common pear genotypes proliferating best on either full- or half-strength (i.e., for the rootstock clone OHF333) MS medium, but on a 10% dilution of MS salts for those of wild pear.

As far as growth regulators are concerned, at least one auxin (NAA) and a cytokinin (BAP) have consistently provided the best results, with the further requirement for a second auxin (IAA) by some genotypes (Williams', Passe Crassane, Old Home), in order to support a sustained division of the protoplast-derived cells. As with most other species, the optimum concentrations for each hormone were genotype-dependent (Table 3).

Irrespective of the source tissue, protoplasts of rootstock genotypes (e.g., wild pear, Old Home, OHF333) generally required a medium richer in organic compounds than those of scion varieties (e.g., Williams', Conference, Passe Crassane, Comice). A similarly differential requirement between scions and rootstocks was verified with apple (Patat-Ochatt 1994), but does not seem to apply for stone fruits (Ochatt 1993).

In terms of osmotica, all common pear protoplast systems studied were cultured in media containing 0.5 M mannitol, where the osmotic pressure was gradually reduced, from the occurrence of first divisions and up to the proliferation of visible microcallus, by additions (every 7–10 days) of an osmoticum-free medium counterpart. Conversely, for wild pear, leaf protoplasts were cultured from day 0 in a medium with 0.45 M of osmoticum, achieved by adding a mixture of five different

Table 3. Optimum protoplast culture medium for various *Pyrus* genotypes

Genotype	Basal medium	Growth regulators	% IPE	% MPE	% FPE
Wild pear	Ammonium-free MS O. 1X, plus sucrose 0.318 M, and 0.033 M each of xylitol, mannitol, sorbitol and inositol*	1 mg/l NAA, 0.4 mg/l BAP	28.21	6.25	1.75
Conference	Ammonium-free MS plus 0.5 M mannitol	2 mg/l NAA, 0.5 mg/l BAP	11.24	4.46	0.97
Williams'	As above	1 mg/l NAA, 1 mg/l IAA, 0.8 mg/l BAP	21.20	7.49	1.89
Passe Crassane	As above	2 mg/l NAA, 0.5 mg/l IAA, 1 mg/l BAP	17.66	2.55	1.41
Old Home	As above[a]	2 mg/l NAA, 2 mg/l IAA, 1 mg/l BAP	20.51	4.78	1.96

[a] The medium was additionally supplemented with a filter-sterilized mixture of organic compounds that consisted of (mg/l): myo-inositol (100), thiamine-HCl (1), pyridoxine-HCl (1), nicotinic acid (1), glycine (2), Ca-panthotenate (0.5), cyanocobalamin (0.005), biotin (0.025), folic acid (0.05) and riboflavin (0.25).

compounds (sucrose, mannitol, xylitol, inosital, sorbitol) (Ochatt and Caso 1986; cf. Table 3).

The timing of developmental responses from the cultured pear protoplasts was fairly similar, regardless of the genotype or source tissue. Thus, protoplasts generally underwent a long lag phase prior to the onset of division, with no less than 3 days before cell wall regeneration and with first division occurring by day 7 of culture at the earliest. Typically, the % IPE (initial plating efficiency) in the optimum medium ranged between 11 and 28%, depending on the genotype and source assessed (Table 3). So far, the fastest-responding genotype was the wild pear, where mesophyll protoplasts gave microcalli of a size suitable for transfer for further growth (i.e., > 100 cells each, 1–2 mm in diameter) after only 3 weeks after isolation (Ochatt and Caso 1986), while this took 80 days for leaf protoplasts of the common pear rootstock cultivar Old Home (Ochatt et al. 1992). For all the other genotypes and sources studied, protoplasts proliferated to the microcallus stage within 50–60 days from isolation. The % FPE (final plating efficiency) was generally 1–2 % of the initially cultured protoplasts (Table 3).

Particular significance was attached to the time required for the proliferation of ten-celled colonies and the percentage of initially cultured protoplasts attaining this stage (% MPE, intermediate plating efficiency; Ochatt and Power 1992), as this was found to be the developmental threshold beyond which most protoplast-derived cell colonies would brown and dye, and when only an optimum medium would support sustained growth. Typical figures for the % MPE are also given in Table 3.

Following the proliferation of microcallus visible to the naked eye, these were either gently pipeted or transferred with a fine forceps onto media devised for their further growth. These media were the same as used for the maintenance of the explant-calli providing protoplasts (i.e., for Conference pear; Ochatt and Power

1988a), or were specifically designed. In this respect, wild pear protoplast-derived calli were grown on an NH_4^+-containing medium where MS salts were raised to half-strength (Ochatt and Caso 1986), whereas for all common pear genotypes, a full-strength MS medium with (Comice, Conference, Williams') or without NH_4^+ ions (Passe Crassane, Old Home, OHF3333) was preferred. Generally, NAA and BAP were the hormones of choice for this stage.

4 Plant Regeneration from Protoplast-Derived Callus

For all the *Pyrus* genotypes studied to date, regeneration responses occurred via organogenesis and, as opposed to results for many herbaceous species, they were strongly dependent on the transfer of callus pieces of at least 200 mg fr. wt. (approx. 125 mm^3) to the regeneration medium. Indeed, if callus portions were smaller than this, they would gradually become loose and ultimately die within 3 weeks of subculturing.

It is interesting to note, that for all genotypes and source tissues, organogenesis occurred only on MS-based media that contained ammonium ions. In addition, differences were apparent between the leaf protoplast-derived calli, depending on the genotype (i.e., wild vs. common pear). Wild pear calli underwent organogenesis on a full-strength MS-based medium, while those of common pear would only do so in a half-strength MS-based medium. Another significant differential requirement for organogenesis from such tissues was that while casein enzymatic hydrolysate and Ca-panthotenate were essential for the induction of shoot bud regeneration from the callus of common pear, they had no detectable effect on those of wild pear. Furthermore, for common pear, on a medium lacking both these compounds, organogenesis did not occur, while in a medium where Ca-panthotenate alone was omitted, this promoted rhizogenesis. This held true for calli derived from protoplasts of either mesophyll (Ochatt and Power 1988b) or embryo callus (Ochatt and Power 1988a) origin. For all those genotypes where caulogenesis did occur, however, BAP was the only cytokinin and IBA the only auxin supporting such organogenic response. Yet, for some genotypes (Williams', Old Home), the regeneration medium had to be further supplemented with gibberellic acid (GA_3). In addition to this, protoplast-derived callus of Old Home regenerated shoot buds only if the medium contained, also, double the concentration of group B vitamins as in the original MS formulation (i.e., 0.2 mg/l thiamine-HCl, 1 mg/l pyridoxine-HCl, 1 mg/l nicotinic acid) (Ochatt et al. 1992).

Also of relevance, in the context of regenerating intact plants from the cultured protoplasts, was the fact that, for Conference pear, shoot bud regeneration could be obtained from rhizogenic protoplast-derived calli (Ochatt and Power 1988a), by adopting an indirect pathway to caulogenesis.

Protoplast-derived calli capable of becoming organogenic underwent a unique sequence of phenotypical changes following their transfer onto the regeneration medium. They were friable and yellowish at the time of transfer, they turned reddish within a week, brownish a week later, and within 1 more week they had sharply darkened (to a coffee-brown color), and smooth, green areas (subsequently developing into shoot buds) appeared thereafter. Conversely, nonregenerating callus portions proliferated directly as green, compact, nodular tissues from the second week following transfer to the regeneration medium. This pattern of responses, which was common to all pear genotypes regenerated so far, has also been observed for most other regenerable top-fruit tree genotypes studied to date.

On average, protoplast-derived pear plants were weaker, and concomitantly more difficult to acclimatize ex vitro, than plants of the same genotype that were obtained by other in vitro methods (e.g., through micropropagation). Nevertheless, once established in soil they were soon capable of resuming normal growth and, with time, they were not distinguishable in terms of vigor or appearance from the mother plants that had provided the original protoplasts.

5 Protoplast Fusion and Somatic Hybridization Studies with *Pyrus* Protoplasts

Articles on somatic hybridization involving *Pyrus* genotypes include a single example (Ochatt et al. 1989). Noteworthy in this article was the fact that it dealt not with *Pyrus-Pyrus* protoplast fusions, but was aimed at producing intergeneric, intersubfamilial somatic hybrid rootstocks. Thus, leaf mesophyll protoplasts of the rootstock form of wild pear, *P. communis* var. *pyraster* L. (*Pomoideae*, 2n = 2x = 34) were chemically fused, following the electroporation of the parental protoplasts as separate populations, with protoplasts isolated from cell suspension cultures of the cherry rootstock Colt, *Prunus avium* × pseudocerasus (*Prunoideae*, 2n = 3x = 24).

In culture, these heterokaryons exhibited a remarkably long lag phase prior to the onset of mitotic divisions (21 days) but, by day 100 of culture, 22 putative hybrid microcalli of about 50 cells each per ml medium were recovered (i.e., equivalent to an heterokaryon plating efficiency = 0.36 %).

For most morphological characteristics assessed, the somatic hybrid plants obtained were intermediate as compared to the wild pear and Colt cherry parents. Further confirmation of hybridity for such plants was provided by somatic chromosome numbers and by assessments of the leaf isozyme banding patterns. More recently, Ochatt and Patat-Ochatt (1990) examined the likely performance of such somatic hybrid clones as rootstocks. These studies provided further confirmation of the hybrid nature of the novel genotypes obtained, after the wild pear and Colt cherry somatic hybridization.

6 Miscellaneous Studies with *Pyrus* Protoplasts

Rech et al. (1987) showed that the application of high-voltage, short-duration electric pulses (electroporation) of freshly isolated protoplasts of several species, including Conference pear, had dramatic effects on their subsequent cultural responses. Of the five genotypes studied in this report, Conference pear protoplasts were those with the largest size (average 43 μm in diameter). In this respect, when 400 μl samples of each protoplast suspension were submitted to three successive exponential pulses (of 250 to 2000 V/cm and with pulse durations of 10 to 50 μs, at 10-s intervals), the pear protoplasts were those most affected by the electrical parameters. The percentage of viable protoplasts decreased with increasing voltage and pulse duration, with extensive membrane lysis and protoplast collapse above 1000 V. Also, a lower voltage with a longer pulse (e.g., 250 V, 50μs) was more detrimental to protoplast viability than higher voltages of short duration (e.g., up to 1000 V, 10 μs). In spite of this, assessments of cell wall regeneration and division of such protoplasts revealed that electromanipulated pear protoplasts had already regenerated a cell wall and entered division by 3 days of culture, while control (nontreated) protoplasts exhibited a lag phase of at least 9 days prior to the onset of division. In addition to stimulating protoplast division, electroporation (up to 1000 V) also significantly increased the plating efficiency, whatever the duration of the pulses applied. Also relevant in this context, were the subsequent (Ochatt, unpubl.) results observed during the callusing stage where, for electroporated, protoplast-derived calli of Conference pear, voltages of 250 V or 500 V favored the largest fresh weight gains (Table 4). Such enhancement of cultural responses, for electroporated protoplasts and protoplast-derived tissues, opened up the way for their subsequent use to foster improved growth responses for various pear protoplasts on their own right (Ochatt, unpubl.), and also for the assessment of such electroporation treatment in the context of somatic hybridization experiments, where the possibility of using it as a selection strategy was explored (cf. Sect. 5; Ochatt et al. 1989).

As discussed in Sect. 1, fire blight, caused by the bacterium *Erwinia amylovora* (Burrill) Winslow et al. is one of the most devastating diseases affecting species of the Pomoideae, including pear and apple. Fire blight has been extensively studied for a long time by several authors, and the severity of the disease is highly dependent on changes in the environmental conditions which, in turn, renders field susceptibility assessments difficult and therefore variable. All previously applied in

Table 4. Growth responses of Conference pear calli following electro-treatment of protoplasts prior to their culture[a]

Voltage	Growth (mg/l) for pulse durations of (μs)				
(V)	10	20	30	40	50
250	3996 ± 288	3583 ± 233	3109 ± 199	2897 ± 203	2667 ± 193
500	3874 ± 215	3621 ± 178	3198 ± 96	3005 ± 105	2745 ± 125
750	3098 ± 183	2754 ± 154	2367 ± 111	2101 ± 88	1886 ± 118
1000	2237 ± 137	1996 ± 56	1577 ± 78	1322 ± 91	1067 ± 135
1250	1425 ± 156	1119 ± 105	982 ± 102	654 ± 47	488 ± 59
1500	1030 ± 99	866 ± 78	579 ± 64	365 ± 39	259 ± 22
1750	665 ± 51	499 ± 62	304 ± 34	197 ± 25	113 ± 9
2000	323 ± 20	265 ± 36	188 ± 108	135 ± 17	110 ± 10

[a] X ± S.D. from three successive subcultures, 20 replicates/treatment. Untreated controls = 825 ± 56 mg fr. wt.

vitro approaches succeeded in separating resistant from susceptible plant genotypes, but were unable to distinguish between cultivars with small differences in their levels of susceptibility to the disease. Hence, more precise and biologically simpler models are needed, especially for the study of the host/pathogen interaction mechanism(s) and, in this respect, the use of protoplasts appeared as a feasible alternative. Against this background, Brisset et al. (1990) have recently shown that a wall-less cell system is capable of reacting to a virulent strain of *E. amylovora*. For this, mesophyll protoplasts isolated from various pear clutivars, exhibiting different degrees of susceptibility to fire blight infection at the whole plant level (e.g., Old Home/resistant, Williams'/susceptible, Passe Crassane/very susceptible), were co-cultured with the virulent wild-type strain CFBP1430 of *E. amylovora*, and with an avirulent transposon mutant of the former (PMV6046). Results as assessed in terms of the effects of bacteria on protoplast viability, the time to the onset of divisions, the percentage of the originally cultivated protoplasts that divided once and of those proliferating to give ten-cell colonies, correlated with field resistance to fire blight of the respective pear genotypes. Thus, no differences were observed for either of the parameters tested, between protoplasts cultured on their own or co-cultured with the avirulent mutant PMV6046. Conversely, when protoplasts were co-cultured with the virulent strain CFBP1430, those of the resistant genotype Old Home responded similarly to the controls, while all parameters were significantly reduced for protoplasts of the susceptible cultivar Williams', and they declined sharply and dramatically for protoplasts of the very susceptible genotype Passe Crassane. Such a model might serve as the basis for an early screening for fire blight resistance among individual trees, produced by either conventional or biotechnological breeding.

As a spin-off from this work, present ongoing research with callus and plants regenerated from such co-cultured protoplasts has provided evidence for the existence of protoclonal variation among the regenerants. In this respect, assessments of the banding profiles for a large number of leaf isozyme systems have revealed that, for Passe Crassane, plants regenerated from protoplasts that had been co-cultured with the virulent strain CFBP1430 possess additional bands for several of the isozymes studied (Ochatt et al. in prep.). Such plants will have to be challenged with bacteria before any possible enhancement of their resistance to fire blight can be ascertained. However, these results highlight the potential usefulness of protoplasts for the study of host/pathogen interaction on a wall-less system, and also the possibility of producing novel genotypes with valuable agronomic characters by exploiting induced (directed) protoclonal variation for pears.

Finally, in a most recent report, Ochatt and Patat-Ochatt (1991) have assessed the effects of cold storage of protoplasts, of many woody plant sources and including various *Pyrus* systems, on the time course of viability and their subsequent competence for growth and (for some genotypes) plant regeneration. Thus, protoplasts were stored (at 4 °C, in the dark) in an osmoticum solution for up to 50 days, with samples taken every 5 days and transferred to culture on the respective optimum medium and conditions for each genotype. The percentage protoplast viability and IPE declined with time of cold storage (gradually up to 20 days and more rapidly thereafter), but absolute data for MPE and FPE were not affected by the duration of the cold-storage period, as compared to the control protoplast populations. Furthermore, no significant differences were observed, in terms of %

FPE, between control protoplasts and those cold-stored for up to 50 days. In fact, if the % MPE and FPE are referred not to the original number of cultured protoplasts but to the % IPE as a 100%, there was a significant increase of both these parameters for cold-stored protoplasts. Interestingly, the duration of the cold storage of protoplasts did not exert any significant effect on the shoot bud regeneration competence of the callus derived from them. It would therefore appear that cold-storage results in an enrichment of the population with viable, growth-competent, regenerable protoplasts while simultaneously eliminating (during the cold-storage period) most of those protoplasts that would, nevertheless, be unable to divide. In turn, this would reduce/negate the leakage, into the medium, of any inimical substances which would adversely affect the subsequent proliferation of the remaining, growth-competent protoplasts cultured in the same dish (environment). Such a strategy could be particularly useful for species where there is a sharp decline in plating efficiency beyond the ten-cell colony stage (i.e., most woody species), or for those genotypes which are very prone to phenolic oxidation, coupled with growth arrest, during the initial protoplast culture stages.

7 Summary and Conclusion

Plants have been regenerated from the cultured protoplasts of different tissue sources and for several pear genotypes. Also, pear protoplasts have been successfully used as partners for somatic hybridization studies, and for the generation of genetic novelty through protoclonal variation. Finally, basic research studies were performed, on the effects of the electro-stimulation and cold storage of pear protoplasts on their subsequent culture and plant regeneration responses. The summation of all these results clearly shows that *Pyrus* species should no longer be considered recalcitrant to such biotechnological breeding approaches.

In this context, and as implied at the onset of this chapter, protoplast technology holds great potential and could, indeed, become a powerful tool for pear breeding. This could apply both to rootstock (through the production of genetic novelty by somatic hybridization) and scion genotypes (through the exploitation of protoclonal variation for an enhanced resistance to either biotic or abiotic stresses).

8 General Protocols

8.1 Protoplast Isolation

1. Use the most fully expanded leaves from axenic shoot cultures, harvested at least 4 weeks post-subculturing.
2. Gently bruise the abaxial surface of leaves with the cutting edge of a scalpel blade and carefully chop them, to approx. 1-mm-wide strips, discarding the central vein.
3. Transfer 1 g fr. wt. of chopped leaf tissue to 10 ml of plasmolyticum (CPW salts medium plus 0.7 M mannitol, 1 % PVP-10 and 5 mM MES, pH 5.6), for 60–90 min.

4. Replace the plasmolyticum with 10 ml of an enzyme mixture containing (w/v) 1 % Cellulase Onozuka R-10, 1% Hemicellulase and 0.1% Pectolyase Y23, dissolved in the plasmolyticum.
5. Digest the tissues overnight (16–20 h), either stationary or with slow (40 rpm) shaking, under low light intensity illumination (500 lx).
6. Sieve the resulting suspension through a 50–80 µm pore-sized mesh, and pellet by centrifugation (10 min, 100 g).
7. Resuspend the pellet in the plasmolyticum and spin twice (as above) for rinsing.
8. Resuspend the pellet, containing protoplasts and debris, in 2 ml of plasmolyticum, layer it on top of 10 ml of CPW salts medium with 21 % (w/v) sucrose and spin the tubes (8–10 min, 80–100 g) for protoplast purification. Protoplasts will gather at the interphase between both solutions.
9. Recover the purified protoplasts in a small, measured volume of CPW salts medium with 0.5 M mannitol and allow them to rest for 1 h, while assessments of yield, viability, and size of protoplasts are to be performed.

8.2 Protoplast Culture

1. Dilute protoplasts to the required initial plating density ($0.5 - 2.5 \times 10^5$/ml depending on the genotype) with culture medium
2. Use an NH_4NO_3-free MS-based medium (at full or half-strength), with 0.5 M mannitol as the osmoticum, and supplemented with NAA (1–2 mg/l) and BAP (0.4–1.0 mg/l). Also examine the effects of adding IAA (0.5–2 mg/l) and casein enzymatic hydrolysate (50–100 mg/l) to the culture medium.
3. Enrich the medium with further organic compounds, if dealing with a rootstock genotype.
4. Keep the protoplasts as liquid layers and place dishes statically under low light intensity illumination.
5. Check cultures for cell wall regeneration and division as from day 7.
6. From day 7–14, gradually reduce the osmotic pressure of the medium by (weekly) additions of the same, mannitol-free medium in a 3–4 : 1 ratio, up to the proliferation of visible microcalli (1–2 mm in diam./approx. 100 cells each).
7. For assessments of the suitability of the medium used, score the percentage of ten-cell colonies produced (% MPE) by approx. day 30.
8. By day 60–80, score the number of microcalli produced (% FPE), as referred to the original number of cultured protoplasts.

8.3 Callus Culture and Plant Regeneration

1. Transfer microcalli 1–2 mm in diam. (day 60–80) to a semi-solid full-strength, NH_4NO_3-containing MS medium, with 2 mg/l NAA and 0.5 mg/l BAP, and keep dishes under the same conditions as used for culturing the axenic shoots that provided protoplasts.
2. Keep calli on the callusing medium until a minimum size of approx. 125 mm^3 is attained, and then transfer intact callus portions to the regeneration medium.
3. For shoot bud regeneration, employ half-strength, NH_4^+-containing MS medium, containing 0.05–0.5 mg/l IBA, 0.5–2 mg/l BAP, 0–0.2 mg/l GA_3, 50 mg/l casein enzymatic hydrolysate and 10–20 mg/l Ca-pantothenate.
4. Caulogenic competence will soon be lost by calli. This is not necessarily an irreversible process and, for some genotypes, addition of minute GA_3 concentrations (to hitherto GA_3-free medium) or an increase in the concentration of group B vitamins may restore caulogenesis.
5. Excise regenerated shoot buds from the callus underneath and transfer to a conventional micropropagation medium (as used for the mother shoots).
6. Once shoots at least 3 cm tall are obtained, transfer them, for 1 week, to a half-strength MS medium with 3 mg/l NAA and 0–0.5 mg/l IBA. Thereafter, transfer such shoots to a similar hormone-free, half-strength MS medium.
7. Regenerated, rooted plants are generally weak and will need great care during the first days following ex vitro transfer. For a detailed description of the steps to be followed, refer to Ochatt et al. (1990).

References

Brisset MN, Ochatt SJ, Paulin JP (1990) Evidence for quantitative responses during co-culture of *Pyrus communis* protoplasts and *Erwinia amylovora*. Plant Cell Rep 9:272–275

Layne REC, Quamme HA (1975) Pears. In: Janick J, Moore JN (eds) Advances in fruit breeding. Purdue University Press, West Lafayette, Indiana: 38–70

Murashige T, Skoog F (1962) A revised medium for rapid growth and bioassays with tobacco tissue cultures. Physiol Plant 15:473–497

Ochatt SJ (1987) Coltura di protoplasti come metodo per il miglioramento genetico nelle piante da frutto. Frutticoltura 49:58–60

Ochatt SJ (1990) Protoplast technology and top-fruit tree breeding. Acta Hortic 280:215–226

Ochatt SJ (1993) Regeneration of plants from protoplasts of some stone fruits (*Prunus* spp.). In: Bajaj YPS (ed) Biotechnology in agriculture and forestry vol 23: Plant protoplasts and genetic engineering IV. Springer Berlin Heidelberg pp 78–96

Ochatt SJ Caso OH (1986) Shoot regeneration from leaf mesophyll protoplasts of wild pear (*Pyrus communis* var. *pyraster* L.). J Plant Physiol 122:243–249

Ochatt SJ, Patat-Ochatt EM (1990) Assessments of graft-compatibility of somatic hybrids, *Pyrus communis* var. *pyraster* L (+) *Prunus avium* × *pseudocerasus*, and species of the subfamilies Pomoideae and Prunoideae, Rosaceae. VII Int Cong Plant Tissue and Cell Culture, Amsterdam: 216

Ochatt SJ, Patat-Ochatt EM (1991) The time-course evolution of viability and competence for proliferation of woody plant protoplasts following cold-storage. VII Int Protoplast Symp. (Uppsala, Sweden). Physiol Plant 82:A16

Ochatt SJ, Power JB (1988a) Rhizogenesis in callus from Conference pear (*Pyrus communis* L.) protoplasts. Plant Cell Tissue Organ Culture 13:159–164

Ochatt SJ, Power JB (1988b) Plant regeneration from mesophyll protoplasts of Williams' Bon Chretien (syn. Bartlett) pear (*Pyrus communis* L.). Plant Cell Rep 7:587–589

Ochatt SJ, Power JB (1992) Fundamental aspects of plant regeneration from cultured cells and protoplasts. In: Moo-Young M, Warren GS, Fowler MW (eds) Comprehensive biotechnology Suppl 2. Pergamon Press, New York: 99–127

Ochatt SJ, Patat-Ochatt EM, Rech EL, Davey MR, Power JB (1989) Somatic hybridization of sexually incompatible top-fruit tree rootstocks, wild pear (*Pyrus communis* var. *pyraster* L.) and Colt cherry (*Prunus avium* × *pseudocerasus*). Theor Appl Genet 78:35–41

Ochatt SJ, Davey MR, Power JB (1990) Tissue culture and top-fruit tree species. In: Walker JM, Pollard J (eds) Methods in molecular biology vol 6: Plant cell culture. The Humana Press, Clifton, New Jersey pp 193–207

Ochatt SJ, Chevreau E, Gallet M (1992) Organogenesis from Passe Crassane and Old Home pear (*Pyrys communis* L.) protoplasts and isoenzymatic trueness-to-type of the regenerated plants. Theor Appl Genet 83:1013–1018

Patat-Ochatt EM (1994) Plant regeneration from protoplasts of apple (*Malus* × *domestica* Borkh.). In: Bajaj YPS (ed) Biotechnology in agriculture and forestry vol 29: Plant protoplasts and genetic engineering. Springer Berlin Heidelberg

Power JB, Davey MR, McLellan MS, Wilson D (1989) Laboratory manual: plant tissue culture. The University of Nottingham

Rech EL, Ochatt SJ, Chand PK, Power JB, Davey MR (1987) Electro-enhancement of division of plant protoplast-derived cells. Protoplasma 141:169–176

Rehder A (1967) Manual of cultivated trees and shrubs, 2nd edn. McMillan, New York

Revilla MA, Ochatt SJ, Doughty D, Power JB (1987) A general strategy for the isolation of leaf mesophyll protoplasts from deciduous fruit and nut tree species. Plant Sci 50:133–137

Van der Zwet T, Keil HL (1979) Fire blight: a bacterial disease of rosaceous plants. USDA Agric Handbook 510:200 pp

Vavrà M, Orel V (1971) Hybridization of pear varieties by Gregor Mendel. Euphytica 20:60–67

Section II
Genetic Transformation

II.1 Transformation in *Arabidopsis thaliana*

R. BILANG and I. POTRYKUS[1]

1 Introduction

Arabidopsis thaliana has attracted the interest of numerous laboratories working in the areas of plant molecular biology, plant development, and genetic engineering. The properties which make this species an attractive tool for plant molecular biology and physiology have been reviewed several times (Rédei 1975; Estelle and Somerville 1986; Meyerowitz 1989). They include small genome size, low content of repetitive DNA, only 5 chromosomes, more than 80 characterized genetic loci, more than 200 RFLP markers, rich mutant collection, short generation time, thousands of seeds per plant, small size allowing growth of many plants in little space. Further progress in the areas mentioned above depends strongly on routine and efficient techniques for transient or stable introduction of genes into *Arabidopsis* cells. *Agrobacterium*-mediated DNA transfer to leaf disks (Lloyd et al. 1986; An et al. 1986; Sheikholeslam and Weeks 1987; Schmidt and Willmitzer 1988), germinating seeds (Feldmann and Marks 1987) or root explants (Valvekens et al. 1988) of various ecotypes of *Arabidopsis* are well established. However, the method of direct gene transfer to protoplasts (Paszkowski et al. 1984; Shillito et al. 1985; Negrutiu et al. 1987) might offer certain advantages compared to *Agrobacterium*-mediated transformation, e.g., for transient expression studies (Fromm and Walbot 1987), in gene-targeting studies (Paszkowski et al. 1988) or in shot-gun cloning approaches aiming at identifying DNA sequences that complement mutant phenotypes. Regeneration of transgenic plants following PEG-mediated gene transfer to protoplasts of various genotypes has been reported (Damm et al. 1989; Karesch et al. 1991a). The present protocol for direct gene transfer to leaf mesophyll protoplasts and regeneration of fertile plants of the ecotype Zürich has been developed on the basis of the published experience with protoplast and cell culture systems of *Arabidopsis* and *Brassica* (e.g., Negrutiu et al. 1975; Gresshoff 1976; Gleba and Hoffmann 1978, 1980; Negrutiu and Jacobs 1978; Xuan and Menczel 1979, 1980; Glimelius 1984; Huang and Yeoman 1984; Feldmann and Marks 1986; Kao and Seguin-Swartz 1987; Damm and Willmitzer 1988), and has been used in our group for more than 2 years.

[1]Institute of Plant Sciences, Swiss Federal Institute of Technology, ETH Zentrum, 8092 Zürich, Switzerland

Biotechnology in Agriculture and Forestry, Vol. 22
Plant Protoplasts and Genetic Engineering III (ed. by Y.P.S. Bajaj)

2 Plant Culture and Protoplast Isolation

A detailed protocol for plant and protoplast culture of ecotype Zürich has been published (Karesch et al. 1991b) and is summarized here. Sterilized seeds were cold-treated (3 days in darkness at 4 °C) and germinated on 1/2 MS-0 medium (Table 1). The plantlets were then grown for 4 to 5 weeks in Plastem boxes (Vaplast AG, Schwarzenburg, Switzerland) on the same medium but under 16 h light (4000 lx) at 22 °C and a night temperature of 16 °C. Good aeration proved to be crucial for the development of the plantlets and was provided by a hole in the lid of the Plastem boxes closed with a rubber sponge stopper.

Protoplasts were isolated from leaves of 4- to 5-week-old plants at the vegetative rosette stage. Leaves were cut into fine slices and pre-incubated for 1 h in a solution containing 0.4 M mannitol and 10 mM $CaCl_2$ (= MaCa-solution). Thereafter, this solution was replaced with enzyme solution containing 1% Cellulase Onozuka R10 and 0.25% Macerozyme R10, both dissolved in 0.4 M mannitol, 6.8 mM $CaCl_2$, pH 5.5. After overnight incubation at room temperature, the protoplasts were separated from undigested tissue in stainless steel sieves of 100 and 50 μm mesh and washed three times in W5 solution (Menczel et al. 1981). Sedimentations were performed in a table top centrifuge (Hettich Universal 1200) for 10 min at 80 *g*. When suspended in W5 at a concentration of 1 to 2×10^6 per ml, protoplasts could be stored at 4 °C in darkness for several hours without loss of competence for transformation.

Table 1. Media used for protoplast culture and regeneration of *Arabidopsis thaliana* ecotype Zürich. (Karesch et al. 1991b)

Medium	Use	Modifications[a]
1/2 MS-0	Axenic shoot growth, shoot and root elongation	Half concentration of MS salts and vitamins, 1.5% sucrose, solidified with 0.8% agar (Difco), pH 6 before autoclaving
MS-PP1	Protoplast culture prior to first divisions	0.4 M glucose, 1 mg/l 2,4-D, 0.15 mg/l kinetin filter sterilized, pH 5.8
MS-PP2	Development of microcolonies	0.2 M glucose, 0.05 mg/l 2,4-D, 2 mg/l NAA 0.15 mg/l kinetin, 0.15 mg/l BAP, pH 5.8; for callus cultures: plus 0.8% agar, pH 6 before autoclaving
MS-S	Shoot induction	3% sucrose, 0.15 mg/l IAA, 5 mg/l 2-iP, solidified with 0.8% agar, pH 6 before autoclaving
MS-R	Root induction	half concentrated MS salts and vitamins, 1.5% sucrose, 1 mg/l NAA, 0.8% agar pH 6 before autoclaving

[a] All media are based on MS—salts (Murashige and Skoog 1962) supplemented with 1 mg/l thiamine HCl, 0.5 mg/l pyridoxine HCl, 0.5 mg/l nicotinic acid, 2 mg/l glycine, 100 mg/l m-inositol.

3 Direct Gene Transfer to Protoplasts

Following the final sedimentation, the protoplasts were resuspended in a solution containing 0.5 M mannitol, 15 mM $MgCl_2$ and 0.1% (w/v) MES (2-[N-morpholino]ethane-sulfonic acid, pH 5.5) at a concentration of 2×10^6 per ml.

The transformation frequency for *Nicotiana* species is known to be dependent on the concentration of metal ions in the transformation mixture (Negrutiu et al. 1987). This correlation was not found to be as obvious for *Arabidopsis* protoplasts (Table 2): variation of the $MgCl_2$-concentration in the above solution between 7 and 60 mM did not significantly alter the absolute transformation frequency (ATF), as determined by the number of antibiotic resistant, hence putatively transformed colonies per 10^6 plated protoplasts 4 weeks after transformation. A decrease was only found with ion concentrations below and above this range. $CaCl_2$ was not an efficient substitute for $MgCl_2$ and yielded only few transformants similar to the control treatments without metal ions.

For transformation, 0.5-ml aliquots of this protoplast suspension were distributed to 12-ml Sterilin tubes, immediately followed by addition of transforming DNA and 0.5 ml of PEG solution. The PEG solution was prepared by dissolving 40% (w/v) polyethylenglycol 4000 (Merck) in 0.4 M mannitol, 100 mM Ca $(NO_3)_2$, and autoclaving. This sterilization method results in a drop of ca. 2 pH units, therefore the pH is adjusted to 8 prior to autoclaving. When the pH was buffered at nonphysiologic levels or the PEG solution was filter sterilized instead of autoclaved and therefore no pH drop occurred, the transformation frequency was markedly reduced. The protoplast/DNA/PEG-mixture was incubated at room temperature, and variation of the incubation time from 5 to 30 min had no detectable effect on the further cell division frequency or transformation frequency.

According to the procedure chosen to select for stably transformed clones (Sect. 4), different selectable marker gene constructs were used in our trans-

Table 2. The influence of the magnesium concentration during transformation on transformation frequency

Concentration of $MgCl_2$ [mM][a]	Absolute transformation frequency[b]
0	74 ± 12.0
7	83 ± 21.5
15	123 ± 33.4
30	111 ± 23.7
60	118 ± 26.3
120	57 ± 9.5

[a] Samples of 10^6 protoplasts were suspended in a solution containing 0.5 M mannitol and the corresponding amount of Mg ions. Each sample was transformed with 10 μg of linearized pSHI943 (Fig. 1) plus 40 μg of calf thymus DNA.

[b] Number of antibiotic-resistant colonies per 10^6 plated protoplasts after 3 weeks of culture in MS-PP2 medium supplemented with 25 μg/ml hygromycin B. Figures represent mean value and standard deviation of six independent repetitions.

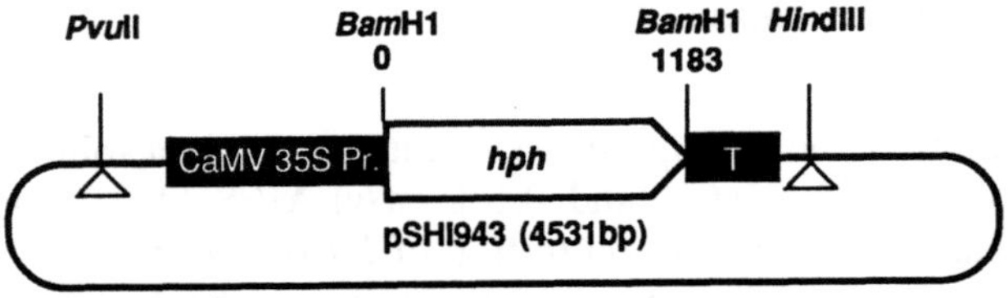

Fig. 1. Plasmid used for transformation experiments. Plasmid pSHI943 (Bilang et al. 1991) carries the gene for hygromycin-phosphotransferase (*hph* Gritz and Davies 1983) flanked by expression signals of the cauliflower mosaic virus 35S transcript. Linearization of double-stranded plasmids was achieved with *Hin*dIII. *Triangles* indicate positions where corresponding single-stranded pTZ-derivatives of pSHI943 were linearized

formation experiments. Some of these constructs, such as pHP28 (Paszkowski et al. 1988), were based on the gene coding for neomycin-phosphotransferase (*nptII*), others, like pSHI943 (Fig. 1), derived from the gene coding for hygromycin-phosphotransferase (*hph*). Plasmid DNA was sterilized in 70% ethanol and dissolved in sterile H_2O at a concentration of 1 μg/μl.

We tested the impact of carrier DNA as well as of the quantity and physical structure of the transforming DNA on transformation frequency. The absolute transformation frequency (ATF) was significantly influenced by the addition of calf thymus DNA (up to 40 μg per sample) as a carrier substrate to the transforming plasmid DNA: treatment of the protoplasts with plasmid DNA alone resulted in a ten times lower ATF. Linearized double-stranded plasmid DNA yielded in general more antibiotic resistant colonies than supercoiled plasmids (Table 3, lanes 1, and 2). Single-stranded pTZ (Mead at al. 1986) – derivatives of the marker gene constructs yielded ATFs comparable to those obtained with equimolar amounts of linear double-stranded plasmids (Table 3, lanes 2, 3, and 4). Single-stranded molecules were also linearized using a method described by Podhajaska and Szybalski (1985). The ATF was not increased by this treatment compared with

Table 3. Impact of the physical structure of the transforming DNA on transformation frequency

Exp.	Structure of DNA[a]					
	1	2	3	4	5	6
a	45[b]	75	60	80	n.d.	n.d.
b	38	90	105	65	n.d.	n.d.
c	90	150	130	78	168	138
d	153	200	174	174	157	212
e	178	178	114	114	72	130

[a] Lane 1, double-stranded circular; lane 2, double-stranded linear; lane 3, coding strand circular, lane 4, noncoding strand circular; lane 5, coding strand linear; lane 6, noncoding strand linear.
[b] Number of antibiotic resistant colonies per 10^6 plated protoplasts after 3 weeks of culture in MS-PP2 medium supplemented with 25 μg/ml hygromycin B. Figures represent mean value of two replicas per experiment, n.d. value not determined.

ATFs obtained with circular single-stranded molecules (Table 3, lanes 3, 4, 5, and 6). Transformation with either the coding or noncoding strand of a marker gene resulted in similar transformation frequencies (Table 3, lanes 3, and 5 vs. lanes 4 and 6). These results are consistent with previously published data obtained with other systems (Furner et al. 1989; Rodenburg et al. 1989) and our experience with mesophyll protoplasts of *Nicotiana tabacum* (Bilang et al. 1992). We therefore conclude that single-stranded DNA was efficiently processed to duplex form after uptake by the protoplasts. Genomic integration patterns of such processed products could not be distinguished from integration patterns obtained after transformation with double-stranded DNA, as revealed by Southern bolt analysis of genomic DNA extracted from antibiotic resistant clones (Fig. 2, lanes a to k). This analysis also reveals that introduced plasmid DNA is often rearranged to a great extent prior to its integration into the genome, a phenomenon that has also been described for mammalian and other plant systems (Lebkowski et al. 1984; Wake et al. 1984; Paszkowski 1990) and applies also to *Agrobacterium*-mediated gene transfer (reviewed by Weising et al. 1988). We therefore tested whether the genomic integration pattern of exogenous DNA and the integration copy number could be controlled by the use of different quantities of transforming plasmid DNA. However, neither transformation frequency (data not shown) nor the copy number of integrated plasmids ranging from 1 to about 12 were correlated with the amount of plasmid DNA used for transformation (Fig. 2, lanes 1 to q).

In summary, it appears that some parameters that affect transformation in other systems (Shillito et al. 1985; Negrutiu et al. 1987) have a similar influence in

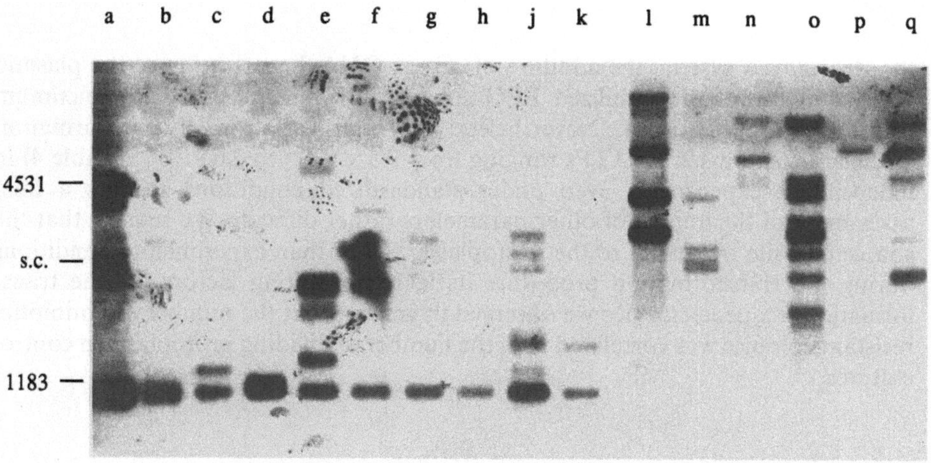

Fig. 2. Southern blot analysis of hygromycin resistant clones obtained after transformation with pSHI943 or its pTZ-derivatives. *Lane a* 15pg of partially *Bam*H1-digested pSHI943 (length of DNA fragments are indicated in bp, *s.c.* (supercoiled plasmid). *Lanes b to q* contain each 5 µg of genomic DNA from independent transformed calli. Configuration of transforming DNA was: *lane a* double-stranded circular; *lanes b, c, l to q* double-stranded linear; *lanes e and f* non-coding strand circular; *lanes g and h* coding strand linear; *lanes j and k* noncoding strand linear. Amount of plasmid DNA used for transformations was per sample: *lanes a to k, m,* and *n* 10 µg of double-stranded/ 5 µg of single-stranded DNA; *lane l* 5 µg; *lanes o and p* 20 µg; *lane q* 30 µg. Genomic DNA was digested with *Bam*H1 (*lanes b to k*) or *Hind*III (*lanes l to q*), separated on a 0.8% agarose gel, blotted onto nylon H-bond membrane and probed with ^{32}P-labeled *Bam*H1-fragment of pSHI943

Table 4. Variation of the transformation frequency

Exp.[a]	Transformation frequency[b]				
	Replicas				Mean
a	70,	72			71
b	98,	122			110
c	40,	66			53
d	40,	46,	74		53
e	114,	141			127
f	148,	157			152
g	62,	120,	149,	153	121
h	139.	179,	188,	229	183
j	250,	283			266
k	95,	213			154
l	66,	133			100
Mean					127 ± 65.4

[a] Samples of 10^6 protoplasts were suspended in 0.5 M mannitol, 15 mM $MgCl_2$, and each transformed with 10 µg of linearized pSHI943 plus 40 µg of calf thymus DNA.
[b] Number of antibiotic resistant colonies per 10^6 plated protoplasts after 3 weeks of culture in MS-PP2 medium supplemented with 25 µg/ml hygromycin B.

the *Arabidopsis* system: the addition of carrier DNA, linearization of the plasmid DNA and use of pH-stabilized PEG solution are recommended for maximum transformation frequency. Nevertheless, the high variation of transformation efficiencies (expressed by ATFs ranging from 0.5×10^{-4} to 2.6×10^{-4}; Table 4) in independent experiments even under standardized conditions renders a final assessment of the impact of other parameters rather difficult. We suggest that the source and the treatment of the protoplasts, rather than experimental conditions during the transformation procedure itself, were limiting factors for the transformation frequencies, since we observed in general that the number of antibiotic-resistant colonies was correlated with the number of dividing protoplasts in control cultures.

4 Selection of Transformed Colonies

After the transformation procedure, the protoplasts were washed once with W5 solution and once with 0.4 M mannitol and then embedded in alginate gels. Plating method and plating density were chosen according to the experimental purposes: plating of protoplasts in thin layers of gels at relatively low densities (2×10^5/ml) allowed the visual screening of protoplast development and the determination of cell division and plating efficiencies (Fig. 3A, B). In contrast, alginate beads containing up to 7×10^5 protoplasts per ml were prepared in large-scale transformation experiments, where subsequent culture of the colonies was performed under selective conditions. Preparation of alginate gels requires that calcium-free protoplast suspensions are allowed to polymerize over a calcium-containing basal layer or in a calcium-containing solution. Therefore, protoplasts suspended at double plating density in 0.5 M mannitol were mixed with the same volume of a solution of 2% (w/v) sodium alginate in 0.4 M mannitol. For the preparation of thin gel layers, aliquots of 1 ml of this mixture were pipeted onto a basal layer of solidified Difco Bacto agar containing 10 mM $CaCl_2$ and 0.4 M mannitol. Following gellification for 30 min, the gellified surface was flooded with 2.5 ml of MaCa-solution, resulting in the release of the gel layers. Alternatively, alginate beads were obtained by dropwise addition of the protoplast/alginate mixture to a gellifying solution (0.4 M mannitol, 50 mM $CaCl_2$).

The plated protoplasts were subjected to a cold-starvation treatment, i.e., incubation for 2 days in darkness at 4 °C. Thereafter, the thin-layer gels were transferred to culture medium MS-PP1 (Table 1), or the gellifying solution was replaced by this medium. Five ml of medium per ml of alginate gel proved to be most suitable for cell proliferation. After 7 more days of culture in darkness at 22 °C, half the volume of medium MS-PP1 was replaced by MS-PP2 (Table 1). Culture was continued in continuous dim light (700 lx), and half the volume of the medium was weekly replaced by fresh MS-PP2 medium. Cell division frequency was usually determined 2 weeks after plating of the protoplasts (Fig. 3B) and reached an average of 5% of cells entering first divisions. Under nonselective culture conditions, the plating effeciency (i.e., the ratio of cells yielding proliferating calli) varied between approx. 0.01% to 5% with an average of approx. 0.5%.

In order to select putative transformed colonies, selective agents were added to MS-PP2 medium 1 week after the first cell divisions occurred. The antibiotics hygromycin B, kanamycin sulfate, and paromomycin suppressed growth of wild-type cells embedded in alginate with different efficiencies. Rapid cell death was caused by hygromycin (25 μg/ml final concentration), whereas the final concentration of kanamycin had to be increased to 200 μg/ml in order to cause complete growth inhibition with no escaping colonies. Low doses of paromomycin are known to suppress cellular growth with high efficiency in other systems (Guerche et al. 1987; Schnorf et al. 1991), but led in the present system only to a slight reduction of callus growth at concentrations of up to 50 μg/ml. Since comparatively lower doses of all these antibiotics were lethal for in vitro-grown *Arabidopsis* seedlings, we suggest that the above observations do not necessarily reflect an enhanced resistance to kanamycin or paromomycin of *Arabidopsis* cells. It is conceivable that physical or chemical properties of the alginate gel matrix interfere with the activity of the antibiotic compounds.

In view of the high selective activity of hygromycin B, most of our transformation experiments were performed with expression vectors carrying the *hph* gene which confers hygromycin-B resistance to transformed clones. Due to the rapid action of hygromycin B, resistant colonies could be detected as soon as 10 days after beginning of selection. After 3 to 4 weeks of culture, antibiotic-resistant colonies were scored, mechanically liberated from the alginate matrix, and transferred onto MS-PP2 plates solidified with 0.8% agar and containing the selective agents at the same concentrations as in liquid culture (Fig. 3C). Molecular analysis

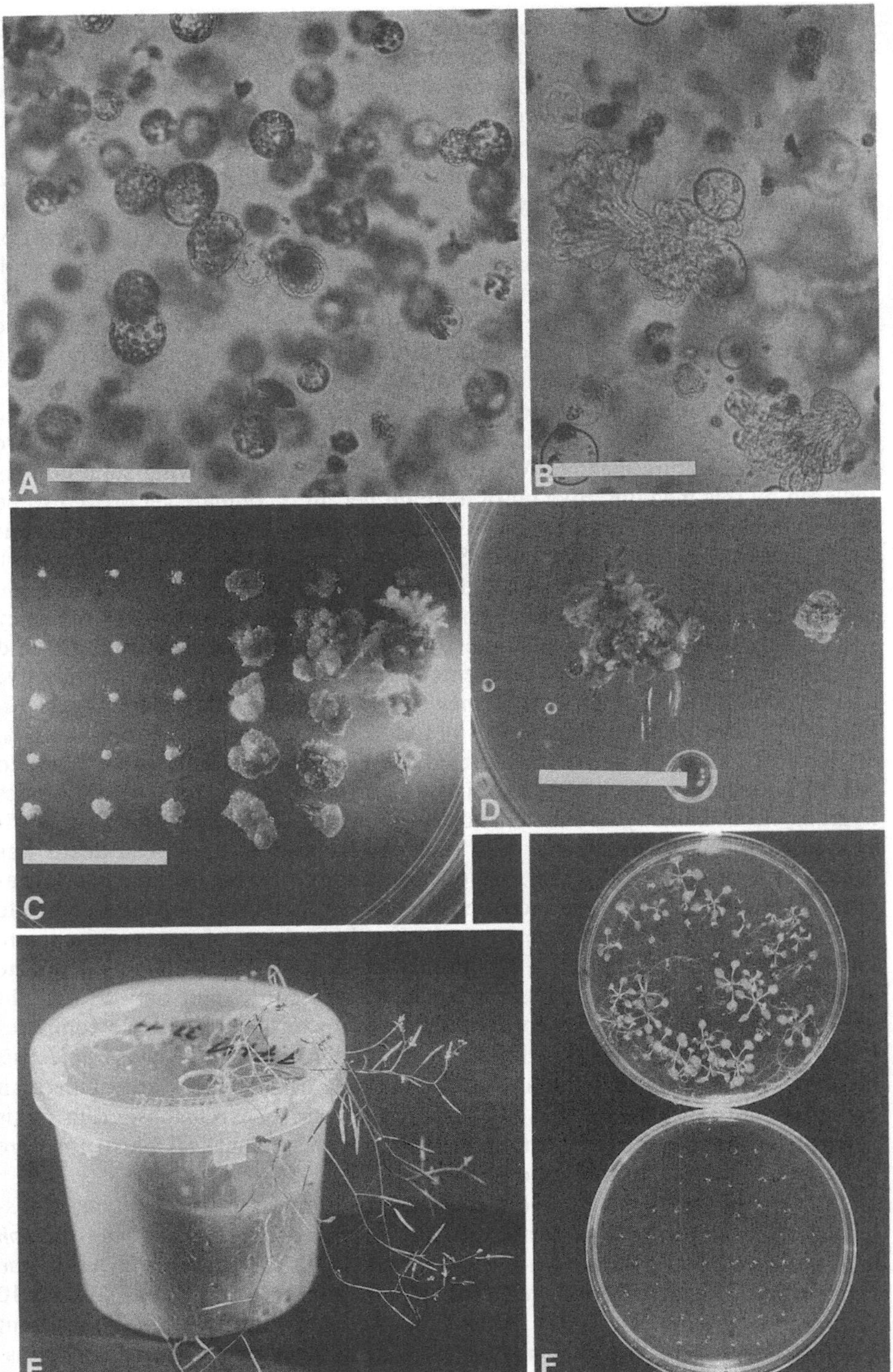
A
B
C
D
E
F

of the putative transformants was performed with genomic DNA isolated from calli grown for 2 to 3 weeks on MS-PP2 plates. All of the more than 100 hygromycin- or kanamycin-resistant colonies that were subjected to Southern blot analysis (examples shown in Fig. 2) turned out to be true transformants with copies of exogenous DNA integrated into the genome.

5 Plant Regeneration

Within 3 weeks of culture, clones reached a size of ca. 5 mm in diameter. At this stage, the development of shoot meristems was induced by transfer to MS-S culture medium (Table 1) and incubation at 20 °C in continuous light (2000 lx). After 10 to 16 days of incubation, shoot meristems could be identified as small dark-green areas. In contrast to protoplast division- and plating efficiencies (Sect. 4), shoot induction frequencies were highly variable and depending on the culture conditions applied to the protoplast-derived calli. Prolonged culture of callus material on MS-PP2 agar plates led to greening and hardening of the entire calli and to the ultimate loss of competence for shooting.

Elongation and differentiation of shoots were supported by reduction of exogenous hormones and good aeration: calli with shoot primordia were transferred to 1/2 MS-0 agar in Plastem boxes as described for axenic shoots (Sect. 1). Incubation was at 20 °C, 16 h 3000 lx, 16 °C at darkness. After 2 weeks, shoot rosettes (Fig. 3D) could be excised from the calli for subsequent rooting. In parallel cultures, non-morphogenic callus material was usually propagated on MS-PP2 agar in order to provide enough material for further molecular analysis.

Rooting of the excised shoots was induced on MS-R medium (Table 1) without selection pressure at 20 °C in continuous dim light for 3 days. Root growth was promoted by subsequent transfer to hormone- and antibiotic-free MS-agar and incubation at 20 °C, 16 h 3000 lx, 16 °C darkness. Rooted shoots were transferred to a mixture of 2 vol of potground and 1 vol of perlite contained in small boxes standing in shallow water in Plastem boxes with three holes in the lid. Adjustment to growth in a less humid atmosphere was achieved by stepwise removal of the stoppers in the lid (Fig. 3E). Flowering and fruit ripening took place in the open air at 20/16 °C, 16 h 3000 lx per day. The fertility of the regenerated plants proved to be very variable: in some of the experiments, all of the independent shoots set seeds, in other circumstances, only sterile siliques were formed. However, these phenomena were apparently not correlated to the shoot culture conditions, since fertile shoots

Fig. 3A–F. Regeneration of fertile transformed *Arabidopsis* plants from protoplasts. **A** Protoplast population embedded in alginate gels 1 day after isolation from leaf tissue. **B** Proliferating protoplast-derived microcalli 10 days after protoplast isolation. *Bars* in **A** and **B** 0.1 mm. **C** Protoplast-derived calli on MS-PP2 plates containing 25 mg/l hygromycin B. *Left part of dish* wild-type calli precultured under nonselective conditions. *Right* transformed, pre-selected calli. *Bar* 2 cm. **D** Late and earlier stage of shoot induction on MS-S plates. *Bar* 1 cm. **E** Fertile transformant after transfer to potground. **F** R1 seedlings of a wild-type regenerant (*bottom*) and a plant transformed with a pSHI943 derivative, grown for 7 days on MS-O plates supplemented with 25 mg/l hygromycin B. (Karesch et al. 1991a) were obtained both by using the system described above as well as in vitro on non-rooted shoots.

Seeds were sterilized as described in Sect. 2 and allowed to germinate on half-concentrated MS agar supplemented, where appropriate, with 25 µg/ml hygromycin B or 200 µg/ml kanamycin sulfate for phenotypic offspring analysis of the regenerated transgenic plants (Fig. 3F). Segregation of the antibiotic-resistant phenotype was not in all cases according to the expected Mendelian rules (Karesch et al. 1991a). Aberrant segregation ratios were shown to be the result of post-meiotic inactivation of the correctly transmitted marker gene (Mittelsten Scheid et al. 1991).

6 Concluding Remarks

The method of direct gene transfer to protoplasts of *Arabidopsis thaliana* as it is described here yielded transformed colonies with reliable frequencies. It has been used to study extrachromosomal intermolecular recombination between co-transformed plasmid molecules (R. Bilang and J. Paszkowski unpubl.). Homologous recombination occurred in up to 50% of the protoplasts that were competent for integration of exogenous DNA, and events occurring at these relatively high frequencies were easily scorable with the present system. Other, less abundant, pathways of DNA metabolism like gene targeting events (Paszkowski et al. 1988) could not be detected in *Arabidopsis* with the present transformation frequency, even in large scale experiments with more than 30×10^6 protoplasts. The value of PEG-mediated gene transfer could therefore be further increased by additional attempts to maximize the transformation frequency. These attempts should include characterization of the protoplast source and optimization of culture conditions for protoplast proliferation.

In view of the difficulties with shoot induction and the nonreliable degree of fertility of the regenerated shoots there is no doubt that the method of choice for the production of transgenic *Arabidopsis* plants is *Agrobacterium*-mediated gene transfer (for references see Sect. 1). The present protocol for direct gene transfer to protoplasts of *Arabidopsis thaliana* seems to be sufficient for transient expression systems and for transformation experiments where growth and selection of protoplast-derived calli can already answer the question of an experiment.

Acknowledgment. The authors wish to thank M. Schrott and Dr. J. Simmonds for careful reading of the manuscript. Figure 3F was kindly provided by Dr. O. Mittelsten Scheid. The results presented here are part of RB's PhD thesis which was accomplished in the laboratory of Dr. J. Paszkowski.

References

An G, Watson BD, Chiang CC (1986) Transformation of tobacco, tomato, potato and *Arabidopsis thaliana* using a binary Ti vector system. Plant Physiol 81:301–305

Bilang R, Iida S, Peterhans A, Potrykus I, Paszkowski J (1991) The 3'-terminal region of the hygromycin-B-resistance gene is important for its activity in *Escherichia coli* and *Nicotiana tabacum.* Gene 100:247–250

Bilang R, Peterhans A, Bogucki A, Paszkowski J (1992) Single-stranded DNA as a recombination substrate in plants assessed by stable and transient recombination assays. Mol Cell Biol 12:329–336

Damm B, Willmitzer L (1988) Regeneration of fertile plants from protoplasts of different *Arabidopsis thaliana* genotypes. Mol Gen Genet 213:15–20

Damm B, Schmidt R, Willmitzer L (1989) Efficient transformation of *Arabidopsis thaliana* using direct gene transfer to protoplasts. Mol Gen Genet 217:6–12

Estelle MA, Somerville C (1986) The mutants of *Arabidopsis.* Trends Genet 2:89–93

Feldmann KA, Marks MD (1986) Rapid and efficient regeneration of plants from explants of *Arabidopsis thaliana.* Plant Sci 47:63–69

Feldmann KA, Marks MD (1987) *Agrobacterium*-mediated transformation of germinating seeds of *Arabidopsis thaliana*: a non-tissue culture approach. Mol Gen Genet 208:1–9

Fromm M, Walbot V (1987) Transient expression of DNA in plant cells. In: Hohn T, Schell J (eds) Plant DNA infectious agents. Springer, Berlin Heidelberg New York, pp 304–310

Furner IJ, Higgins ES, Berrington AW (1989) Single-stranded DNA transforms plant protoplasts. Mol Gen Genet 220:65–68

Gleba YY, Hofmann F (1978) Hybrid cell lines *Arabidopsis thaliana* + *Brassica campestris*: no evidence for specific chromosome elimination. Mol Gen Genet 165:257–264

Gleba YY, Hofmann F (1980) "Arabidobrassica": a novel plant obtained by protoplast fusion. Planta 149:112–117

Glimelius K (1984) High growth rate and regeneration capacity of hypocotyl protoplasts in some Brassicaceae. Physiol Plant 61:38–44

Gresshoff PM (1976) Protoplast and callus regeneration of *Arabidopsis thaliana.* Arabidopsis Inf Serv 13:211–214

Gritz L, Davies J (1983) Plasmid-encoded hygromycin B resistance: the sequence of hygromycin B phosphotransferase gene and its expression in *E. coli* and in *S. cerevisiae.* Gene 25:179–188

Guerche P, Bellini C, Le Moullec JM, Caboche M (1987) Use of a transient expression assay for the optimization of direct gene transfer into tobacco mesophyll protoplasts by electroporation. Biochimie 69:621–628

Huang BC, Yeoman MM (1984) Callus proliferation and morphogenesis in tissue cultures of *Arabidopsis thaliana* L. Plant Sci Lett 33:353–363

Kao HM, Seguin-Swartz G (1987) Study of factors affecting culture of *Brassica napus* L. and *B. juncea* Coss. mesophyll protoplasts. Plant Cell Tissue Organ Cult 10:79–90

Karesch H, Bilang R, Mittelstem Scheid O, Potrykus I (1991a) Direct gene transfer to protoplasts of *Arabidopsis thaliana.* Plant Cell Rep 9:571–574

Karesch H, Bilang R, Potrykus I (1991b) *Arabidopsis thaliana*: protocol for plant regeneration from protoplasts. Plant Cell Rep 9:575–578

Lebkowski JS, DuBridge RB, Antell EA, Greisen KS, Calos MP (1984) Transfected DNA is mutated in monkey, mouse, and human cells. Mol Cell Biol 4:1951–1960

Lloyd AM, Barnason AR, Rogers SG, Byrne MC, Fraley RT, Horsch RB (1986) Transformation of *Arabidopsis thaliana* with *Agrobacterium tumefaciens.* Science 324:464–466

Mead DA, Szczesna-Skorupa E, Kemper B (1986) Single-stranded DNA "blue" T/ promoter plasmids: a versatile tandem promoter system for cloning and protein engineering. Prot Eng 1:67–74

Menczel L, Nagy F, Kiss ZR, Maliga P (1981) Streptomycin resistant and sensitive somatic hybrids of *Nicotiana tabacum* + *N. knightiana*: correlation of resistance to *N. tabacum* plastids. Theor Appl Genet 59:191–195

Meyerowitz EM (1989) *Arabidopsis,* a useful weed. Cell 56:263–569

Mittelsten Scheid O, Paszkowski J, Potrykus I (1991) Reversible inactivation of a transgene in *Arabidopsis thaliana.* Mol Gen Genet 228:104–112

Murashige T, Skoog F (1962) A revised medium for rapid growth and bioassays with tobacco tissue cultures. Physiol Plant 15:473–497

Negrutiu I, Jacobs M (1978) Factors which enhance in vitro morphogenesis of *Arabidopsis thaliana*. Z Pflanzenphysiol 90:423–430
Negrutiu I, Beeftink F, Jacobs M (1975) *Arabidopsis thaliana*. as a model system in somatic cell genetics. Plant Sci Lett 5:293–304
Negrutiu I, Shillito R, Potrykus I, Biasini G, Sala F (1987) Hybrid genes in the analysis of transformation conditions I. Setting up a simple method for direct gene transfer in plant protoplasts. Plant Mol Biol 8:363–373
Paszkowski J (1990) Fate of foreigh DNA introduced to plant cells. In: Setlow JK (ed) Genetic engineering, principles and methods, vol 11. Plenum Press, New York, pp 145–157
Paszkowski J, Shillito RD, Saul MW, Mandák V, Hohn T, Hohn B, Potrykus I (1984) Direct gene transfer to plants. EMBO J 3:2717–2722
Paszkowski J, Baur M, Bogucki A, Potrykus I (1988) Gene targeting in plants. EMBO J 7:4021–4026
Podhajaska AJ, Szybalski W (1985) Conversion of the FokI endonuclease to a universal restriction enzyme: cleavage of phage M13mp7 DNA at predetermined sites. Gene 40:175–182
Rédei GP (1975) *Arabidopsis* as a genetic tool. Annu Rev Genet 9:111–127
Rodenburg KW, de Groot MJA, Schilperoort RA, Hooykaas PJJ (1989) Single-stranded DNA used as an efficient new vehicle for transformation of plant protoplasts. Plant Mol Biol 13:711–719
Schmidt R, Willmitzer L (1988) High efficiency *Agrobacterium tumefaciens*-mediated transformation of *Arabidopsis thaliana* leaf and cotyledon explants. Plant Cell Rep 7:583–586
Schnorf M, Neuhaus-Url G, Galli A, Iida S, Potrykus I, Neuhaus G (1991) An improved approach for transformation of plant cells by microinjection: molecular and genetic analysis. Trans Res 1:23–30
Sheikholeslam SN, Weeks DP (1987) Acetosyringone promotes high efficiency transformation of *Arabidopsis thaliana* explants by *Agrobacterium tumefaciens*. Plant Mol Biol 8:291–298
Shillito RD, Saul MW, Paszkowski J, Potrykus I (1985) High efficiency direct gene transfer to plants. Biotechnology 3:1099–1103
Valvekens D, Van Montagu M, Lijsebettens M (1988) *Agrobacterium tumefaciens*-mediated transformation of *Arabidopsis thaliana* root explants by using kanamycin selection. Proc Natl Acad Sci USA 85:5536–5540
Wake CT, Gudewicz T, Porter T, White A, Wilson JH (1984) How damaged is the biologically active subpopulation of transfected DNA? Mol Cell Biol 4:387–398
Weising K, Schell J, Kahl G (1988) Foreign genes in plants: transfer, structure, expression, and applications. Annu Rev Genet 22: 421–477
Xuan LT, Menczel L (1979) Improved protoplast culture in *Arabidopsis thaliana*. Arabidopsis Inf Serv 16:82–83
Xuan LT, Menczel L (1980) Improved protoplast culture and plant regeneration from protoplast-derived callus in *Arabidopsis thaliana*. Z Pflanzenphysiol 96:77–80

II.2 Transformation in *Atropa belladonna*

YOSHIHITO SUZUKI[1], YURIKO KURIOKA[2], TAKESHI OGASAWARA[2], and HIROSHI KAMADA[2]

1 Introduction

Atropa belladonna is native to dry area from southwestern Europe to West Asia, and has been cultivated as an important source of the pharmacologically useful alkaloids atropine, hyoscyamine, and scopolamine. These alkaloids as plant secondary metabolites cannot be easily artificially synthesized and must be extracted from plants which are not adapted to agriculture. Production of alkaloids by in vitro cultured undifferentiated cells as an alternative to field cultivation is also difficult (Bajaj and Simola 1991), because the undifferentiated cells tend to be genetically unstable in culture and the selection of cell lines producing the alkaloids at a high level has not been achieved. In plants, alkaloids are biosynthesized in the roots and translocated to the leaves. Production of the alkaloids using root culture has been explored as a new type of in vitro culture. Root culture was facilitated by the transformation by *Agrobacterium rhizogenes*. Actually, in some reports on transformation of *A. belladonna*, the main purpose is to obtain not transformed plants but hairy roots for the study of alkaloid production (Kamada et al. 1986; Jung and Tepfer 1987; Walton et al. 1990).

There are several reports describing transformation of *A. belladonna* by *Agrobacterium* (Table 1). In some of them, wild-type Ti or Ri plasmids are used (Kamada et al. 1986; Ondrej and Protiva 1987; Ondrej et al. 1987; Jung and Tepfer 1987; Walton et al. 1990), and neomycin phosphotransferase II gene (Beck et al. 1982), hygromycin phosphotransferase gene (in preparation) and the *rolC* gene locating on Ri T-DNA are the only foreign genes introduced into *A. belladonna* (Ondrej and Vlasák 1987; Mathews et al. 1990; Kurioka et al. 1992). The information on methods applied to transformation of *A. belladonna* along the general procedure of plant transformation by using *Agrobacterium* is presented here.

2 Preparation of *Agrobacterium*

There are two categories of transformation vectors based on *Agrobacterium*, cointegrating vectors and binary vectors. There are some reports of transferring the

[1] Faculty of Agriculture, University of Tokyo, Yayoi 1-1-1, Bunkyo-ku, Tokyo 113, Japan
[2] Gene Experiment Center, University of Tsukuba, Tsukuba, Ibaraki 305, Japan

Biotechnology in Agriculture and Forestry, Vol. 22
Plant Protoplasts and Genetic Engineering III (ed. by Y.P.S. Bajaj)

Table 1. Summary of *Agrobacterium*-mediated transformation of *Atropa belladonna*

Agrobacterium and vector gene	Material used for inoculation	Selection method	Tissue obtained	Reference
A. rhizogenes 15834	Seedling	Hormone autotrophy	Hairy root, plant	Kamada et al. (1986)
A. tumefaciens B6S3	Seedling	Hormone autotrophy	Tumor, plant	Ondrej and Protiva (1987)
A. tumefaciens C58	Seedling	Hormone autotrophy	Tumor, plant	Ondrej and Protiva (1987)
A. tumefaciens Bo542	Seedling	Hormone autotrophy	Tumor, plant	Ondrej and Protiva (1987)
A. tumefaciens C58C1 (pRiA4b)	Seedling	Hormone autotrophy	Hairy root, plant	Ondrej and Protiva (1987)
A. rhizogenes 8196	Seedling	Hormone autotrophy	Tumor	Ondrej and Protiva (1987)
A. rhizogenes 8196	Seedling	Hormone autotrophy	Tumor,	Ondrej and Vlasák (1987)
A. tumefaciens Bo542	Seedling	Hormone autotrophy	Tumor, plant	Ondrej and Vlasák (1987)
A. tumefaciens Bo542 (pGA472)	Seedling	Kanamycin resistance	Tumor, plant	Ondrej and Vlasák (1987)
A. rubi ATCC13335	Seedling	Hormone autotrophy	Tumor, plant	Ondrej et al. (1987)
A. rhizogenes A4	Stem segment	Hormone autotrophy	Hairy root, plant	Jung and Tepfer (1987)
A. rhizogenes 8196	Stem segment	Hormone autotrophy	Hairy root, plant	Jung and Tepfer (1987)
A. rhizogenes LBA9402	Leaf disc	Hormone autotrophy	Hairy root	Walton et al. (1990)
A. tumefaciens C58C1 (pGV3850::1103)	Leaf disc	Kanamycin resistance	Callus, shoot	Mathews et al. (1990)
A. tumefaciens LBA4404 (pSK223)	Leaf disc	Kanamycin resistance	Plant	Kurioka et al. (1992)
A. tumefaciens R1000 (pRiA4b)	Leaf disc	Hormone autotrophy	Hairy root	Kamada et al. (unpubl.)
A. tumefaciens LBA4404 (pRiA4b + pBI300)	Leaf disc	Hygromycin resistance	Hairy root	Kamada et al. (unpubl.)
A. rhizogenes (pRi15834, pARK5)	Leaf disc	Herbicide resistance	Hairy root	Saito et al. (1992)

kanamycin-resistant gene (neomycin phosphotransferase II gene, NPT II gene) (Beck et al. 1982) as a foreign gene into *A. belladonna* by using both binary vector pGA472 (An et al. 1985; Ondrej and Vlasák 1987) and cointegrating vector pGV3580::1103 (Mathews et al. 1990).

Plasmids pSK233 and pBI300 constructed by us were based on a general purpose binary vector pBI121 (Jefferson et al. 1987) (Fig. 1), which is now commercially available. The plasmid pSK223 contains a *rolC* gene under a control of CaMV 35S promoter (Fig. 2). The *rolC* is one of the genes locating on Ri T-DNA. The GUS and NOS-ter were excised at the smaI and EcoRI sites and replaced by the *rolC* gene of pRiA4b. We also constructed a plasmid pBI300 (Fig. 2) using pBI121, in which the GUS gene was replaced by the hygromycin phosphotransferase gene (van den Elzen et al. 1985; Waldron et al. 1985). Either of these vectors was transferred from *E. coli* to a host *Agrobacterium* LBA4404 with the aid of *E. coli* HB101 harboring a helper plasmid (pRK2013), giving *Agrobacterium* possessing all conditions necessary for transformation.

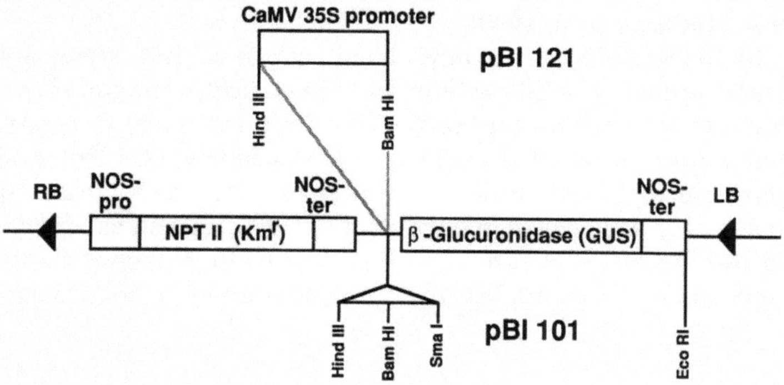

Fig. 1. Structure of pBI101 and pBI121. T-DNA region of pBI101 contains a kanamycin selectable marker, polylinker cloning sites, GUS and nopaline synthase polyadenylation site. In pBI121, a CaMV 35S promoter is ligated into Hind III-Bam HI sites of pBI101.

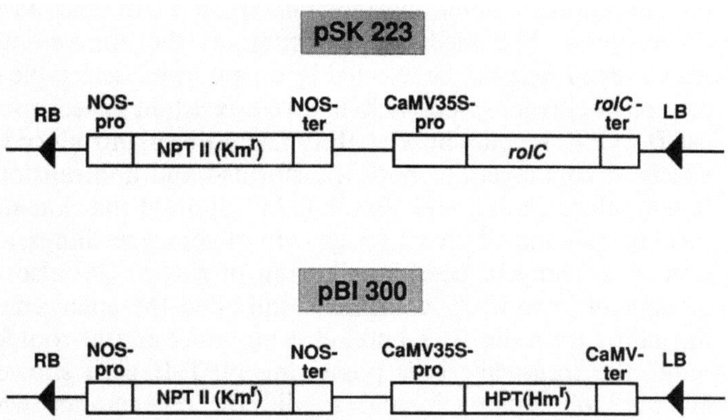

Fig. 2. Structure of pSK223 and pBI300

3 Transformation

When one desires to obtain transformed tissues as crown galls or hairy roots by introducing entire Ti or Ri T-DNA, it is possible to use intact seedlings for transformation by directly infecting bacteria on hypocotyls or stems with a needle. Induced tumors or adventitious roots at infection site are excised and transferred onto solid medium. The excised tissues can grow autonomously on the medium without phytohormones. This method has been successfully applied to transformation of *A. belladonna* (Kamada et al. 1986; Ondrej and Protiva 1987; Ondrej and Vlasák 1987; Ondrej et al. 1987).

Explants must be used for transformation when the transformed tissues do not have morphological phenotypes and need to be selected by an antibiotic selectable marker or an enzyme activity marker given to the transformed cells. Two basic approaches have been used to obtain transgenic plants using explants; cocultivation of regenerating protoplasts and the leaf disc method. The former has not been applied to *A. belladonna*, and the latter has been extensively used (Jung and Tepfer 1987; Mathews et al. 1990).

In the leaf disc procedure, axenic explants (leaf, stem) are cocultured with *Agrobacterium* on solid medium for 2 days. Acetosyringone is a factor from plants which activates the *vir* gene responsible for transfer and integration of T-DNA into plant chromosomes (Bolten et al. 1986; Machida et al. 1986). Addition of acetosyringone during cocultivation has been reported to increase transformation frequency of *A. belladonna* (Mathews et al. 1990) as well as *Arabidopsis* (Sheikholeslam and Weeks 1987). A nurse culture which also enhances transformation in some plants has not been applied to transformation of *A. belladonna.*

4 Elimination of Bacteria and Selection of Transformed Cells

After cocultivation with *Agrobacterium*, the explants are transferred to regeneration/selection medium. This medium contains carbenicillin or claforan to eliminate the bacteria. The medium also contains the appropriate antibiotic when the transformed cells can be selected by an antibiotic selectable marker gene integrated into plant chromosomes. A kynamycin-resistant gene, neomycin phosphotransferase II (NPT II) gene (Beck et al. 1982), has been introduced into *A. belladonna* and effects of kanamycin to both transformed and nontransformed tissues have been investigated. Ondrej and Vlasák (1987) showed that kanamycin at 25 mg/l in the medium was already toxic for growth of young seedlings, and a higher concentration of kanamycin caused yellowing of leaves. We also obtained a result that 50 mg/l of kanamycin completely inhibited the enlargement of leaf discs, callus formation from the discs, and following shoot and/or root formation. On the other hand, the transgenic cells possessing NPT II gene showed complete growth at 100 mg/l of kanamycin and so did the regenerants, which demonstrated the efficiency of NPT II gene as a selectable marker for obtaining transgenic plants of *A. belladonna.*

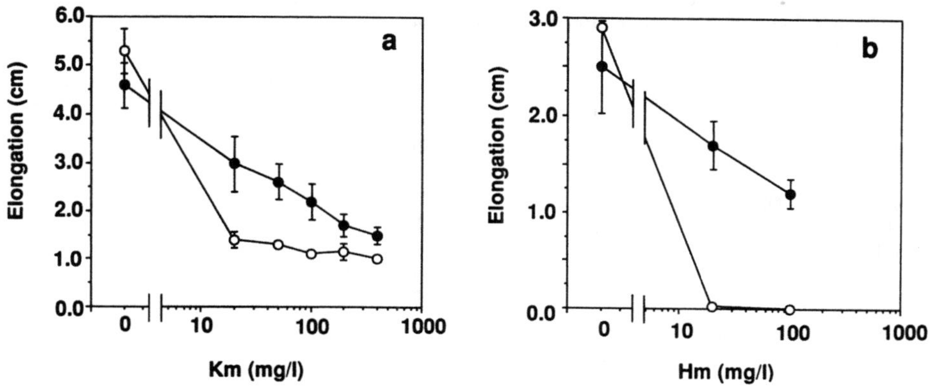

Fig. 3. a The effects of kanamycin on the elongation of hairy roots carrying pRiA4b T-DNA (—○—) and those carrying both pRiA4b and pBI121 T-DNA (—●—). **b** The effects of hygromycin on the elongation of hairy roots carrying pRiA4b T-DNA (—○—) and those carrying both pRiA4b and pBI300 T-DNA (—●—)

Although elongation of the hairy roots of *A. belladonna* transformed by Ri plasmid A4b, which do not have the NPT II gene, is suppressed by kanamycin at 20 mg/l, they still show growth and even the concentration at 400 mg/l could not stop their growth (Fig. 3a). This shows that the NPT II gene cannot be an appropriate selectable marker when one desires to introduce another gene into hairy roots. We examined hygromycin phosphotransferase (HPT) gene as an alternative antibiotic selectable marker, which is encoding resistance to hygromycin (Waldron et al. 1985; van den Elzen et al. 1985). The efficiency of this marker has been demonstrated in a number of plant species (Lloyd et al. 1986). A plasmid vector pBI300 (Fig. 2) was constructed by excising the GUS gene of pBI121 and replaced by the HPT gene. The hairy roots obtained by introduction of both A4b T-DNA and pBI300 T-DNA showed fairly good growth at 20 mg/l of hygromycin, while the hairy roots transformed only by A4b T-DNA stopped elongation at the same concentration of hygromycin (Fig. 3b). This shows the efficiency of the HPT gene as a selectable marker in retransformation of hairy roots of *A. belladonna*.

5 Regeneration of Plants

In general, the regeneration frequency of plants from callus or other tissues is mostly affected by the presence of phytohormones, auxin, and cytokinin in the medium. It has been reported that the presence of benzyladenine (BA) at 3 mg/l in the medium increases the frequency of the shoot formation from *A. belladonna* hairy roots transformed by *A. tumefaciens* C58C1 harboring a Ri plasmid A4b (Ondrej and Protiva 1987). Addition of naphthalene acetic acid (NAA, 1 μM) in addition to BA (6 μM) has also been reported to promote callus formation from leaf explants and following shoot differentiation (Mathews et al. 1990). We

Table 2. Effects of NAA and BA on shoot formation in *A. belladonna*

	BA (mg/l)			
NAA (mg/l)	0	0.1	1	10
a) From leaf segments				
0	0[a]	0	9	18
0.1	0	10	2	11
1	0	0	11	39
10	0	0	0	7
b) From stem segments				
0	0	0	0	4
0.1	6	3	10	0
1	0	0	12	52
10	0	0	0	33

[a] Regeneration frequency was revealed by the percentage of the explants with shoot formation. Leaf and stem segments were cultured for 3 weeks on the MS plates containing various concentrations of NAA and BA. The number of the segments with shoot formation was counted.

examined the effects of NAA and BA on regeneration frequency from leaf and stem segments of *A. belladonna.* The results are shown in Table 2. The numbers show the percentages of the segments with shoots. A combination of BA 10 mg/l and NAA 1 mg/l showed the highest shoot formation frequency both from leaf and stem segments. Considering the transformation frequency, 52% for stem segments and of 39% for leaf segments are not high enough to obtain transformed plants. When 80 leaf discs were inoculated with *Agrobacterium* harboring pSK223, only a few shoots differentiated out of calli on the kanamycin selection medium, although calli were formed on all of the leaf segments.

6 Transgenic *A. belladonna*

The transgenic *A. belladonna* regenerated from crown galls or cane galls are reported to be deformed (short and thick shoots, shortened internodes, narrow and small leaves) (Ondrej and Protiva 1987; Ondrej et al. 1987). These symptoms decreased gradually during subcultivation, which shows that this phenotypic alteration is not always stable. The transformed *A. belladonna* regenerated from hairy roots exhibit characteristic features, such as wrinkled leaves, shortened internodal length, and decrease of apical dominancy (Fig. 4) (Kamada et al. 1986). These phenotypic features of the transformant are stable under successive subcultivation and have also been reported for other plant species (Tepfer 1984; Ooms et al. 1985).

The transformants with NPT II gene and HPT gene showed resistance to kanamycin and hygromycin, respectively, as described above.

Fig. 4. **A** Untransformed *A. belladonna.* **B** *A. belladonna* regenerated from hairy roots transformed with *A. tumefaciens* strain R1000 (pRiA4b). It has wrinkled leaves, shortened internodal length, and decreased apical dominancy. **C** *A. belladonna* regenerated from callus transformed with *A. tumefaciens* strain LBA4404 (pSK223). It shows extensive flowering and extremely decreased apical dominancy

The transformants in which CaMV 35S pro + *rolC* gene was introduced showed extensive flowering even in the culture bottle under subcultivation, while normal plants nor transformants with wild-type Ri T-DNA do not show flowering under the same condition. The transformants also exhibited the extreme decrease of apical dominancy. They have a shoot branching and a flower bud at almost every internodal base (Fig. 4).

7 Conclusion

Agrobacterium-mediated transformation has been shown to be practical to *A. belladonna* and that NPT II gene is an appropriate selectable marker. HPT gene was also shown to be efficient on retransformation of the hairy roots. The most desirable improvement to obtain transgenic *A. belladonna* is to increase regeneration frequency from transformed cells. Throughout the procedure mentioned above, including inoculation of plant tissues with *Agrobacterium*, transformation of plant cells, and regeneration, young seedlings seem to provide better material than those subcultured for a long period in culture bottles, although it is not clear which step of the procedure is affected by the age of the plant.

8 Protocols for Transformation

8.1 Plasmids

The constructed binary plasmids, such as pSK223 or pBI300, were conjugatively transferred by triparental mating using pRK2013 as a helper plasmid from *E. coli* to *Agrobacterium* such as LBA4404 or R1000.

8.2 Plant Transformation (an example in pSK223)

1. Inoculate *A. tumefaciens* (pAL4404 + pSK223) in 5 ml of LB broth medium containing 100 μg/ml of kanamycin and incubate at 28 °C for 24 h with vigorous shaking.
2. Transfer culture of *A. tumefaciens* into a plastic dish (ø 6 cm).
3. Cut leaves of axenic plants into squares approximately 5 × 5 mm.
4. Float leaf discs in the bacterial culture for a few seconds.
5. Eliminate excess *A. tumefaciens* with the sterilized Kimwipe.
6. Transfer leaf discs on a Murashige and Skoog's (MS) medium (ø 9 cm plastic dish) and then incubate the plate in a culture room (3000 lx, 16-h light/8-h dark, 25 °C).
7. After 2 to 3 days, transfer leaf discs on a MS plate containing Claforan (0.5 g/l)[3], BA (10 mg/l), and NAA (1 mg/l).
8. After 1 week, transfer leaf discs to the fresh MS plate with Claforan and incubate the plate in a culture room. Repeat this step until callus formation.
9. After 3 or 4 weeks, transfer leaf discs to a MS plate containing Claforan (0.5 g/l), BA (10 mg/l), NAA (1 mg/l) and kanamycin (0.1 g/l)[4]. Repeat this step once a week until shoot formation.

References

An G, Watson BD, Stachel S, Gordon MP, Nester EW (1985) New cloning vehicles for transformation of higher plants. EMBO J 4:277–284

Bajaj YPS, Simola LK (1991) *Atropa belladonna* L.: In vitro culture, regeneration of plants, cryopreservation, and the production of tropane alkaloids. In: Bajaj YPS (ed) Biotechnology in agriculture and forestry, vol 15. Medicinal and aromatic plants III. Springer, Berlin Heidelberg New York, pp 1–23

Beck E, Ludwig G, Auerswald EA, Reiss B, Schaller H (1982) Nucleotide sequence and exact localization of the neomycin phosphotransferase gene from transposon Tn5. Gene 19:327–336

Bolten GW, Nester EW, Gordon MP (1986) Plant phenolic compounds induce expression of *Agrobacterium tumefaciens* loci needed for virulence. Science 232:983–985

[3] To eliminate *Agrobacterium* from the infected tissue.

[4] To select Km-resistant plant tissues with NPTII gene integrated into the chromosome by *Agrobacterium* infection.

Jefferson RA, Kavanagh TA, Bevan MW (1987) GUS fusions: β-glucuronidase as a sensitive and versatile gene fusion marker in higher plants. EMBO J 6:3901–3907

Jung G, Tepfer D (1987) Use of genetic transformation by the Ri T-DNA of *Agrobacterium rhizogenes* to stimulate biomass and tropane alkaloid production in *Atropa belladonna* and *Calystegia sepium* roots grown in vitro. Plant Sci 50:145–151

Kamada H, Okamura N, Satake M, Harada H, Shimomura K (1986) Alkaloid production by hairy root cultures in *Atropa belladonna.* Plant Cell Rep 5:239–242

Kurioka Y, Suzuki Y, Kamada H, Harada H (1992) Promotion of flowering and morphological alterations in *Atropa belladonna* transformed with a CaMV 35S-*rolC* chimeric gene of the Ri plasmid. Plant Cell Rep 12:1–6

Lloyd A, Barnason A, Rogers SG, Byrne M, Fraley RT, Horsch RB (1986) Transformation of *Arabidopsis thaliana* with *Agrobacterium tumefaciens.* Science 234:464–466

Machida Y, Usami S, Yamamoto A, Niwa Y, Takebe I (1986) Plant-inducible recombination between the 25bp border sequences of T-DNA in *Agrobacterium tumefaciens.* Mol Gen Genet 204:374–382

Mathews H. Bharathan N, Litz RE, Narayanan KR, Rao PS, Bhatia CR (1990) The promotion of *Agrobacterium*-mediated transformation in *Atropa belladonna* L. by acetosyringone. J. Plant Physiol 136:404–409

Ondrej M, Protiva J (1987) In vitro culture of crown gall and hairy root tumors of *Atropa belladonna*: differentiation and alkaloid production. Biol Plant 29:241–246

Ondrej M, Valsák J (1987) Expression of kanamycin resistance introduced by *Agrobacterium* binary vector into *Nicotiana tabacum* and *Atropa belladonna.* Biol Plant 29:161–166

Ondrej M, Matousek J, Vlasák J (1987) Differentiation of transformed plants from tumors induced by *Agrobacterium rubi* ATCC 13335. J. Plant Physiol 126:397–407

Ooms G, Karp A, Burrell MM, Twell D, Roberts J (1985) Genetic modification of potato development using Ri T-DNA. Theor Appl Genet 70:440–446

Saito K, Yamazaki M, Anzai H, Yoneyama, Murakoshi I (1992) Transgenic herbicide-resistant *Atropa belladonna* using an Ri binary vector and inheritance of the transgenic trait. Plant Cell Rep 11:219–224

Sheikholeslam SN, Weeks DP (1987) Acetosyringone promotes high efficiency transformation of *Arabidopsis thaliana* explants by *Agrobacterium tumefaciens.* Plant Mol Biol 8:291–298

Tepfer DA (1984) Transformation of several species of higher plants by *Agrobacterium rhizogenes*: sexual transmission of the transformed genotype and phenotype. Cell 37:959–967

van den Elzen P, Townsend J, Lee KY, Bedbrook J (1985) A chimeric hygromycin resistance gene as a selectable marker in plant cells. Plant Mol Biol 5:299–302

Waldron C, Murphy EB, Roberts JL, Gustafson GD, Armour SL, Malcolm SK (1985) Resistance to hygromycin-B. Plant Mol Biol 5:103–108

Walton NJ, Robins RJ, Peerless ACJ (1990) Enzymes of N-methylputrescine biosynthesis in relation to hyoscyamine formation in transformed root cultures of *Datura stramonium* and *Atropa belladonna.* Planta 182:136–141

II.3 Transformation in *Catharanthus* Species (Madagascar Periwinkle)

C. David[1] and J. Tempé[1,2]

1 Introduction

Secondary metabolite production by plant cell cultures has been the subject of many investigations. Indole alkaloids such as ajmalicine, an antihypertensive drug, or vinblastine and vincristin, two antitumor compounds produced by *Catharanthus roseus*, are very important in the pharmaceutical industry. Cell cultures from the same plant species have been propagated in several laboratories since they are expected to produce physiologically potent indole alkaloids (Carew 1975). Undifferentiated cell lines producing high yields of ajmalicine, or serpentine, that can be easily converted into ajmalicine by reduction, have been obtained by selection (Zenk et al. 1977).

However, a problem for the industrial exploitation of such cultures has been the variability in the productivity (Deus-Neumann and Zenk 1984) attributed to somaclonal variation, described as the result of genetic instability of such cultures (Larkin and Scowcroft 1981; Evans and Sharp 1986; Bajaj 1990). Since somaclonal variation has often been correlated with the undifferentiated state, a possible solution to this problem could be to use differentiated tissues such as shoots or roots. Root and/or shoot cultures of different species were reported to exhibit secondary metabolite composition similar to that of parent plant organs. Because of their structural simplicity, root cultures would appear to be easier to handle and more suitable than shoot cultures for large-scale industrial production in liquid medium. Furthermore, the contribution of root to secondary metabolism in the whole plant is well known, but root cultures are still little utilized for the production of plant chemical. Such cultures are indeed difficult to establish and to maintain, which may have limited their use. The finding that hairy roots can easily be propagated in vitro, with high growth rates and massive increase of the biomass, has therefore stirred much interest. A number of papers describing obtaining hairy root cultures and their properties have thus appeared in the past few years. Measurements of secondary metabolite production by hairy root cultures indicate that productivity of these is within the range of that of roots isolated from whole plants. These cultures, that produce an important biomass, make it possible to obtain compounds that occur in native roots in trace only (Davioud et al. 1989a). The variability observed from one line to another (Mano et al. 1986; Jaziri et al. 1988) and within different cultures of the same line is certainly much less than for

[1] Institut des Sciences Végétales, CNRS, Avenue de la Terrasse, 91198 Gif sur Yvette, France
[2] Institut National Agronomique Paris Grignon, 75231 Paris, France

Biotechnology in Agriculture and Forestry, Vol. 22
Plant Protoplasts and Genetic Engineering III (ed. by Y.P.S. Bajaj)

undifferentiated subcultures. Root cultures are indeed genetically more stable than cell cultures. A fundamental feature, the clonal origin of root meristems (David et al. 1984; Van Sluys and Tempé 1989), may be important, since root cultures derived from a single root tip will be genetically homogeneous. This, associated with genetic stability of meristems, should ensure that physiological and biochemical properties remain constant over long periods of time: hairy root cultures isolated and cultivated in vitro for 8 years were perfectly stable over this period, by criteria of opine production and T-DNA structure (David and Tempé 1987). On the other hand, when individual root tips are not isolated, the resulting cultures are not of clonal origin, allowing for changes in proportions of the several root lines present in the culture. Such changes may explain variation in composition as observed for opine synthesis by various authors (Tepfer 1984; Hamill et al. 1986; Kamada et al. 1986).

Other important points to keep in mind are the synthesis site of the secondary metabolites studied, and the fact that a compound accumulates in a root does not imply that such an organ is also the site of synthesis. Therefore hairy root cultures should be suitable only for those alkaloids which are synthesized within the root and do not in principle apply to production of metabolites synthesized in the aerial parts of the plant, such as vinblastine in *C. roseus.* In this respect, the finding that vinblastine is present in callus and hairy root cultures of this species (Parr et al. 1988) is surprising and would need to be confirmed. In this chapter, we report work done on *Catharanthus* species, a plant of major interest for the pharmaceutical industry.

2 Importance of *Catharanthus*

The Madagascar periwinkle is a pantropical ornamental plant of the family Apocynaceae. The botanical nomenclature of this plant was previously somewhat confused and it has been called *Ammocalis rosea, Catharanthus roseus, Lochnera rosea,* and more frequently *Vinca rosea.* The correct botanical name for this plant is *Catharanthus roseus* G. Don (Stearn 1966) and this prevails now. Since the plant was known as *Vinca rosea* when its alkaloids were being studied, several of these have received vernacular names that recall this previous denomination.

This plant was first studied because of its supposed hypoglycemic properties. The discovery of the antitumor activity of *Catharanthus* alkaloids represents one of the most significant additions of natural products to cancer therapy. Vinblastine was discovered simultaneously by two groups, and this discovery constituted a great event in alkaloids chemistry. A few years latter, vincristine was isolated from *C. roseus* (Svoboda 1961). Three groups of alkaloids were identified, among which the first, including vinblastine and vincristine, is the most important. Vinblastine has also been isolated from the minor species, *C. trichophyllus* (Atta-ur-Rahman et al. 1983).

It should be noted that the alkaloids of *C. roseus* containing two different building blocks, such an indole and dihydroindole, were sometimes referred to as dimeric indole alkaloids: it is more accurate to use the term binary or bisindole alkaloids, since chemically these are not dimers of two equal subunits.

C. roseus produces more than 100 indole alkaloids (Scott 1970), many of which are pharmacologically active. The chemical structures of 14 of them are still unknown (Blasko and Cordell 1990). Since the amounts of these alkaloids in the plant are extremely low and the commercial production of plants is unreliable, attempts were made to extract these products from plant tissue cultures.

Catharanthus alkaloids are used for treating both malignant and nonmalignant diseases, such as platelet disorders. Their antimitotic properties, due to inhibition of microtubule assembly, make them indispensable in cancer chemotherapy. The specific binding site on tubulin of vinblastine or vincristine is distinct from the target site of other natural products, such as colchicine. As microtubular function is intracellular, the alkaloids have to enter cells and remain within them to be effective. In vitro, the cytotoxic effects of the drugs require a minimal exposure time approximately equal to the doubling time of the cells, usually 12 to 72 h, at a concentration of 10 nM to 1 μM. Cell types with short doubling times tend to be more sensitive than those that divide slowly but, surprisingly, neuronal cells are extremely sensitive.

The most important clinical use of bisindole alkaloids is in the treatment of leukemias and lymphomas. In this respect, the activities of vinblastine and vincristine appear to be quite different in their use: vinblastine has a striking efficiency in Hodgkin's disease and vincristine has curative effects in childhood lymphocytic leukemia.

3 Genetic Transformation of *Catharanthus* Species

3.1 Studies of *Catharanthus roseus* Crown Gall

To analyze the role of *Agrobacterium tumefaciens* in crown gall formation, an experimental method was used that allowed the selective thermal killing of the bacteria at any desired time following their inoculation to the host (Braun 1974). *C. roseus* was utilized, since this species is very resistant to high temperature, e.g., 46 °C, sufficient to kill *Agrobacterium* but not the plant. From these experiments, it was concluded that tumor induction, now known as a genetic transformation process, occurs as early as 34 h after the bacteria are introduced into a host.

Later, the rapid proliferation of crown gall cells was considered as a method to obtain a rapid increase of biomass and to study secondary metabolites (De Cleene 1980). Alkaloid production of habituated and tumorous cell suspension cultures of *C. roseus* were analyzed (Eilert et al. 1987a): these secondary metabolites no longer accumulated after transfer of the cultures in an auxin-containing medium of two enzymatic activities involved in the indole alkaloid synthesis pathway, tryptophane decarboxylase and strictosidine synthase, dropped considerably (Eilert et al. 1987b). Similarly, crown gall culture of *Atropa belladonna* did not produce alkaloids at all (Ondrej and Protiva 1987). These results can be related to the undifferentiated state of crown gall cells, that were thus found unsuitable for secondary metabolite production.

3.2 A Model for Organ Culture, the Hairy Root

Fast-growing hairy root cultures can easily be obtained and have been established for a wide range of dicotyledonous plants (Mugnier 1988; Tempé and Casse-Delbart 1989). Their potential for in vitro production of secondary metabolites has been investigated and hairy root cultures were established for many different plants. Hairy roots were found to have qualitatively the same composition as parent plant roots, with similar or higher yields.

Hairy roots are induced upon inoculation of *Agrobacterium rhizogenes* to many dicotyledonous plants (for a review, see Birot et al. 1987, and Tempé and Casse-Delbart 1989). Virulent strains of this bacterium contain plasmids called Ri (Root-inducing) responsible for the hairy root disease, resulting in abundant root production at the site of inoculation on susceptible hosts (Riker et al. 1930). The molecular basis for hairy root is the transfer to the plant genome of specific DNA fragment(s), called T-DNA for transferred DNA, originally located in the T-region of Ri plasmids (Chilton et al. 1982). Hairy root proliferation results from T-DNA integration and expression in the plant cells.

To study indole alkaloid production, *C. roseus* G. Don and *C. trichophyllus* (Bak.) Pich. hairy root cultures were isolated from plants inoculated with different *A. rhizogenes* strains. Hairy root lines of *C. roseus* were established in vitro on solid culture medium after the inoculation with *A. rhizogenes* (Fig. 1). Spontaneous regenerants were obtained (Brillanceau 1986; Brillanceau et al. 1989). Unfortunately, all the lines studied exhibited a rather poor adaptation to liquid culture

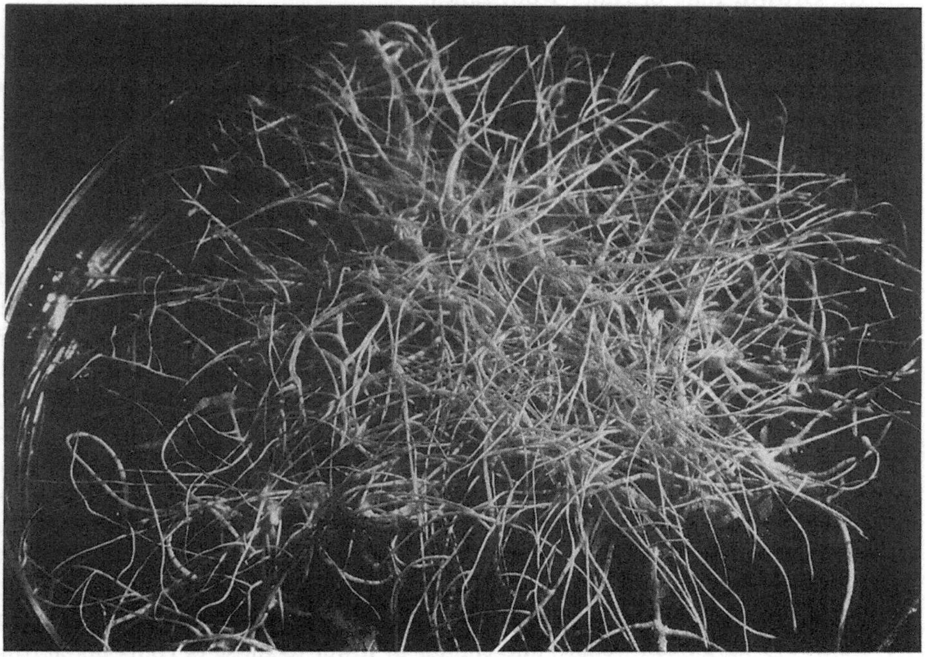

Fig. 1. *Catharanthus roseus* hairy root culture on solid culture medium, 4 weeks after transfer

conditions, under which they grew as a mixture of dissociated cells, cell aggregates, and roots. This heterogeneous material was found to be unsuitable for large-scale liquid culture. One hairy root line evolved rapidly as a dedifferentiated culture with spontaneous bud formation which developed in vitro as plantlets after cutting. The plants exhibited a very abnormal phenotype and did not survive more than a few months after transfer to the greenhouse. The presence in this line of at least six inserts of TL-DNA (Huffman et al. 1984; Jouanin 1984) may be responsible for the phenotype and the very poor adaptation to the greenhouse. In another line, the TL and TR-regions (Huffman et al. 1984; Jouanin 1984) were transferred as one fragment, which is very unusual for hairy root lines.

3.3 Genetic Transformation of *Catharanthus trichophyllus* by *Agrobacterium rhizogenes*

Following the difficulties with *C. roseus*, studies were undertaken with the minor species *C. trichophyllus*, whose hairy roots exhibit a very good adaptation to liquid culture conditions. The alkaloid content of in vitro-cultivated normal roots, hairy roots, and of roots from greenhouse-grown plants were compared (Davioud et al. 1989b). Five new compounds, belonging to a new series of yohimbine type bases, were identified (Davioud et al. 1989a).

3.3.1 Production of Indole Alkaloids by in Vitro Root Cultures from C. trichophyllus

3.3.1.1 Growth Rate of Hairy Root Lines

One axenic normal root culture and six hairy root lines, derived from independent root tips, were established from one inoculated *C. trichophyllus* plant with *A. rhizogenes* agropine-type strain 15834. The growth rate of normal and hairy root lines was compared: growth was measured as relative weight increase over a 40-day period. Five of the selected lines exhibited a typical hairy root phenotype, i.e., fast-growing and highly branched roots (Fig. 2), whereas one had a very reduced growth rate and spontaneously formed small calli.

The transformed nature of each hairy root line was checked by analysis of the opine content and T-DNA hybridization (Table 1). Mannopine and agropine, used as chemical markers of the presence of TR-DNA (De Paolis et al., 1985), were detected in extracts from two transformed root lines. To assess the presence of T-DNA in *C. trichophyllus* roots, Southern blots were hybridized against T-DNA specific probes (Shen et al. 1990). The phenotype of the slow-growing hairy root line was suggestive of the expression of T-DNA-borne auxin synthesis genes, and physiological studies indicated that at least one of the two T-DNA genes involved in auxin synthesis was expressed. Amounts of alkaloids production by this line (Table 1) were consistently lower (average: 160 μg/g fresh weight) than the amounts produced by the other hairy root lines (averages: 507 and 627 μg/g fresh weight), that were similar to those of normal roots obtained in the same conditions (average: 679 μg/g fresh weight) and similar or lower to those of roots taken from greenhouse grown plants (respectively 589, 497, 1006, and 1522 μg/g fresh weight

Table 1. Characteristics and alkaloid content of greenhouse-grown plant roots, in vitro normal roots and three hairy root lines from *C. trichophyllus*. For plant roots, average of four greenhouse-grown plants and for root cultures, average of three different 5-week-old cultures at 5-week intervals (10^{-3} mg/g fr. wt) (After Davioud et al. 1989b and Shen et al. 1990)

Roots	Phenotype	Opine	T-DNA	1	2	3	4	5	6	12	14	18	T
Normal greenhouse-grown plant root	Normal	–	–	130	142	33	221	109	55	51	150	27	903
Normal root line	Normal	–	–	158	161	225	63	2	14	3	55	nd	679
T4	Hairy root	–	TL	129	64	91	56	9	47	12	98	nd	507
T5	Hairy root	+	TL + TR	183	153	134	66	tr	14	2	75	nd	627
T6	Tumors	+	TR	71	9	7	9	2	8	3	53	nd	160

1: Horhammericine, 2: Vindolinine, 3: Lochnericine, 4: Echitovenine, 5: Ajmalicine, 6: Tabersonine, 12: Serpentine, 14: Dimethoxyanthraserpine, 18: Tetrahydroalstonine.

nd: not detected; tr: trace amounts (< 5).

T: total of alkaloid (1–6, 12, 14, 18) amounts.

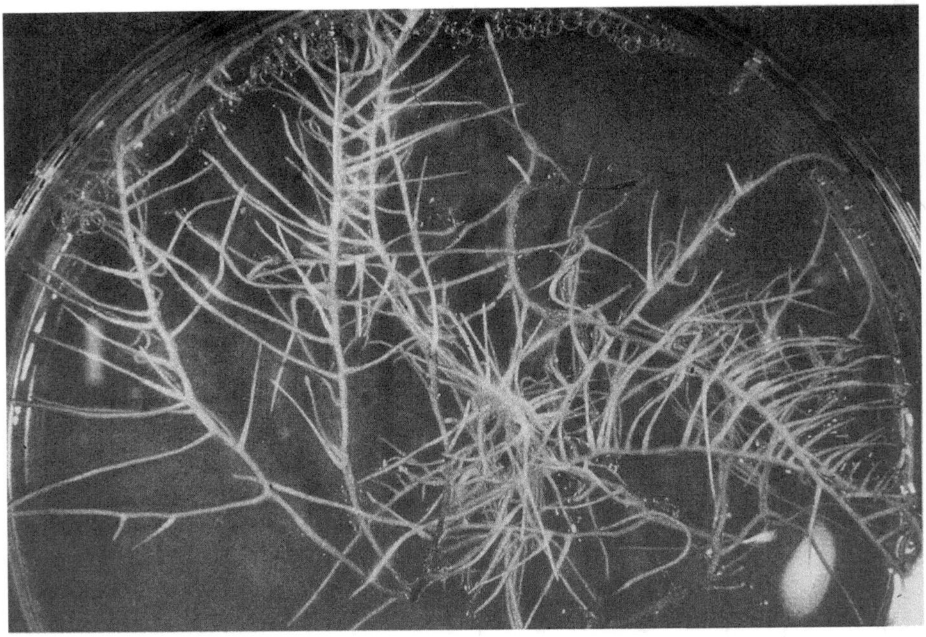

Fig. 2. *Catharanthus trichophyllus* hairy root culture on solid culture medium, 2 weeks after transfer

for the roots of four different greenhouse-grown plants). Since both alkaloids and auxin are products of tryptophane metabolism, this observation might be a consequence of limited availability of tryptophane, due to its transformation to indole acetic acid.

3.3.1.2 Indole Alkaloid Production by Hairy Root Fermenter Cultures

A rapidly growing hairy root line was selected for fermenter culture (Fig. 3) to examine its growth rate and alkaloid content. The results of two bioreactor runs were analyzed (Davioud et al. 1989b): the fresh weight of this root line increased from an original inoculum (10 g each) by 200 and 250 times over 6.5-week and 9.5-week culture periods, respectively. The alkaloids were retained almost entirely in the root tissue, less than 0.01 g/l being released in the medium, and extraction of 350 g (freeze-dried) hairy root gave 3.22 g of crude alkaloids. The alkaloid content of roots of four greenhouse-grown plants showed variable compositions when examined. To avoid genetic heterogeneity between different plants, the alkaloid content of 5-week-old cultures, derived from the same parent plant were analyzed. *C. trichophyllus* plant roots and cultured roots exhibit similar alkaloid patterns. They produce a rather broad alkaloid spectrum, usually found in the roots of this species. Almost all the biogenetic pathway for the monoterpenoid indole alkaloids is therefore functional in roots from this species. Members of the *Iboga* type representing the last step of the indole alkaloid biosynthesis evolution in the Apocynaceae (Scott 1970) are, however, not observed.

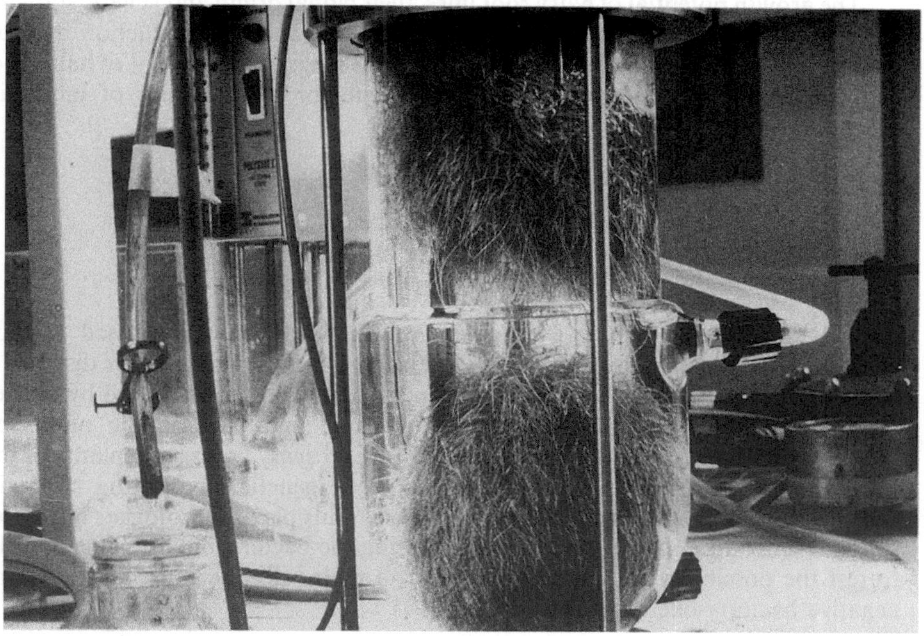

Fig. 3. Six-week-old culture of the *Catharanthus trichophyllus* hairy root line shown in Fig. 2 obtained in a 2-l fermenter

In addition to already described alkaloids, five compounds were identified that had not been found previously: they were characterized as anthraserpine derivatives. The presence of these in normal root culture confirms that transformation does not modify the pattern of alkaloid biosynthesis in *C. trichophyllus*. It was suggested that the expression of hairy root T-DNA has no direct influence on secondary metabolite production, although it may have an indirect one by controlling the growth rate, as observed for one hairy root line. Other authors reported a positive influence of growth rate on productivity of secondary metabolites (Jung and Tepfer 1987).

4 Potential of Hairy Root Cultures

In spite of research conducted over more than 30 years (Routien et al. 1956), the commercial feasibility of secondary metabolite production by in vitro culture has still to be demonstrated (Constabel 1990). Only a handful of species have been used industrially (Fontanel and Tabata 1987) and shikonine production, an example of the large-scale application of plant cell cultures, has now totally stopped for economic reasons. This naphtoquinone pigment was obtained from cell culture of *Lithospermum erythrorhizon* (Tabata and Fujita 1985) and commercialized by the Japanese company Mitsui Petrochemical, a group of seven private companies: 0.4 kg shikonine was produced in 1 day in a 750 l fermenter.

The growth potential of hairy root lines, their clonal origin and genetic stability, and the complete differentiation of root tissue warrant the production of root-specific biochemicals at substantial levels. At this stage, the potential of hairy root cultures for biomass production, a prerequisite for development of industrial processes, would appear to be satisfactory.

5 Transformation of *C. roseus* and Plant Pathology

A field in which transgenic *Catharanthus roseus* plant could be used is plant pathology. This species is quite often used as an experimental host for diseases of cultivated plants. This is the case, for instance, of *Citrus* diseases caused by viruses, mycoplasma-like organisms (MLO), spiroplasma, and bacteria-like organisms (BLO), which can be readily be transmitted to *C. roseus*. Transgenic plants of this species could be used to check the potential of genetic engineering strategies designed to protect *Citrus* crops against disease. Thus plant pathologists have been trying to introduce in *C. roseus* a gene encoding the bactericidal lisozyme in order to test the possibility to control "greening", a major disease caused by a Gram-negative bacteria-like organism (Bedin 1991).

6 Genetic Engineering with Genes Involved in Tropane Alkaloids Synthesis

It is now possible to direct the synthesis of enzymes in living cells, to control the levels produced and, in the near future, to modify their stability and specificity. Up till now the prospects of genetic manipulation for industrial application remain rather limited, the limitations being due more to lack of knowledge of biological processes than to the ability to engineer organisms. For what concerns the biosynthesis of secondary metabolites, applications are clearly possible, and important results have been already obtained. Thus, it has been possible to modify pigment production in petunia flowers, either by antisense messenger RNA strategies (Van der Krol et al. 1988) or by channeling a pigment precursor to another pathway by introducing a gene encoding the synthesis of a key enzyme (Meyer et al. 1987).

The same type of manipulation applies to the biosynthesis of any other compound. However, it implies that the genes encoding the enzymes involved in these biosyntheses are available, and also knowledge of metabolic pathways and regulation of the synthesis of precursors and related products. Ideally, once the genes and the knowledge are both available, it should be possible to modify existing pathways and eventually to introduce a complete pathway in a plant that had been selected as being particularly adapted to extractive processes. It is clear that there is still a long way to go: to this end, investigations on biosynthetic pathways are an obvious prerequisite (Leete 1989; Pfitzner and Zenk 1989).

The purification and characterization of two key enzymes involved in the biosynthesis of indole alkaloids, tryptophane decarboxylase and strictosidine synthase, and the isolation of complementary DNA clones (cDNA), were achieved. The first enzyme is responsible for the decarboxylation of tryptophan to form tryptamine, and the latter allows the condensation of stereospecifically tryptamine with secologanin to form strictosidine.

The alkaloidal glucoside strictosidine was identified as a biosynthetic precursor of monoterpenoid indole alkaloids. The enzyme strictosidine synthase has been characterized (Stöckigt and Zenk 1977) and purified (Treimer and Zenk 1979a, b) of indole alkaloid-producing plants. The cDNA clone for strictosidine synthase from *Rauvolfia serpentina* (Kutchan et al. 1988) was the first obtained for an enzyme involved in alkaloid metabolism, and its nucleotide sequence is one of the very few examples of an enzyme involved in secondary plant metabolism. This cDNA has been expressed in an enzymatically active form in *Escherichia coli* (Kutchan 1989).

A tryptophane decarboxylase cDNA clone was isolated from seedlings of *C. roseus* (De Luca et al. 1989). Transgenic tobacco plants expressing this gene were obtained: they synthesize and accumulate high levels of tryptamine without apparent adverse effects (Songstad et al. 1990).

Thus, in *C. roseus*, overexpression of these enzymes could result in increasing the flow towards the synthesis of vinblastine and vincristine, which would open a new field in the study of alkaloid biosynthesis and its application to biotechnology. It should however be realized that, because our knowledge is limited, unexpected problems could be encountered such as toxic effects of high concentrations of alkaloids.

It should be possible to combine genetic engineering for improved alkaloid production with the potential of hairy root cultures as a source of raw material for extraction. Despite the fact that we still have much to learn, it is already possible to perform some experiments with the genes and the techniques available. In fact, one could combine the advantages of hairy root culture and the possibility to engineer *Catharanthus* species with genes such as tryptophane decarboxylase and strictosidine synthase. It is indeed possible to obtain in one step transformants carrying both hairy root T-DNA and other genes such as those involved in alkaloid biosynthesis, by a cotransformation strategy (Depicker et al. 1985; Petit et al. 1986). Transformants, selected on their morphogenetic potential to grow as roots, could be used as hairy root cultures for alkaloid production.

7 Conclusion

In spite of many years of active research and progress on the use of in vitro methods to produce raw material for the extraction of secondary metabolites there is, as far we know, no active industrial development based on these methods. This is due to the low productivity or the variability of yield which makes in vitro-grown material uncompetitive when compared to other sources. Similarly, the hopes that crown gall or hairy root cultures could perform better than normal material have not been substantiated.

However, with the development of transformation techniques and the ongoing research on secondary metabolites, we see genetic engineering as probably the most powerful tool to increase the productivity of *Catharanthus* genotypes. Although this improvement may not be straightforward, because of the complexity of the metabolic pathways involved or because overproduction could be deleterious to the engineered material, or for other unforeseen reasons, there is reasonable hope that productivity will be improved by genetic engineering.

Whether these improvements will lead to industrial development, and under which form – extraction from whole plants grown in the field or from in vitro-cultured material – will depend on economic considerations which cannot be evaluated a priori.

References

Atta-ur-Rahman, Bashir M, Hafeez M, Perveen N, Fatima J, Mistry AN (1983) A rapid procedure for the isolation of catharanthine, vindoline and vinblastine. Planta Med 47:246–247

Bajaj YPS (1990) Somaclonal variation – Origin, induction, cryopreservation, and implications in plant breeding. In: Bajaj YPS (ed) Biotechnology in agriculture and forestry, vol 11. Somaclonal variation in crop improvement I. Springer, Berlin Heidelberg New York, pp 3–48

Bedin (1991) Construction de vecteurs du gène du lysozyme: étude d'un système modèle et expression dans des plantes transgéniques. Thèse de l'Université Bordeaux II

Birot AM, Boucher D, Casse-Delbart F, Durand-Tardif M, Jouanin L, Pautot V, Robaglia C, Tepfer D, Tepfer M, Tourneur J, Vilaine F (1987) Studies and uses of the Ri plasmids of *Agrobacterium rhizogenes*. Plant Physiol Biochem 25:323–335

Blasko G, Cordell G (1990) Isolation, structure elucidation and biosynthesis of the bisindole alkaloids of *Catharanthus*. In: Brossi A, Suffness M (eds) The alkaloids, vol 37. Antitumor bisindole alkaloids from *Catharanthus roseus*. Academic Press, San Diego, California, pp 1–76.

Braun AC (1947) Thermal studies on the factors responsible for tumor initiation in crown gall. Am J Bot 34:234–240

Brillanceau MH (1986) Etude chimique des alcaloïdes de deux espèces du genre *Guettarda*. Culture in vitro de racines transformées par *Agrobacterium rhizogenes*. Thèse de l'Université Paris-sud, Centre de Chatenay-Malabry

Brillanceau MH, David C, Tempé J (1989) Transformation of *Catharanthus roseus* G. Don by *Agrobacterium rhizogenes*. Plant Cell Rep 8:63–66

Carew DP (1975) Tissue culture studies of *Catharanthus roseus*. In: Taylor WI, Farnsworth NR (eds) The *Catharanthus* alkaloids. Marcel Dekker, New York, p 193

Chilton MD, Tepfer DA, Petit A, David C, Casse-Delbart F, Tempé J (1982) *Agrobacterium rhizogenes* T-DNA into the genomes of the host plant root cells. Nature 295:432–434

Constabel CP (1990) Medicinal plant biotechnology. Planta Med 56:421–425

Constabel CP, Towers GHN (1989) Thiarubine accumulation in hairy root cultures of *Chaenactis douglasii*. J Plant Physiol 133:67–70

David C, Tempé J (1987) Segregation of T-DNA copies in the progeny of a regenerant plant from a mannopine-positive hairy root line. Plant Mol Biol 9:585–592

David C, Chilton MD, Tempé J (1984) Conservation of T-DNA in plants regenerated from hairy root cultures. Bio/Technol 2:73–76

Davioud E, Kan C, Quirion JC, Das BC, Husson HP (1989a) *Epiallo*-yohimbine derivatives isolated from in vitro hairy root cultures of *Catharanthus trichophyllus*. Phytochemistry 28:1383–1387

Davioud E, Kan C, Hamon J. Tempe J, Husson HP (1989b) Production of indole alkaloids by in vitro root cultures from *Catharanthus trichophyllus*. Phytochemistry 28:2675–2680

De Cleene (1980) A possibility for increasing the production of physiologically active substances by in vitro plant tumour cultures. J Speculations Sci Technol 3:353–356

De Luca V, Marineau C, Brisson N (1989) Molecular cloning and analysis of cDNA encoding a plant tryptophan decarboxylase: comparison with animal dopa decarboxylase. Proc Natl Acad Sci USA 86:2582–2586

De Paolis A, Mauro ML, Pomponi M, Cardarelli M, Spano L, Costantino P (1985) Localization of agropine-synthesizing functions in the TR region of the root inducing plasmid of *Agrobacterium rhizogenes.* Plasmid 13:1–7

Depicker A, Herman L, Jacobs A, Schell J, van Montagu M (1985) Frequencies of simultaneous transformation with different T-DNAs and their relevance to the *Agrobacterium*/cell interaction. Mol Gen Genet 210:477–484

Deus-Neumann B, Zenk MH (1984) Instability of indole alkaloid production in *Catharanthus roseus* cell suspension cultures. Plant Med 50:427–431

Eilert U, De Luca V, Kurz WGW, Constabel F (1987a) Alkaloids formation by habituated and tumorous cell suspension cultures of *Catharanthus roseus.* Plant Cell Rep 6:271–274

Eilert U, De Luca V, Constabel F Kurz WGW (1987b) Elicitor-mediated induction of tryptophan decarboxylase and strictosidine synthase activities in cell suspension cultures of *Catharanthus roseus.* Arch Biochem Biophys 254:491–497

Evans DA, Sharp WR (1986) Applications of somaclonal variation. Bio/Technol 4:528–532

Fontanel A, Tabata M (1987) Production of secondary metabolites by plant tissue and cell cultures. Present aspects and prospects. In: Nestec LTD (ed) Nestlé Research News, Vevey, Switzerland, pp 92–103

Hamill JD, Parr AJ, Robins RJ, Rhodes MJC (1986) Secondary product formation by culture of *Beta vulgaris* and *Nicotiana rustica* transformed with *Agrobacterium rhizogenes.* Plant Cell Rep. 5:111–114

Huffman GA, White FF, Gordon MP, Nester GW (1984) Hairy root inducing plasmid: physical map and homology to tumor inducing plasmids. J Bacteriol 157:269–276

Jaziri M, Legros M, Homes J, Vanhaelen M, (1988) Tropane alkaloids production by hairy root cultures of *Datura stramonium* and *Hyoscyamus niger.* Phytochemistry 27:419–420

Jouanin L (1984) Restriction map of an agropine-type Ri plasmid and its homologies with Ti plasmids. Plasmid 12:91–102

Jung G, Tepfer D (1987) Use of genetic transformation by the Ri T-DNA of *Agrobacterium rhizogenes* to stimulate biomass and tropane alkaloid production in *Atropa belladonna* and *Calystegia sepium* roots grown in vitro. Plant Sci 50:145–151

Kamada H, Okamura N, Satake M, Harada H, Shimomura K (1986) Alkaloid production by hairy root cultures in *Atropa belladonna.* Plant Cell Rep 5:239–242

Kutchan TM (1989) Expression of enzymatically active cloned strictosidine synthase from the higher plant *Rauvolfia serpentina* in *Escherichia coli.* FEBs Lett 257:127–130

Kutchan TM, Hampp N, Lottspeich F, Beyreuter K, Zenk MH (1988) The cDNA clone for strictosidine synthase from *Rauvolfia serpentina.* DNA sequence determination and expression in *Escherichia coli.* FEBS Lett 237:40–44

Larkin PJ, Scowcroft WR (1981) Somaclonal variation. A novel source of variability from cell cultures. Theor Appl Genet 60:197–214

Leete E (1989) Recent developments in the biosynthesis of the tropane alkaloids. Planta Med 56:339–352

Mano Y, Nabeshima S, Matsui C, Ohkawa H (1986) Production of tropane alkaloids by hairy root cultures of *Scopolia japonica.* Agric Biol Chem 50:2715–2722

Meyer P, Heidmann I, Forkmann G, Saedler H (1987) A new *Petunia* flower colour generated by transformation of a mutant with a maize gene. Nature 330:677–678

Mugnier J (1988) Establishment of new axenic hairy root lines by inoculation with *Agrobacterium rhizogenes.* Plant Cell Rep 7:9–12

Ondrej M, Protiva J (1987) In vitro culture of crown gall and hairy root tumors of *Atropa belladonna*: differentiation and alkaloid production. Biol Plant 29:241–246

Parr AJ, Peerless ACJ, Hamill JB, Walton NJ, Robins RJ, Rhodes MJC (1988) Alkaloid production by transformed root cultures of *Catharanthus roseus.* Plant Cell Rep 7:309–312

Petit A, Berkaloff A, Tempé J (1986) Multiple transformation of plant cells by *Agrobacterium* may be responsible for the complex organization of T-DNA in crown gall and hairy root. Mol Gen Genet 202:388–393

Pfitzner U, Zenk MH (1989) Homogeneous strictosidine synthase isoenzymes from cell suspension cultures of *Catharanthus roseus.* Planta Med 55:525–530

Riker AJ, Banfield WM, Wright WH, Keitt GW, Sagen HE (1930) Studies on infectious hairy root of nursery apple trees. J. Agric Res (Washington DC) 41:507–540

Routien JB, Tenafly NJ, Nickell LG (1956) Cultivation of plant tissue. US Patent Office (chas. Pfizer and Co.) No 2.747.334

Scott AI (1970) Biosynthesis of the indole alkaloids. Acc Chem Res 3:151–157

Shen WH, Davioud E, David C, Barbier-Brygoo H, Tempé, Guern J (1990) High sensitivity to auxin is a common feature of hairy root. Plant Physiol 94:554–560

Songstad DD, De Luca V, Brisson N, Kurz WGW, Nessler CL (1990) High levels of tryptamine accumulation in transgenic tobacco expressing tryptophane decarboxylase. Plant Physiol 94:1410–1413

Stearn WT (1966) *Catharanthus roseus*, the correct name for the Madagascar periwinkle. Llyodia 29:196–200

Stöckigty, Zenk MH (1977) Isovincoside (strictosidine), the key intermediate in the enzymatic formation of indole alkaloids. FEBS Lett 79:233–237

Svoboda GH (1961) Alkaloids of *Vinca rosea* (*Catharanthus roseus*). IX. Extraction and characterization of leurosidine and leurocristine. Llyodia 24:173–178

Tabata M, Fujita Y (1985) Production of shikonin by plant cell cultures. In: Zaitlin M, Day PR, Hollaender A (eds) Biotechnology in plant science. Academic Press, Orlando, Florida, pp 217–218

Tempé J, Casse-Delbart F (1989) Plant gene vectors and genetic transformation. *Agrobacterium* Ri plasmids. In: Schell J, Vasil IK (eds) Cell culture and somatic cell genetics of plants: the molecular biology of nuclear genes, vol 6. Academic Press, San Diego, California, pp 25–49

Tepfer D (1984) Transformation of several species of higher plants by *Agrobacterium rhizogenes*: Sexual transmission of the transformed genotype and phenotype. Cell 37:959–967

Treimer J, Zenk MH (1979a) Purification and properties of strictosidine synthase, the key enzyme in indole alkaloid formation. Eur J Biochem 101:225–233

Treimer J, Zenk MH (1979b) Strictosidine synthase from cell cultures of Apocynaceae plants. Febs Lett 97:159–162

Van der Krol AR, Lenting PE, Veenstra J, van der Meer I, Koes RE, Gerats AGM, Mol JNM, Stuitje AR (1988) An anti-sense chalcone synthase gene in transgenic plants inhibits flower pigmentation. Nature 333:866–869

Van Sluys M, Tempé J (1989) Behavior of the maize transposable element *Activator* in *Daucus carota.* Mol Gen Genet 219:313–319

Zenk MH, El-Shagi H, Arens H, Stöckigt J, Weiler EW, Deus B (1977) Formation of the indole alkaloids serpentine and ajmalicine in cell suspension cultures of *Catharanthus roseus.* In: Barz W, Reinhard E, Zenk MH (eds) Plant tissue culture and its bio-technological application, Springer, Berlin Heidelberg New York, pp 27–44

II.4 Transformation in *Datura* Species

P. CHRISTEN[1] and M.F. ROBERTS[2]

1 Introduction

One of the main reasons why there are so few commercial processes which utilize plant cell cultures on a large scale is the cost effectiveness of the maintenance and processing of sterile cultures versus yield. Poor secondary metabolite biosynthesis and culture instability have prevented commercialization of many plant culture systems. Secondary products are frequently produced in cell cultures in lower yield than in the parent plant; however, there are more than 30 examples where the yield of secondary compounds from plant cell cultures is equal to or in excess of that produced by the parent plant (Phillipson 1990). Providing selection for high-yielding cells can be made and stable cultures produced, commercial exploitation may be considered, providing that the market cost of the compound produced allows the process to be cost-effective.

It has been considered that, in many instances, poor yield in undifferentiated tissues relates to the fact that growth without differentiation is incompatible with the expression of a secondary metabolic pathway. Hence, there has been an interest in the production of differentiated cultures, i.e., shoot and root cultures, the most recent development being the production of transformed root cultures using *Agrobacterium rhizogenes* (hairy roots). In this chapter, work on *Datura* species is discussed.

2 Tropane Alkaloid Production

The tropane alkaloids scopolamine (also known as hyoscine) and atropine are medicinally important compounds because of their action on the autonomic nervous system. Atropine, a racemic mixture of (+) and (−) hyoscyamine, is generally utilized commercially although only the (−) form is pharmacologically active. Synthetic production of these compounds is more costly than extraction from natural sources and cannot compete in price. Commercial sources of these

[1] Dépt. de Pharmacognosie, Ecole de Pharmacie, Université de Genève, 30, Quai E.-Ansermet, 1211 Genève 4, Switzerland
[2] Dept. of Pharmacognosy, The School of Pharmacy, University of London, London WC1N 1AX, United Kingdom

Biotechnology in Agriculture and Forestry, Vol. 22
Plant Protoplasts and Genetic Engineering III (ed. by Y.P.S. Bajaj)

alkaloids are species belonging to such genera as *Atropa*, *Datura*, *Duboisia*, and *Hyoscyamus* of the Solanaceae.

2.1 The Importance of *Datura* Species as a Commercial Source of Tropane Alkaloids

A number of medicinal preparations utilize scopolamine as scopolamine-N-butyl bromide as the preferred parasympatholytic agent because hyoscyamine has a stimulant action on the central nervous system. It is therefore commercially useful to have suitable species containing principally scopolamine without or with a smaller proportion of hyoscyamine. In this connection, the genus *Datura* has received some attention with respect to alkaloid production (see Petri and Bajaj 1989).

In Wettstein's classification of the Solanaceae (Wettstein 1895), the tribe Datureae comprises four genera: *Datura*, *Dyssochroma*, *Solandra*, and *Trianaea*. The genus *Datura* is divided into four sections: *Stramonium*, *Dutra*, *Ceratocaulis*, and *Brugmansia*. Some species are herbaceous, some aquatic, and others arborescent (Safford 1921). All species of the genus so far investigated chemically contain the ornithine-derived alkaloids of the hygrine and tropane groups. The aerial parts of all plants contain scopolamine and/or hyoscyamine as principal alkaloids with smaller amounts of derivatives of these bases. The roots contain, in addition, a large number of esters formed from tropane-3,6-diol and tropane-3,6,7-triol. The known alkaloids from the genus were reviewed by Evans (1979).

The major commercial sources of scopolamine are *Datura metel*, a herbaceous *Datura*, and two species of the Australian genus *Duboisia* (*D. myoporoides* and *D. leichhardtii*). During the past 15 years, a South American tree *Datura*, *Datura sanguinea*, extensively cultivated in Ecuador, has emerged as an alternative economic source of scopolamine. Various forms of the plant exist and selected races are used commercially. Tissue culture clonal propagation technique has also been used for the propagation of thousands of *D. sanguinea* plants from a single desirable individual (Evans 1990).

An investigation of *Datura candida*, another arborescent *Datura* species, and various cultivars reported scopolamine as the major alkaloid with a yield of between 0.10 and 0.33% dry wt. (Bristol et al. 1969). Griffin (1976) has cited the possible commercial value of the Australian peach flowered form of *D. candida*.

2.2 *Datura* Hybrids

Hybridization followed by selection can be valuable in studying the inheritance of specific chemical constituents and offers the possibility of producing new chemical races superior to either of the parents. Species used commercially for their secondary metabolites are often improved by plant breeding and the use of either natural or artificial hybrids is important.

With *Datura*, most work has been reported on the inheritance of scopolamine and hyoscyamine in the herbaceous hybrids (Evans 1979) and to a lesser extent with the tree daturas (El-Dabbas and Evans 1982; El-Imam and Evans 1990).

The work initiated by Romeike (1962) with the herbaceous hybrid between *Datura stramonium* and *Datura ferox* showed that the F_1 generation of the cross is characterized by scopolamine dominance and in the F_2 generation an apparent 3:1 segregation into scopolamine and predominantly hyoscyamine types occurs. The author stabilized the desirable F_1 characteristics by tetraploid formation and Evans (1979) produced by further hybridization and selection high scopolamine-containing plants.

The hybrid formed by crossing *Datura stramonium* var. *godronii* × *D. discolor* has also been studied (Al-Yahya and Evans 1975). *D. stramonium* varieties, at the time of flowering and early fruiting, contain in the aerial parts the alkaloids scopolamine and hyoscyamine in the ratio 1:3, *D. discolor*, on the other hand, contains principally scopolamine with a smaller proportion of hyoscyamine, ratio about 6:1. The total alkaloid content of the aerial parts of the hybrid created from these two species averaged 0.12% dry wt., somewhat lower than that of either parent. The scopolamine/hyoscyamine ratio (about 1:2) lies between that of the two parents, being closer to that of *D. stramonium*.

Concerning the tree daturas, El-Dabbas and Evans (1982) reported on the alkaloid content of the F_1 generation obtained by crossing *Datura candida*, a white-flowered tree *Datura* with *Datura aurea*, a closely related yellow-flowered form and that of the reciprocal cross. F_1 progeny possessed alkaloid contents of up to 0.72% dry wt., the amount being superior to those of the parents: *D. candida* 0.25 and *D. aurea* 0.66% dry wt. Recently, Griffin (1992) reported a F_1 progeny of a Queensland peach-flowered form of *Datura candida*.

Another potentially useful tree *Datura* hybrid involved crosses between *Datura suaveolens* and *Datura candida* (El-Imam and Evans 1990); whereas the parents contained 0.14 and 0.47% dry wt. scopolamine in the aerial parts, respectively, individual nonuniform F_1 hybrids ranged in content from 0.32 to 0.87% dry wt. The same authors reported also on the hybrids involving crosses between *Datura candida* and an orange-pink-flowered cultivar of the same species. Alkaloid yield of up to 1.5% dry wt. in the leaves was higher than in either of the parents.

3 Transformation

Callus and suspension cultures of various *Solanaceae* species have been studied as an alternative source for the production of hyoscyamine and scopolamine (Hay et al. 1988). The content of the alkaloids in such cultures remains much lower than that found in original plants despite the use of cell screening techniques in an attempt to improve productivity. Since the root is the site of tropane alkaloid biosynthesis and since the formation of secondary metabolites is associated with tissue organization, in vitro root cultures have been studied. The productivity is broadly similar to that of roots from the parent plants, but the slow growth of these cultures is a major drawback to their commercial use in alkaloid production.

Root cultures genetically transformed with *Agrobacterium rhizogenes* provide an alternative to the whole plant in the production of root derived secondary metabolites. Hairy root cultures have received considerable attention during the past decade and have opened broad perspectives for the production of pharmaceuticals. In order to overcome the problems and restrictions that are inherent to undifferentiated plant cell cultures, organized tissue cultures such as hairy roots represent a reliable alternative as a source of secondary metabolites. This type of culture has a number of features which make them valuable for studying the biosynthesis of plant secondary products with the aim of manipulating the expression of the pathways. These include: growth rate faster than the corresponding untransformed roots or cell suspension cultures; stable biosynthesis of secondary products characteristic of the roots of the species involved; inherent chromosome stability. This stability is associated with the maintenance of the organized state since exogenous auxin treatments which induce disorganization of the transformed roots lead to a rapid loss of product formation.

Within the Solanaceae, production of tropane alkaloids by hairy root cultures has been reported for several genera including *Atropa* (Kamada et al. 1986; Jung and Tepfer 1987; Walton et al. 1990), *Datura* (Flores et al. 1987; Payne et al. 1987; Jaziri et al. 1988; Christen et al. 1989; Hilton and Rhodes 1990; Parr et al. 1990; Walton et al. 1990), *Hyoscyamus* (Flores et al. 1987; Jaziri et al. 1988), *Scopolia* (Mano et al. 1986; Nabeshima et al. 1986, see also Zhang, Chap. II. 18 this vol.), and *Duboisia* (Knopp et al. 1988; Mano et al. 1989). In considering *Datura*, many species have been transformed by a variety of strains of *Agrobacterium* (Table 1). It should be noted, however, that most of the species investigated in this manner have been herbaceous *Datura*.

Table 1. *Datura* species transformed with *Agrobacterium rhizogenes*

Herbaceous *Datura* species	References
Datura chlorantha	Knopp et al. (1988)
D. ferox	Knopp et al. (1988); Parr et al. (1990)
D. fastuosa	Parr et al. (1990)
D. fastuosa var. *violacea*	Knopp et al. (1988)
D. innoxia	Knopp et al. (1988); Parr et al. (1990); Shimomura et al. (1991)
D. metel	Knopp et al. (1988)
D. meteloides	Knopp et al. (1988)
D. quercifolia	Knopp et al. (1988); Parr et al. (1990)
D. rosei	Knopp et al. (1988)
D. stramonium	Knopp et al. (1988); Parr et al. (1990), Hilton and Rhodes (1990); Jaziri et al. (1988)
D. stramonium var. *inermis*	Knopp et al. (1988)
D. stramonium var. *stramonium*	Payne et al. (1987); Knopp et al. (1988), Munier (1988)
D. stramonium var. *tatula*	Knopp et al. (1988)
D. wrightii	Parr et al. (1990)
Arborescent *Datura* species	
D. sanguinea	Knopp et al. (1988); Parr et al. (1990)
Arborescent *Datura* hybrids	
D. candida × *D. candida* var. Flintham Hall	Christen et al. (1989, 1990)
D. candida × *D. aurea*	Robins et al. (1990)

In order to produce highly productive transformed root cultures, it is essential to use plant material selected for its interesting alkaloid spectrum or with a high alkaloid content relative to other plants of the same species or genus. The selection of plants showing desirable biosynthetic characteristics should be an important first step in obtaining hairy root cultures (Parr et al. 1988).

3.1 The *Agrobacterium* Transformation

Agrobacterium tumefaciens and *A. rhizogenes* are commonly occurring bacteria widely distributed in the soil. The infection of a plant which has been wounded naturally or artificially by *A. tumefaciens* results in tumor production, while a similar infection by *A. rhizogenes* generally induces adventitious root formation in the wounded tissue on dicotyledonous plants. The respective plant diseases are called crown gall and hairy root diseases. Only a very few monocotyledonous plants have been reported to be even weakly susceptible to crown gall and hairy root diseases (Hooykaas 1989). The reason for this limit in host range is not completely understood.

The molecular mechanisms underlying crown gall and hairy root induction are similar: transfer of a portion of a bacterial plasmid to plant cells. The virulence plasmid of *A. tumefaciens* has been called the Ti (Tumor-inducing) plasmid, while that of *A. rhizogenes* has been named the Ri (Root-inducing) plasmid. At the infection site, plant cells are transformed by a small fragment of DNA (T-DNA: 14–22 kb) from Ti or Ri plasmid. T-DNA may be transferred and integrated either as one or as two separate DNA fragments [T_L (left) and T_R (right)]. Thus genetic material is transferred from plasmids of *Agrobacterium* to plant hosts and these transferred genes are expressed in host cells. Once T-DNA is integrated into the plant nucleus, some genes encode synthesis of enzymes which catalyze reactions leading to the production of the plant growth factors auxins and cytokinins in transformed cells. Therefore crown gall and hairy root cells can grow in vitro in the absence of the bacterium in tissue culture media lacking exogenous supplies of phytohormones in contrast to normal plant cells.

Other genes on T-DNA code for the production within hosts of enzymes involved in the production of compounds called opines which are either condensates of an amino acid and a sugar or phosphorylated sugar derivatives. The type of opines synthesized is dependent on the strain of *Agrobacterium* that initiated the infection. Opines seem to serve as nutrients for the *Agrobacteria* (Clare 1990) and also to create a chemical environment favorable for growth and propagation of the pathogenic agent (Petit et al. 1983). It has been demonstrated (Spencer and Towers 1988) that *Agrobacteria* are attracted to hosts across chemical gradients of phenolic compounds released by injured plant cells. One specific highly active compound in this respect has been identified as acetosyringone.

3.2 Hairy Root Cultures of Arborescent Daturas

Hairy root cultures of two hybrid daturas have been developed and investigated. Both have their origins in the earlier investigations of *Datura* hybrids made by Dr.

W.C. Evans at Nottingham University (El-Iman and Evans 1990). Hairy root cultures of the tree *Datura* hybrid involving *Datura candida* × *D. candida* var. Flintham Hall (an orange-pink cultivar) (A) have been established and the alkaloid content determined (Christen et al. 1989). Another hybrid, *Datura candida* × *D. aurea* (B) has been similarly investigated (Robins et al. 1990). In the case of hybrid A, sterile seedlings were established and transformed root cultures initiated by wounding stems of 18-day-old sterile plantlets with *Agrobacterium rhizogenes* strain ATCC 15834. For the production of hairy roots from hybrid B, detached leaves were taken from greenhouse-grown plantlets, sterilized and inoculated with *A. rhizogenes* strain LBA 9402. One to 2 weeks after inoculation, hairy roots emerged at infection sites. Tissue bearing transformed roots was excised and, for hybrid A, transferred onto hormone-free MS medium (Murashige and Skoog 1962) with 1% agar, 3% sucrose, and containing 500 mg/l ampicillin to remove living *Agrobacterium*. After elimination of the bacteria, the transformed roots were cultivated on hormone-free MS medium with 3% sucrose. The fatest-growing roots were subcultured into fresh medium at 2-week intervals. The elimination of *Agrobacterium* may sometimes be difficult. A number of different approaches have proved successful in overcoming this problem:

1. Treatment with an antibiotic or a combination of antibiotics.
2. Subculturing the root tips for a short time in medium containing high agar concentration.
3. Culturing hairy roots in liquid medium at 40 °C for a few hours.

Generally, it is possible to establish transformed roots in axenic culture after subculturing a few times at 2- or 3-week intervals in liquid media containing antibiotics.

About 0.5 g (fresh weight) of the hairy roots were inoculated into 80 ml liquid MS medium containing 3% sucrose in 250-ml Erlenmeyer flasks. They were cultivated on a rotary shaker at 100 rpm in darkness at 25 °C over a period of 1 month. Transformed root cultures were characterized by a high degree of lateral branching, an absence of geotropism, profusion of root hairs, and a variable growth rate. The root clones developed from one plant showed a wide variation in alkaloid content varying from 0.12 to 1.6% dry wt. The fastest-growing roots were subcultured on a regular basis and showed the same growth characteristics after 1 year of culture.

The culture of hairy roots from hybrid B varied somewhat in that B5 salt (Gamborg et al. 1968) and vitamin base was used and roots were immediately placed in liquid media initially containing ampicillin.

One characteristic feature of the transformed roots is that they synthesize opines. They were identified by paper electrophoresis (Whatman 3MM) essentially according to the method of Petit et al. (1983). After drying, the electrophoretogram was stained with alkaline silver nitrate reagent (Ellis et al. 1984). Detection of T-DNA by Southern hybridization may also be used to confirm that root cultures have been transformed, and this is a useful alternative since opine production is not of general occurrence and the expression of opine genes in hairy root tissue has been shown in certain circumstances to be unstable with time (Rhodes et al. 1987).

Analysis of tropane alkaloids: the hairy roots, harvested after 1 month of culture, were lyophilized, then powdered, and extracted essentially according to

Payne et al. (1987). Leaf, stem, and root tissues of untransformed 5-month-old pot-grown plants were analyzed by the same procedure for comparison of the alkaloid content (Christen et al. 1989, hybrid A).

The alkaloids hyoscyamine and scopolamine may be quantified either by HPLC or by capillary GLC with a nitrogen detector. Quantification by HPLC used the method of Kamada et al. (1986) (Christen et al. 1989) on a μ-Bondapak C-18 column (30 cm × 3.9 mm i.d.; 10 μm) (Waters, USA) fitted with a Guard-Pak μ-Bondapak C-18 precolumn module. The solvent system used was 10 mM sodium 1-heptanesulfonate in H_2O (adjusted to pH 4.0 with acetic acid): methanol 6:4 and the flow rate was 1 ml/min. The alkaloids were detected by UV at 215 nm and the quantification was calculated by the use of standard curves fitted by linear regression.

After 1 month of culture, the scopolamine and hyoscyamine content in the hairy roots of hybrid A was compared with that of the untransformed plants (Table 2). Scopolamine was the principal alkaloid and the concentration of scopolamine and hyoscyamine in the hairy roots was 0.68% dry wt. By comparison, the aerial parts and the roots of 5-month-old pot-grown plants contained 0.425 and 0.258% dry wt. respectively. Alkaloid concentration was increased by a factor of 1.6 and 2.6 relative to the aerial parts and to the roots of the normal plants. The ratio scopolamine/hyoscyamine was about 5:1 in the hairy roots, similar to that in the parent plant, which indicates that Ri T-DNA does not have a major qualitative effect on the alkaloid composition. In hairy roots from hybrid B the ratio hyoscyamine/scopolamine was 5:2 (Robins et al. 1990).

In addition, the alkaloid extracts from hairy roots from both hybrids, A and B, were analyzed by GLC and GC/MS. Nineteen tropane alkaloids comprising 73% of the peaks detected have been identified in hairy roots from hybrid A and some 37 in hairy roots of B (Table 3). Details on the extraction procedure and GLC and GC/MS have been published recently (Christen et al. 1990; Robins et al. 1990). The GLC analysis for hairy roots from hybrid A was performed with a 15-m fused silica capillary column coated with the methylsilicone phase DB-1 (J & W Scientific) and a Perkin-Elmer 8700 gas chromatograph fitted with two detectors: a flame

Table 2. Comparison of alkaloid content of *Datura candida* × *D. candida* var. Flintham Hall hybrid hairy root cultures and normal untransformed plants. (Christen et al. 1989)

	Alkaloids (% dry wt.)		
	Hairy roots[a]	Untransformed plants[b]	
		Aerial parts	Roots
Scopolamine	0.570[c]	0.360	0.220
Hyoscyamine	0.110	0.065	0.038
Scopolamine + hyoscyamine	0.680	0.425	0.258
Ratio S/H	5.2:1	5.5:1	5.8:1

[a] Hairy roots cultured for 1 month in liquid MS medium.
[b] Five-month-old pot-grown plants.
[c] The alkaloids were quantified by HPLC. Scopolamine and hyoscyamine values are the sum of the alkaloid in the hairy roots and in the medium.

ionization detector (FID) and a more specific and more sensitive nitrogen phosphorus detector (NPD). The derivatization of alkaloids was omitted without notable problems of adsorption. GC/MS analysis was carried out with the same column using a Hewlett-Packard 5890A instrument in combination with a VG 12-250 mass data system. Chemical ionization mass spectra were recorded on the same instrument using ammonia as reagent gas. The different constituents were identified by comparison of their retention times with those of authentic samples and of their mass spectral data with those published (Witte et al. 1987; Lounasmaa 1988). In the absence of NMR data (A) the stereochemistry of substituents at carbon 3 and 6 could only be determined by GLC with alkaloids for which reference material was available. An estimate of the relative percentage of each alkaloid was carried out by computerized peak area integration. In the case of hairy roots from hybrid B, NMR data was available for apoatropine (Robins et al. 1990).

Scopolamine has been identified as the main compound in hairy roots of hybrid A accounting for about 17% of the alkaloid mixture. Tropine (13.8%), 3-acetoxy-6-hydroxytropane (4.1%), hyoscyamine (3.4%), apohyoscyamine (5.3%), and 6-hydroxyhyoscyamine (4.8%) are also at a relatively high amount in the mixture. Furthermore, 16 alkaloids are present at levels exceeding 1% of the total alkaloids. Scopolamine, tropine, and hyoscyamine account for about 47% of the alkaloids detected. Among the 19 alkaloids detected, 17 are known compounds, but only 5 components in *Datura candida* and 18 components in the cultivar have been identified previously (Table 3).

Three apo-derivatives have been identified from the alkaloid mixture: 6-hydroxyapohyoscyamine, apohyoscyamine and aposcopolamine. These three compounds are at least partially produced artifactually during GC analysis, since there is an easy thermal loss of water from tropic acid derivatives leading to the apo-compounds.

In hairy root cultures of hybrid B, hyoscyamine in the total alkaloids was about 29%, whereas scopolamine was 18%, aposcopolamine 11%, and apohyoscyamine 10%. The remaining compounds found in the hairy roots of hybrid B were mainly esters formed from tropane-3,6-diol and tropane-3,6,7-triol. The predominant esterifying acids are tropic and tiglic acids (Table 3).

Among the 19 alkaloids detected for hairy roots of hybrid A and the 37 for hybrid B, three compounds have never been reported in the literature before (Fig. 1). These are 3-hydroxy-6-propionyloxytropane (A), 3-hydroxy-6-butyryloxytropane (A), and 3-(hydroxyacetoxy)tropane (B) (Christen et al. 1990; Robins et al. 1990).

The results of Christen et al. (1989) and those of Robins et al. (1990) demonstrate that the alkaloids produced by the hairy root cultures are typical of those found in many species of *Datura* (Evans 1979). The transformed roots produce tropane alkaloids which do not differ significantly from those produced by the parent plants. Scopolamine, identified as the major alkaloid in the hairy root cultures of hybrid A, is also the principal alkaloid in the noninfected plants.

The analysis of the spent liquid medium from hairy root cultures of hybrid A revealed that the release of tropane alkaloids was very low (about 3% of the total). This result resembles that found by Payne et al. (1987) with the hairy roots of *Datura stramonium.*

Table 3. Tropane alkaloids identified in transformed root cultures of two hybrids: *Datura candida* × *D. candida* var. Flintham Hall (A) and *D. candida* × *D. aurea* (B)

Alkaloids	M^+	Hybrid A μM/g fr. wt.	Hybrid B nM/g fr. wt.
Hygrine[a]	141	–	220
Tropinone	139	–	10
Tropine[a]	141	54	270
Pseudotropine[a]	141	11	10
Oscine[a]	155	5	–
3α-Acetoxytropane[a]	183	7	160
3β-Acetoxytropane	183	–	2
3,6-Dihydroxytropane	157	–	Trace
3-(Hydroxyacetoxy)tropane[b]	199	–	5
3-Acetoxy-6-hydroxytropane	199	11	15
N-Methylpyrrolidinylhygrine A	224	Trace	10
N-Methylpyrrolidinylhygrine B	224	–	10
3α-Tigloyloxytropane	223	–	100
3β-Tigloyloxytropane (tigloidine)[a]	223	7	3
Cuscohygrine	224	–	95
3-Hydroxy-6-propionyloxytropane[b]	213	8	–
3-Hydroxy-6-(2-methylbutyryloxy)tropane[c]	241	3	8
3-Tigloyloxy-6-hydroxytropane	239	5	245
3-Hydroxy-6-tigloyloxytropane	239	–	65
3α-Hydroxy-6β-tigloyloxytropane[a]	239	4	–
3-Tigloyloxy-6-acetoxytropane	281	–	95
3α-Tigloyloxy-6β-acetoxytropane[a]	281	2	–
3-Phenylacetoxytropane	259	–	10
3-Tigloyloxy-6-propionyloxytropane	295	–	10
3-Hydroxy-6-butyryloxytropane[b, c]	227	3	–
3-Tigloyloxy-6-isobutyryloxytropane[c]	309	–	5
apohyoscyamine[a,d]	271	11	550
6-Hydroxyapohyoscyamine	287	Trace	ca. 40
3-Phenylacetoxy-6,7-epoxytropane[d]	273	–	110
3α-Tigloyloxy-6β-(2-methylbutyryloxy)tropane[c]	323	–	3
3-Tigloyloxy-6-(methylbutyryloxy)tropane[c]	323	–	5
Aposcopolamine[a, d]	285	7	580
3-Phenylacetoxy-6-hydroxytropane	275	–	Trace
Littorine	289	–	ca. 50
Hyoscyamine[a]	289	7	1580
3α,6β-Ditigloyloxytropanenape	321	–	ca. 120
3-Tigloyloxy-6-isovaleroyloxy-7-hydroxytropane[c]	339	–	ca. 5
Scopolamine[a]	303	32	995
3α,6β-Ditigloyloxy-7β-hydroxytropane	337	–	15
(?7β)-Hydroxyhyoscyamine	305	–	20
6β-Hydroxyhyoscyamine[a]	305	9	20
Unknown	414	–	15

[a] Comparison with authentic samples.
[b] New alkaloids.
[c] C_4 or C_5 alkyl isomer not positively identified.
[d] Includes some alkaloid produced artifactually.

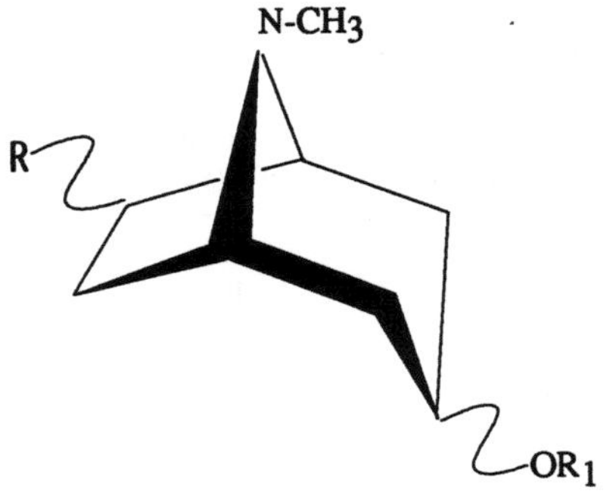

$R = OCOC_2H_5, R_1 = H$ (Hybrid A)
$R = OCOC_3H_7, R_1 = H$ (Hybrid A)
$R = H, R_1 = OCCH_2OH$ (Hybrid B)

Fig. 1. New tropane alkaloids identified from hairy root cultures of the hybrids *Datura candida* × *D. candida* var. Flintham Hall and *D. candida* × *D. aurea*

In many solanaceous species, it is generally considered that the biosynthesis of tropane alkaloids takes place in the roots but mainly accumulate in the aerial parts of the plant. In hairy root cultures, there is no alkaloid transport to aerial parts for storage and this apparent lack of storage facility may contribute to the release of the alkaloids into the growth medium.

In order to establish continuous production of tropane alkaloids, it is essential that they are released into the growth medium. Compounds may be released by means of pH manipulation or more generally by altering the medium composition. The permeabilization of cells can be altered by the addition of various chemicals such as DMSO, antibiotics, or fungal elicitors. The excretion may also be stimulated by continually removing the compounds from the medium by adsorption on ion exchangers, thus relieving any feedback mechanisms that may occur.

Concerning the biochemistry of the pathway leading to tropane alkaloids, most of the experiments to date have utilized whole plants or untransformed root cultures (Leete 1990; Yamada et al. 1990). However, recently a series of papers have been published in which transformed roots of *Datura stramonium* have been used to investigate the regulation of the earlier part of the biosynthetic pathway to the tropanes, in particular the effect of the metabolic flux of intermediates on the two key enzymes ornithine decarboxylase and arginine decarboxylase. The investigation of hairy roots from hybrid B (Robins et al. 1990) indicated that of the six enzymes involved in the pathway to hyoscyamine and scopolamine, ornithine decarboxylase and arginine decarboxylase are present in potentially flux-limiting levels, in contrast to those enzymes assayed which act further down the pathway. In *Datura stramonium* hairy roots, Robins et al. (1991a) reported on a series of experiments aimed at determining the principal internal factors that control the flux into hyoscyamine. It appeared that the regulation of tropane formation is complex. Arginine decarboxylase is the more important activity for hyoscyamine formation. The suppression of this enzyme resulted in a severe decrease in free and conjugated putrescine and in the putrescine-derived intermediates of hyoscyamine biosynthesis (Robins et al. 1991b). An additional limitation to flux involving the supply of activated acids for condensation with tropine to form the identified tropoyl and tigloyl derivatives is also indicated from the observed effect of feeding free acids to hairy root cultures. The relative contribution to flux limitation caused by these two interacting phenomena is inferred from an analysis of the changing relative levels of metabolic intermediates and end product as the culture matures (Robins et al. 1990).

Important research efforts are being directed towards an understanding of the biosynthesis of tropane alkaloids. Robins et al. (1991c) demonstrated that transformed roots treated with exogenous plant growth regulators lost their competence in tropane alkaloid biosynthesis. Transformed root cultures of *Datura stramonium* have been used to purify enzymes involved in the formation of 3α- and 3β-acetoxytropanes (Robins et al. 1991d). Two distinct acyltransferases have been identified; one acetylates tropine (tropan-3α-ol) using acetyl-coenzyme A as cosubstrate, and the other enzyme acetylates pseudotropine (tropan-3β-ol) with the same cosubstrate.

Of particular interest is the formation of scopolamine by oxidative transformation of hyoscyamine via the intermediate 6β-hydroxyhyoscyamine. These reactions require the presence of the enzymes hyoscyamine 6β-hydroxylase (Hashimoto and Yamada 1986; Hashimoto et al. 1991) and 6β-hydroxyhyoscyamine epoxidase (Hashimoto et al. 1989). 6β-hydroxyhyoscyamine has been unambiguously identified in the alkaloid mixture from the hairy root cultures of the hybrids A and B (Christen et al. 1990; Robins et al. 1990).

3.3 Hairy Root Culture of Herbaceous *Datura*

Another work concerning the variation in tropane alkaloid accumulation within different solanaceous species has been reported by Parr et al. (1990). Hairy root cultures of six herbaceous species of *Datura* were established (Table 4). The plants were selected on the basis of their high petiolar alkaloid levels, for example all *Datura stramonium* cultures examined produced a similar spectrum of alkaloids with no scopolamine and with hyoscyamine accounting for about 80% of the total alkaloids. The best line produced a level of 1053 μg/g of hyoscyamine (Tables 4, 5).

Payne et al. (1987) established hairy root cultures of *Datura stramonium* and studied the effects of medium composition and pH on hyoscyamine production.

Table 4. Alkaloid content of transformed root cultures of herbaceous *Datura* species. (Parr et al. 1990)

	μg/g fr. wt.	
Species	Hyoscyamine	Scopolamine
D. stramonium	1053	0
D. ferox	849	43
D. innoxia	486	110
D. wrightii	820	15
D. fastuosa	561	9
D. quercifolia	419	Trace

Table 5. Comparison of petiole levels of hyoscyamine and scopolamine with the transformed root cultures developed from *Datura stramonium*. (Parr et al. 1990)

No. of plants	μg/g fr. wt.		
	Petioles	Hairy root cultures	
	Hyosyamine + scopolamine	Hyoscyamine	Scopolamine
2	1001–1350	870–1053	0
3	801–1000	887–896	0
9	600–800	617–847	0–Trace
4	80–200	663–901	0–Trace

Values for the root cultures are the sum of intracellular and extracellular alkaloid and have been averaged over two culture cycles.

They concluded that B5 medium with 3–5% sucrose is satisfactory for promoting growth of hairy root cultures. A medium containing 5% sucrose and decreased levels of phosphate (0.1–0.5 mM) and nitrate (10–20 mM) increased overall hyoscyamine yield. Furthermore, cultures initially buffered to pH 5.0–5.5 (50 mM MES) grew and produced alkaloids as well as or better than unbuffered controls. Manipulating the pH of the culture medium failed to stimulate the release of hyoscyamine into the medium.

Hilton and Rhodes (1990) examined the growth and hyoscyamine production of transformed roots of *Datura stramonium* in 14-1 fermenters on media containing half or full strength B5 as nutrient medium and at three different temperatures: 25, 30 and 35 °C. Roots grown on half-strength B5 medium with 3% sucrose had a higher dry matter content and a higher hyoscyamine content than roots grown on full strength B5 medium with the same level of sucrose. Growth at 30 °C was faster than at either 25 or 35 °C. The final hyoscyamine concentrations attained, about 0.5 mg/g wet weight, were similar at 25 and 30 °C but some 40% lower at 35 °C. The highest production rate of 8.2 mg hyoscyamine/l/day was obtained with roots grown at 30 °C. In continuous fermentation at 25 °C, the release of hyoscyamine into the culture medium was less than 0.5% w/w of the total but was up to sevenfold higher in fermenters operated at 30 and 35 °C. The authors suggest that the higher temperatures may have increased the permeability of the cells to hyoscyamine, allowing increased transport into the medium.

Ohkawa et al. (1989), examined the effects of several plant growth regulators on the hairy root growth of *Datura innoxia*. Treatment with gibberellic acid (GA_3) at concentrations between 10 ng/l and 1 mg/l accelerated the fresh weight increase of the hairy roots. Elongation and lateral branching were also enhanced by treatment with GA_3. Nothing is said, however, about the influence of the plant growth regulators on the production of alkaloids.

The influence of biotic and abiotic elicitors on product formation in *Datura stramonium* root cultures has been studied recently (Furze et al. 1991). Transformed roots treated with copper or cadmium salts showed an accumulation of sesquiterpenoid phytoalexins lubimin and 3-hydroxylubimin. On the other hand, the biosynthesis of tropane alkaloids was not significantly stimulated. However, a rapid and, in some instances, reversible release of alkaloids was observed.

4 Summary and Conclusions

By comparison with cell cultures, hairy root cultures can be established more easily. Furthermore, the genetic stability and the higher productivity make this technique worthy of consideration for commercial sources of secondary metabolites.

The alkaloid spectra of the two tree datura hairy root cultures described here are dominated by hyoscyamine and scopolamine. In addition, several other minor alkaloids were detected by GC-MS in both cultures and three new compounds identified. Concerning the transformed root cultures of *Datura candida* × *D. candida* var. Flintham Hall, the total amount of scopolamine (principal alkaloid) and hyoscyamine reached a level of 0.68% dry wt., which is 1.6 and 2.6 times higher than in the aerial parts and in the roots of the nontransformed 5-month-old plants, respectively. However, many steps remain to be improved:

1. Detailed information concerning the understanding of the complete pathway and its in vivo regulation will undoubtedly play a significant role in the establishment of hairy root cultures for commercial production.
2. Improving the yield in batch process by employing selected, high-producing clones and optimized culture conditions appears to be more readily accomplished with hairy roots than with cell cultures.
3. For production on an industrial scale, adopting a continuous on-line, nondestructive harvest procedure in which the secondary metabolites are released into the growth medium throughout the life of the culture rather than a batch process.

Although the use of *Agrobacterium* is an interesting technique capable of transferring new genes to a wide variety of plant species, its practical use is limited to those species susceptible to crown gall and hairy root diseases and, apparently, for the production of secondary metabolites biosynthesized in the roots of normal plants.

References

Al-Yahya M, Evans WC (1975) Alkaloids of the F_1 hybrid of *Datura stramonium* × *D. discolor*. J Pharm Pharmacol 27:87p (Suppl)

Bristol ML, Evans WC, Lampard JF (1969) The alkaloids of the genus *Datura*, section *Brugmansia*. Part VI. Tree *Datura* drugs (*Datura candida* cvs.) of the Colombian Sibundoy. J Nat Prod (Lloydia) 32:123–130

Christen P, Roberts MF, Phillipson JD, Evans WC (1989) High-yield production of tropane alkaloids by hairy-root cultures of a *Datura candida* hybrid. Plant Cell Rep 8:75–77

Christen P, Roberts MF, Phillipson JD, Evans WC (1990) Alkaloids of hairy root cultures of a *Datura candida* hybrid. Plant Cell Rep 9:101–104

Clare BG (1990) *Agrobacterium* in plant disease, biological disease control and plant genetic engineering. Sci Progr London 74:1–13

El-Dabbas SW, Evans WC (1982) Alkaloids of the genus *Datura*, section *Brugmansia*. X. Alkaloid content of *Datura* hybrids. Planta Med 44:184–185

El-Imam YMA, Evans WC (1990) Alkaloids of a *Datura candida* cultivar, *D. aurea* and various hybrids. Fitoterapia 61:148–152

Ellis JG, Ryder MH, Tate ME (1984) *Agrobacterium tumefaciens* T_R-DNA encodes a pathway for agropine biosynthesis. Mol Gen Genet 195:466–473

Evans WC (1979) Tropane alkaloids of the Solanaceae. In: Hawkes JG, Lester RN, Skelding AD (eds) The biology and taxonomy of the Solanaceae. Academic Press, New York, pp 241–254

Evans WC (1990) *Datura*, a commercial source of hyoscine. Pharm J 244:651–653

Flores HE, Hoy MW, Pickard JJ (1987) Secondary metabolites from root cultures. Trends Biotechnol 5:64–69

Furze JM, Rhodes MJC, Parr AJ, Robins RJ, Withehead IM, Threlfall DR (1991) Abiotic factors elicit sesquiterpenoid phytoalexin production but not alkaloid production in transformed root cultures of *Datura stramonium.* Plant Cell Rep. 10:111–114

Gamborg OL, Miller RA, Ojima K (1968) Nutrient requirements of suspension cultures of soybean root cells. Exp Cell Res 50:151–158

Griffin WJ (1976) Agronomic evaluation of *Datura candida* – a new source of hyoscine. Econ Bot 30:361–369

Griffin WJ (1992) Alkaloids of a *Datura candida* cultivar. Phytochemistry 31:367–368

Hashimoto T, Yamada Y (1986) Hyoscyamine 6β-hydroxylase, a 2-oxoglutarate-dependant dioxygenase, in alkaloid producing root cultures. Plant Physiol 81:619–625

Hashimoto T, Kohno J, Yamada Y (1989) 6β-Hydroxyhyoscyamine epoxidase from cultured roots of *Hyoscyamus niger.* Phytochemistry 28:1077–1082

Hashimoto T, Hayashi A, Amano Y, Fohno J, Iwanari H, Usuda S, Yamada Y (1991) Hyoscyamine 6β-hydroxylase, an enzyme involved in tropane alkaloid biosynthesis, is localized at the pericycle of the root. J Biol Chem 266:4648–4653

Hay CA, Anderson LA, Roberts MF, Phillipson JD (1988) Alkaloid production by plant cell cultures. In: Mizrahi A (ed) Biotechnology in agriculture, advances in biotechnological processes, vol 9. Alan R. Riss, New York, pp 97–140

Hilton MG, Rhodes MJC (1990) Growth and hyoscyamine production of hairy root cultures of *Datura stramonium* in a modified stirred tank reactor. Appl Microbiol Biotechnol 33:132–138

Hooykaas PJJ (1989) Transformation of plant cells via *Agrobacterium.* Plant Mol Biol 13:327–336

Jaziri M, Legros M, Homes J, Vanhaelen M (1988) Tropane alkaloids production by hairy root cultures of *Datura stramonium* and *Hyoscyamus niger.* Phytochemistry 27:419–420

Jung G, Tepfer D (1987) Use of genetic transformation by the Ri T-DNA of *Agrobacterium rhizogenes* to stimulate biomass and tropane alkaloid production in *Atropa belladonna* and *Calystegia sepium* roots grown in vitro. Plant Sci 50:145–151

Kamada H, Okamura N, Satake M, Harada H, Shimomura K (1986) Alkaloid production by hairy root cultures in *Atropa belladonna.* Plant Cell Rep 5:239–242

Knopp E, Strauss A, Wehrli W (1988) Root induction on several Solanaceae species by *Agrobacterium rhizogenes* and the determination of root tropane alkaloid content. Plant Cell Rep 7:590–593

Leete E (1990) Recent developments in the biosynthesis of tropane alkaloids. Planta Med 56:339–352

Lounasmaa M (1988) The tropane alkaloids. In: Brossi A (ed) The alkaloids, vol 33. Academic Press, New York, pp 1–81

Mano Y, Nabeshima S, Matsui C, Ohkawa H (1986) Production of tropane alkaloids by hairy root cultures of *Scopolia japonica.* Agric Biol Chem 50:2715–2722

Mano Y, Ohkawa H, Yamada Y (1989) Production of tropane alkaloids by hairy root cultures of *Duboisia leichhardtii* transformed by *Agrobacterium rhizogenes.* Plant Sci 59:191–201

Munier J (1988) Establishment of new axenic hairy root lines by inoculation with *Agrobacterium rhizogenes.* Plant Cell Rep 7:9–12

Murashige T, Skoog F (1962) A revised medium for rapid growth and bioassays with tobacco tissue cultures. Physiol Plant 15:473–497

Nabeshima S, Mano Y, Ohkawa H (1986) Production of tropane alkaloids by hairy root cultures of *Scopolia japonica.* Symbiosis 2:11–18

Ohkawa H, Kamada H, Sudo H, Harada H (1989) Effects of gibberellic acid on hairy root growth in *Datura innoxia.* J Plant Physiol 134:633–636

Parr AJ, Hamill JD, Payne J, Rhodes MJC, Robins RJ, Walton NJ (1988) Exploiting variation in tropane alkaloid production by solanaceous plants. In: Robins RJ, Rhodes MJC (eds) Manipulating secondary metabolism in culture. Cambridge University Press, Cambridge, pp 101–107

Parr AJ, Payne J, Eagles J, Chapman BT, Robins RJ, Rhodes MJC (1990) Variation in tropane alkaloid accumulation within the Solanaceae and strategies for its exploitation. Phytochemistry 29:2545–2550

Payne J, Hamill JD, Robins RJ, Rhodes MJC (1987) Production of hyoscyamine by hairy root cultures of *Datura stramonium*. Planta Med 53:474–478

Petit A, David C, Dahl GA, Ellis JG, Guyon P, Casse-Delbart F, Tempé J (1983) Further extension of the opine concept: plasmids in *Agrobacterium rhizogenes* cooperate for opine degradation. Mol Gen Genet 190:204–214

Petri G, Bajaj YPS (1989) *Datura* spp.: In vitro regeneration and the production of tropanes. In: Bajaj YPS (ed) Biotechnology in agriculture and forestry, vol 7. Medicinal and aromatic plants II. Springer, Berlin Heidelberg New York, pp 135–161

Phillipson JD (1990) Plants as sources of valuable products. In: Charlwood BV, Rhodes MJC (eds) Secondary products from plant tissue culture. Clarendon Press, Oxford, pp 1–21

Rhodes MJC, Robins RJ, Hamill JD, Parr AJ, Walton NJ (1987) Secondary product formation using *Agrobacterium rhizogenes*-transformed hairy root culture. IAPTC Newslett 53:2–15

Robins RJ, Parr AJ, Payne J, Walton NJ, Rhodes MJC (1990) Factors regulating tropane-alkaloid production in a transformed root culture of a *Datura candida* × *D. aurea* hybrid. Planta 181:414–422

Robins RJ, Parr AJ, Bent EG, Rhodes MJC (1991a) Studies on the biosynthesis of tropane alkaloids in *Datura stramonium* L. transformed root cultures. 1. The kinetics of alkaloid production and the influence of feeding intermediate metabolites. Planta 183:185–195

Robins RJ, Parr AJ, Walton NJ (1991b) Studies on the biosynthesis of tropane alkaloids in *Datura stramonium* L. transformed root cultures. 2. On the relative contributions of L-arginine and L-ornithine to the formation of the tropane ring. Planta 183:196–201

Robins RJ, Bent EG, Rhodes MJC (1991c) Studies on the biosynthesis of tropane alkaloids by *Datura stramonium* L. transformed root cultures. Planta 185:385–390

Robins RJ, Bachmann P, Robinson T, Rhodes MJC, Yamada Y (1991d) The formation of 3α- and 3β-acetoxytropanes by *Datura stramonium* transformed root cultures involves two acetyl-CoA-dependent acyltransferases. FEBS Lett. 292:293–297

Romeike A (1962) Weitere Versuche zur Züchtung einer scopolaminreichen *Datura*-Hybride. Kulturpflanze 10:140–148

Safford WE (1921) Synopsis of the genus *Datura*. J Wash Acad Sci 11:173–189

Scragg AH (1990) Fermentation systems for plant cells. In: Charlwood BV, Rhodes MJC (eds) Secondary products from plant tissue culture. Clarendon Press, Oxford, pp 243–263

Spencer PA, Towers GHN (1988) Specificity of signal compounds detected by *Agrobacterium tumefaciens*. Phytochemistry 27:2781–2785

Shimomura K, Sauerwein M, Ishimaru K (1991) Tropane alkaloids in the adventitious and hairy root cultures of solanaceous plants. Phytochemistry 30:2275–2278

Walton NJ, Robins RJ, Peerless ACJ (1990) Enzymes of N-methylputrescine biosynthesis in relation to hyoscyamine formation in transformed root cultures of *Datura stramonium* and *Atropa belladonna*. Planta 182:136–141

Wettstein R (1895) Solanaceae. In: Engler A, Prantl K (eds) Die natürlichen Pflanzenfamilien. Englemann, Leipzig, IV (3b) pp 4–38

Witte L, Müller K, Arfmann H-A (1987) Investigation of the alkaloid pattern of *Datura innoxia* plants by capillary gas-liquid chromatography-mass spectrometry. Planta Med 53:192–197

Yamada Y, Hashimoto T, Endo T, Yukimune Y, Kohno J, Hamaguchi N, Dräger B (1990) Biochemistry of alkaloid production in vitro. In: Charlwood BV, Rhodes MJC (eds) Secondary products from plant tissue culture. Clarendon Press, Oxford, pp 227–242

II.5 Transformation in *Daucus carota* L. (Carrot)

R. Bower and R.G. Birch[1]

1 Introduction

Carrot (*Daucus carota*) has, for two main reasons, been the subject of a variety of genetic transformation studies. As with most food crops, the prospect exists for increased yield and production efficiency by introduction of agronomically useful genes into commercial cultivars. Such genes can potentially confer a range of desirable traits including resistance to pathogens, insects, drought, salt stress, and herbicides.

Carrot has also proven extremely valuable as a model dicot species for basic research purposes. In this context its major attribute is its amenability to culture and manipulation in vitro. In particular carrot has been used as a model to study the genetics and physiology of somatic embryogenesis (Steward et al. 1975; Choi et al. 1987; Wilde et al. 1988; Thomas et al. 1989b).

Isolation, culture, and regeneration of carrot protoplasts have also been extensively studied (see Sect. 4) and the relative ease with which these procedures are accomplished in carrot has particularly facilitated experiments based on gene transfer by electroporation.

2 Stable Genetic Transformation of Carrot

Carrot was one of the first crop species transformed using *Agrobacterium tumefaciens* (Scott and Draper 1987). Because of the simplicity and efficiency of this procedure it is the stable transformation method of choice. Electroporation of carrot protoplasts has also allowed recovery of transformed carrot somatic embryos (Langridge et al. 1985) which should be regenerable into plants. However, production of transformed carrot plants by electroporation is not as simple as *A. tumefaciens*-mediated transformation. Because of the greater time required in tissue culture, it may also generate undesirable somaclonal variation.

Both *A. tumefaciens* and *A. rhizogenes* have been used as vehicles for introduction of genes into carrot. Genetic transformation by *A. rhizogenes* results in characteristic phenotypic variations in regenerated plants which have generally precluded its use for introduction of agronomically useful genes into commercial cultivars, although it still has applications as a research tool. Tepfer (1985) used

[1] Department of Botany, The University of Queensland, Brisbane, 4072, Australia

Biotechnology in Agriculture and Forestry, Vol. 22
Plant Protoplasts and Genetic Engineering III (ed. by Y.P.S. Bajaj)

inoculation of carrot roots to produce transformed tissue from which plants were regenerated via somatic embryogenesis. These plants exhibited altered root and leaf morphology and reduced seed set, but allowed the quantification of rates of transfer of introduced genes through successive generations. Transformation by *A. rhizogenes* was also used to introduce the maize *Ac* element into carrot (van Sluys et al. 1987; van Sluys and Tempe 1989). Plants regenerated after transformation had the characteristic hairy root phenotype, although this did not impede study of *Ac* activity.

Two systems have been described for *A. tumefaciens*-mediated gene transfer into carrot. Cells from an embryogenic cell suspension of *D. carota* cv. Early Nantes were cultured for 5–7 days over a layer of feeder cells before inoculation and co-cultivation with *A. tumefaciens* strain C58C1 carrying the plasmid pGV3850::1103 (Scott and Draper 1987). The plasmid encodes the neomycin phosphotransferase (*npt*II) gene driven by the nopaline synthase (NOS) promoter. Transformed calli were selected on 100 µg/ml kanamycin and used to re-establish cell suspensions from which transformed somatic embryos were recovered and regenerated into plants. Transformation efficiency per plating unit was estimated to be 62–73%.

An alternative method entailed slicing hypocotyl sections from 1-week-old carrot seedlings and pre-culturing these in darkness for 2 days prior to co-cultivation with *A. tumefaciens* strain LBA 4404 carrying CaMV35S promoter-*npt* II construct on a binary vector (Thomas et al. 1989a). No transformation occurred in the absence of the pre-culture treatment. Selection for resistance to 100 µg/ml kanamycin was applied after 2–3 days co-cultivation. After 6 weeks, transformed calli up to 1 cm diameter were used to re-establish suspension cultures. Transformed plants were regenerated from these via somatic embryogenesis. Testing of the method on four commercial cultivars yielded transformation efficiencies per initial explant portion of 0.9–5.8%, which are sufficient for practical genetic engineering purposes.

3 Applications of Electroporation of Carrot Protoplasts

Electroporation of plant protoplasts involves the reversible permeabilization of the plasmalemma by exposure to a short, high voltage electric pulse, allowing entry of macromolecules, including DNA, into the cell. The ability to introduce foreign DNA into plant cells by this method has proven effective both for the stable genetic transformation of a range of crop species and as a tool for investigation of aspects of cell and molecular biology. Polyethylene glycol treatment of protoplasts also facilitates DNA uptake, but generally at a lower efficiency than electroporation. Carrot protoplasts have proved a very convenient tool for experiments requiring direct gene transfer into plant cells.

3.1 Investigation of the Control of Gene Expression

Understanding the effects of genetic elements such as promoters, enhancer elements, and introns which influence transcriptional or translational efficiencies has been enormously increased by the ability to introduce, by electroporation, such sequences coupled with reporter genes into plant cells (Ow et al. 1987; Planckaert and Walbot 1989; Last et al. 1990; Rathus et al. in prep.). Use of electroporation for these purposes has the advantage that results are obtained within 48 h. Further, interpretation of these results is not complicated by effects such as position of integration of introduced genes into the host genome, variation in copy number, and co-suppression phenomena, in contrast to stable transformants (Fluhr et al. 1986; Napoli et al. 1990).

Because of their ease of isolation and efficiency of electroporation, carrot protoplasts are well suited to such experiments. Ow et al. (1987) electroporated carrot protoplasts with a series of partial deletions of the CaMV35S promoter in conjunction with the firefly luciferase reporter gene and identified at least three regions essential for promoter activity. Electroporation of carrot protoplasts was also used to compare the functions of a maize zein promoter and the mannopine synthase promoter (Boston et al. 1987). Sequences introduced into carrot cells by the method of Krens et al. (1982) enabled investigation of the effects of internal AUG codons on initiation of translation (Putterill and Gardner 1989). Analysis of RNA transcripts in electroporated carrot protoplasts has been used in developing methodology for detection of transcription products such as those from the octopine synthase (*ocs*) *npt*II, chloramphenicol acetyl transferase (*cat*), *Bacillus thuringiensis* crystal protein, and alfalfa mosaic virus coat protein genes (Murray et al. 1990). Electroporation of luciferase mRNA into carrot, tobacco, rice, and maize protoplasts enabled analysis of the effects of a translational enhancer sequence from tobacco mosaic virus and of a minimal polyadenylation tail on translational efficiency in plant cells (Gallie et al. 1989). We have used electroporation of carrot and sugarcane protoplasts as model dicot and monocot species to compare expression from constructs containing combinations of promoters, the OCS enhancer element and the maize alcohol dehydrogenase (Adh) Intron 1 (Rathus et al. in prep.).

3.2 Investigation of Aspects of Cell Biology

A variety of cellular processes can be investigated through their responses to introduced DNA. Thomas et al. (1989b) electroporated carrot protoplasts with the *cat* reporter gene then applied concentrations of 5-bromodeoxyuridine to assess whether inhibition of transcription or translation by this compound was the mechanism by which it prevented normal development of somatic embryos. There has been speculation that plant cells, like some microbial cells, may need to be in a state of physiological competence for uptake, expression, or integration of foreign genes. A study using electroporation of carrot revealed that only a subpopulation of protoplasts isolated from a uniform cell suspension is competent for uptake and expression of a foreign gene (Bower and Birch 1990). The results also revealed the basis for increased transient reporter gene expression following electroporation of higher DNA concentrations into protoplasts. Use of a combination of histochemical and fluorometric GUS assays allowed differentiation between increases due to a higher proportion of expressing protoplasts and increases due to increased expression per expressing protoplasts (Fig. 1).

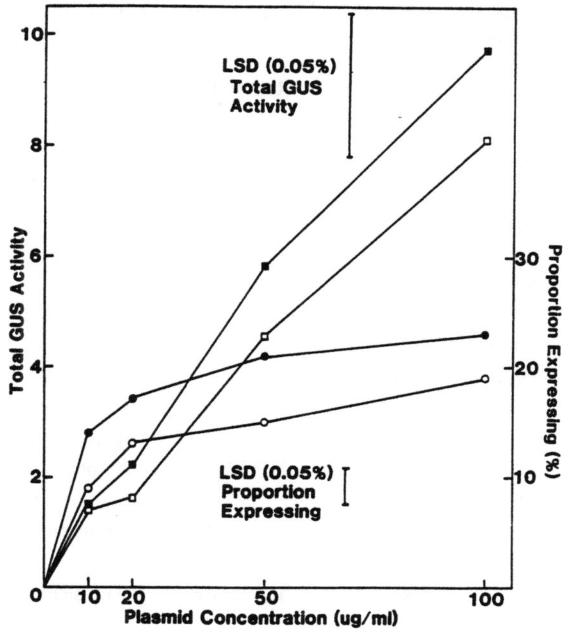

Fig. 1. Effect of carrier DNA and increasing plasmid concentration of total GUS activity (nM methylumbelliferone/min/mg protein) and on the proportion of carrot protoplasts expressing GUS activity. (Bower and Birch 1990). Total GUS activity: ■ + carrier DNA; □ – carrier DNA; proportion expressing: ● + carrier DNA; ○ – carrier DNA

4 Isolation and Culture of Carrot Protoplasts

An effective system for isolation and culture of protoplasts is critical to the success of experiments involving electroporation. The system must produce reliable, high yields of viable protoplasts, capable of high levels of gene expression following electroporation. Numerous protocols exist for isolation and culture of carrot protoplasts (Gosch et al. 1975; Mathews and Cress 1981; Nomura et al. 1982; Boston et al. 1987; Langridge et al. 1987; Nea and Bates 1987; Putterill and Gardner 1989; Shea et al. 1989). Differences in carrot cell lines and protoplasts isolated from these may partly account for the variation in "optimal" conditions reported for electroporation of carrot protoplasts.

Protoplasts used for electroporation experiments in our laboratory are isolated as described in Sect. 6.2. The ease of isolation and relative robustness of carrot protoplasts during subsequent manipulations are two of the major advantages of using carrot as a model dicot species.

5 Electroporation Parameters

Many variables have been shown to influence electroporation efficiency, including voltage, duration, and spacing of electrical pulses, buffer composition, temperature, and DNA concentration and form. Partially as a result of their interactive nature, different combinations of electroporation conditions appear capable of efficient

Table 1. Electroporation conditions used for stable and transient genetic transformation of carrot protoplasts

Electroporation conditions	Reporter gene	Type of expression	Reference
875 V/cm, 54 ms, 1 pulse	CAT	Transient	Fromm et al. (1985)
3.8 kV/cm, 6 μs 6 pulses 0.2 s intervals	Ti plasmid	Stable	Langridge et al. (1985)
550 V/cm, 50 ms 1 pulse	CAT	Transient	Boston et al. (1987)
220 V/cm, 60 ms 1 pulse	CAT	Transient	Fromm et al. (1985)
200 V, 1000 μF or 300 V, 510 μF 1 pulse	CAT NPT II	Transient Stable	Hauptman et al. (1987)
5 kV/cm, 90 μs 8 pulses	TMV RNA	Transient	Langridge et al. (1987)
875 V/cm, 54 ms 1 pulse	Luciferase	Transient	Ow et al. (1987)
750 V/cm, 8 ms 1 pulse	CAT	Transient	Bates et al. (1988)
270 V/cm, 10 ms 1550 μF, 1 pulse	Luciferase GUS	Transient	Gallie et al. (1989)
875 V/cm, 75 ms 750 μF, 1 pulse	CAT	Transient	Thomas et al. (1989a, b)
530 V/cm, 1000 μF 10 ms, 4 pulses 0.2-s intervals	GUS	Transient	Bower and Birch (1990)

transformation of protoplasts of a particular species. This is well illustrated by the diverse set of electroporation protocols published for transformation of carrot protoplasts (Table 1). The following is a summary of some of the main parameters shown to influence electroporation efficiency and the levels of these parameters found optimal for carrot protoplasts.

5.1 Pulse Characteristics

The degree of permeabilization of the protoplast plasmalemma depends primarily on the characteristics of the electric pulse used. Pulse variables include electric field strength, duration, shape, number, and spacing. Higher degrees of permeabilization result in high levels of DNA uptake but reduce protoplast viability. Pulses are usually either square or exponentially decaying, with square pulses resulting in a greater degree of permeabilization (Fromm et al. 1987). The decay characteristics of exponentially decaying pulses can be altered by capacitor size and the conductivity of the electroporation buffer. Pulse field strength at which a membrane is permeabilized is inversely proportional to the protoplast diameter (Nagata 1989). Thus protoplasts of smaller diameter require higher field strengths for effective gene transfer.

Varied combinations of pulse characteristics have been used to achieve the desired degree of permeabilization, which may vary according to whether transient or stable transformation is the goal. Pulse characteristics used for electroporation of carrot protoplasts usually fall into two broad categories:

1. Short high voltage pulses (5–100 μs, 1–5 kV/cm) applied in a low salt medium (Langridge et al. 1985, 1987).
2. Long low voltage pulses (1–100 ms, 200–800 V/cm) applied in a high salt medium (Fromm et al. 1985, 1987; Boston et al. 1987; Bates et al. 1988; Bower and Birch 1990).

A wide range of intermediate condition has also been used.

In general long, low voltage pulses are applied either singly or in low numbers (Fromm et al. 1985; Hauptman et al. 1987; Thomas et al. 1989a), while short, high voltage pulses are used as a battery of three to ten pulses (Langridge et al. 1985, 1987). It has been suggested that long, low voltage pulses may result in larger membrane pores (Bates et al. 1988).

We have found delivery of either four square pulses of 530 V/cm or six square pulses of 500 V/cm, 10 ms duration, and 0.2 s spacing to give reliable, high levels of expression in carrot protoplasts (Fig. 2). Use of these starting conditions should enable rapid optimization for use in other laboratories

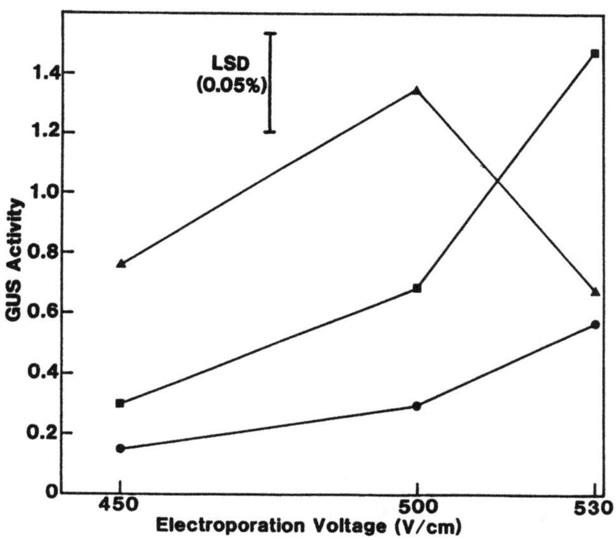

Fig. 2. Effect of electroporation voltage and pulse number on GUS activity (nM methylumbelliferone/min/mg protein) in carrot protoplasts. (unpubl.). ● 2 pulses; ■ 4 pulses, ▲ 6 pulses

5.2 Electroporation Buffer Characteristics

Buffer characteristics vary considerably but most are based on a nonionic osmotic buffer, a pH buffer, and a divalent cation such as calcium or magnesium to assist in maintenance of protoplast membrane integrity. Typically the nonionic osmolarity is adjusted to 0.2 M with mannitol or sorbitol, the pH is 7.2–9.0, ca. 4 mM calcium or magnesium chloride is used to stabilize the plasmalemma and conductivity is adjusted with sodium or magnesium chloride.

Fromm et al. (1985) found variation in calcium chloride concentration around an optimum of 4 mM to substantially alter expression levels in electroporated carrot protoplasts. Some groups report that addition of 6–13% polyethylene glycol to the electroporation buffer increases electroporation efficiency (Boston et al. 1987; Murray et al. 1990). Selection of an electroporation buffer shown to be effective by groups using similar electric pulse characteristics is advisable, as buffer characteristics, particularly conductivity, affect the kinetics of pulse delivery (Fromm et al. 1985).

5.3 Temperature

Most electroporations are performed at 0–4 °C using pre-chilled electroporation solutions (Fromm et al. 1985; Langridge et al. 1987) because electroporation-induced pores remain open for longer at lower temperatures (Nagata 1989). In addition, when high voltages are used, localized heating effects may occur, resulting in decreased protoplast viability.

An exception to the use of chilled electroporation solutions was reported by Bates et al. (1988), who electroporated carrot protoplasts on ice or at room temperature and found increased transient expression in protoplasts electroporated at room temperature. No explanation was offered, but it is possible that the state of the plasmalemma at room temperature rendered it more susceptible to permeabilization by the electroporation conditions used. Both temperatures are convenient to use and can readily be tested for any new system.

5.4 DNA Concentration

Increasing DNA concentration in the electroporation buffer results in increased transient expression levels and stable transformation frequencies, although the range of concentrations for which this holds varies considerably in different electroporation systems. In our electroporation system a linear increase occurred in transient expression of the GUS reporter gene over a concentration range of 10–100 μg/ml plasmid DNA in the electroporation buffer (Fig. 1) and 20 μg/ml is routinely used.

5.5 Carrier DNA

Addition of carrier DNA such as sonicated calf thymus DNA or salmon sperm DNA to plasmid DNA in the electroporation solution has been shown to increase both stable and transient transformation efficiency (Shillito et al. 1985), but the mechanism of action is unknown. It has been suggested (Jones et al. 1987) that carrier DNA acts to saturate endonuclease activity in electroporated protoplasts. Addition of 50 μg/ml carrier DNA was found approximately to double transient CAT expression in electroporated carrot protoplasts (Bates et al. 1988), although its use was reported to decrease expression levels in another study (Boston et al. 1987).

Some groups routinely use carrier DNA concentrations of around 50–60 μg/ml (Langridge et al. 1987; Murray et al. 1990). Our results have shown an approximate doubling of GUS transient expression with addition of 50 μg/ml carrier DNA and no further increase after addition of up to 100 μg/ml, but the degree of increase varied considerably between experiments and we do not routinely use carrier DNA.

6 Protocol for the Establishment of an Electroporation System

Due to inevitable differences in laboratory practices and facilities, protocols usually require a degree of modification and optimisation for maximum effectiveness. The following is a brief summary of the steps involved in establishment and operation of a system for transient GUS expression experiments using electroporation of carrot protoplasts. We have found this system to be very reliable over a period of several years.

6.1 Callus and Cell Suspension Cultures

Establish callus from root discs on MS nutrients (Murashige and Skoog 1962) plus 1 mg/l 2,4-dichlorophenoxyacetic acid (2,4-D), incubate at 28 °C in darkness, and subculture at 28-day intervals. Establish cell suspension from callus cultures by adding 2 g of friable callus to 40 ml MS medium plus 0.2 mg/l 2,4-D in 100 ml flasks. These are shaken at 100 rpm on an orbital shaker at 28 °C in darkness and subcultured weekly by transfer of 10 ml of suspension culture into 30 ml of fresh medium.

6.2 Protoplast Isolation

Take an actively growing cell suspension 3 days after subculture, allow cells to settle and replace half the medium with fresh medium. After 24 h further growth, harvest cells (400 *g*, 5 min). Resuspend each 3 ml of cells in 10 ml of isolation medium containing 2% Cellulase (Onozuka R-10) and 1% Macerozyme (Onozuka R-10) in isolation buffer (3 mM morpholinoethanesulphonic acid, 0.2 mM NaH_2PO_4, 7 mM $CaCl_2$, 0.7 M mannitol, pH 5.6). Incubate 16 h at 28 °C and 60 rpm in 9-cm Petri dishes containing 13 ml of cell suspension. Collect cells by centrifugation (70 *g*, 5 min), wash twice in wash medium (0.37 M glucose, 1.5 mM $CaCl_2$, 0.05% morpholinoethanesulphonic acid, pH 5.6), filter through 100 μm and 44 μm nylon screens and count using a hemocytometer. Protoplasts are stored in wash medium, on ice, for no longer than 1 h prior to use in experiments.

6.3 Electroporation

Resuspend protoplasts at 2×10^6/ml in Hepes buffered saline (Fromm et al. 1985) immediately prior to use. Add plasmid DNA (e.g., 20 g/ml), mix, incubate 10 min on ice. Resuspend gently and electroporate on ice, in 1 ml aliquots (four pulses, 530 V/cm, 10 ms duration, 0.2 s intervals). Incubate on ice 10 min, dilute into 9 ml culture medium (Langridge et al. 1985) in 9-cm Petri dishes and incubate at 28 °C in darkness.

6.4 Transient GUS Expression Assays

After 24–28 h culture, harvest protoplasts (70 *g*, 5 min) and resuspend in chilled GUS extraction buffer (Jefferson et al. 1987). Sonicate on ice, centrifuge thoroughly and remove the supernatant for fluorometric GUS assays (Jefferson et al. 1987), Histochemical GUS assays are performed on intact protoplasts as described (Gallie et al. 1989).

7 Conclusion

Two effective systems exist for transformation of carrot. Stable transformation is most readily achieved using *Agrobacterium*-mediated transformation methods. These are suitable for introduction of agronomically important genes into commercial cultivars and for analysis of patterns of gene expression in intact plants.

Electroporation of carrot protoplasts is a convenient and powerful tool for transient gene expression experiments aimed at investigation of control of gene expression and other aspects of cellular and molecular biology. Although some electroporation parameters are likely to require optimization in each laboratory, depending on the electroporation apparatus and perhaps the carrot cultivar used, the system is highly reproducible and excellent for quantitative studies of gene expression.

References

Bates GW, Piastuch W, Riggs CD, Rabussay D (1988) Electroporation for DNA delivery to plant cells. Plant Cell Tissue and Organ Cult 12:213–218

Boston RS, Becwar MR, Ryan RD, Goldsborough PB, Larkins BA, Hodges TK (1987) Expression from heterologous promoters in electroporated carrot protoplasts. Plant Physiol 83:742–746

Bower R, Birch RG (1990) Competence for gene transfer by electroporation in a sub-population of protoplasts from a uniform carrot cell suspension culture. Plant Cell Rep 9:386–389

Choi JH, Lui L-S, Borkird C, Sung ZR (1987) Cloning of genes developmentally regulated during plant embryogenesis. Proc Natl Acad Sci USA 84:1906–1910

Fluhr R, Kuhlemeier C, Nagy F, Chua N (1986) Organ-specific and light induced expression of plant genes. Science 232:1106–1112

Fromm M, Loverine P, Taylor LP, Walbot V (1985) Expression of genes transferred into monocot and dicot plant cells by electroporation. Proc Natl Acad Sci USA 82:5824–5828

Fromm M, Callis J, Taylor LP, Walbot V (1987) Electroporation of DNA and RNA into plant protoplasts. Methods Enzymol 153:351–366

Gallie DR, Lucas WJ, Walbot V (1989) Visualizing mRNA expression in plant protoplasts: factors influencing efficient mRNA uptake and translation. Plant Cell 1:301–311

Gosch G, Bajaj YPS, Reinert J (1975) Isolation, culture and fusion studies on protoplasts from different species. Protoplasma 85:327–336

Hauptman RM, Ozias-Akins P, Vasil V, Tabaeizadeh Z, Rogers SG, Horsch RB, Vasil IK, Fraley RT

(1987) Transient expression of electroporated DNA in monocotyledenous and dicotyledenous species. Plant Cell Rep 6:265–270
Jefferson RA, Kavanagh TA, Bevan MW (1987) GUS fusions: β-glucuronidase as a sensitive and versatile gene fusion marker in higher plants. EMBO J 6:3901–3907
Jones H, Tempelaar MJ, Jones MGK (1987) Recent advances in plant electroporation. Oxford Surv Plant Mol Cell Biol 4:347–357
Krens FA, Molandijk L, Wullens GJ, Schilperoort RA (1982) In vitro transformation of plant protoplasts with Ti plasmid DNA. Nature 296:72–74
Langridge WHR, Li BJ, Szalay AA (1985) Electric field mediated stable transformation of carrot protoplasts with naked DNA. Plant Cell Rep 4:355–359
Langridge WHR, Li BJ, Szalay AA (1987) Uptake of DNA and RNA into cells mediated by electroporation. Methods Enzymol 153:336–350
Last DI, Brettell RIS, Chamberlain DA, Chaudhury AM, Larkin PJ, Marsh EL, Peacock WJ, Dennis ES (1990) pEmu: and improved promoter for gene expression in cereal cells. Theor Appl Genet 81:581–588
Mathews BF, Cress DE (1981) Liposome mediated delivery of DNA to carrot protoplasts. Planta 153:90–94
Murashige T, Skoog F (1962) A revised medium for rapid growth and bioassays with tobacco tissue cultures. Physiol Plant 15:473–497
Murray EE, Buchholz WG, Bowen B (1990) Direct analysis of RNA transcripts in electroporated carrot protoplasts. Plant Cell Rep 6:337–340
Nagata T (1989) Cell biological aspects of gene delivery into plant protoplasts by electroporation. Int Rev Cytol 116:229–255
Napoli C, Lemieux C, Jorgensen R (1990) Introduction of a chalcone synthase gene into petunia results in reversible co-suppression of homologous genes in trans. Plant Cell 2:279–289
Nea LJ, Bates GW (1987) Factors affecting carrot electroporation efficiency. Plant Cell Rep 6:337–340
Nomura K, Nitta T, Fukuei K, Fujimura T, Komamine A (1982) Isolation and characterisation of protoplasts from carrot somatic embryos. Proc 5th Int Congr Plant Tissue and Cell Cult. Maruzen Tokyo, pp 587–588
Ow DW, Jacobs JD, Howell SH (1987) Functional regions of the cauliflower mosaic virus 35S RNA promoter determined by use of the firefly luciferase gene as a reporter of promoter activity. Proc Natl Acad Sci USA 84:4870–4874
Planckaert F, Walbot V (1989) Transient gene expression after electroporation of protoplasts derived from embryogenic maize callus. Plant Cell Rep 8:144–147
Putterill JJ, Gardner RC (1989) Initiation of translation of the β-glucuronidase gene at internal AUG codons in plant cells. Plant Sci 62:199–205
Rathus C, Bower R, Birch RG Evaluation of promoter, intron and enhancer effects on transient gene expression in sugarcane and carrot protoplasts (in prep)
Scott RJ, Draper J (1987) Transformation of carrot tissues derived from proembryonic suspension cells: a useful model system for gene expression studies in plants. Plant Mol Biol 8:265–274
Shea EM, Gibeaut DM, Carpita NC (1989) Structural analysis of cell walls regenerated by carrot protoplasts. Planta 179:293–308
Shillito RD, Saul MW, Paszkowski J, Muller M, Potrykus I (1985) High efficiency direct gene transfer to plants. Bio/Technol 3:1099–1103
Steward FC, Israel HW, Mott RL, Wilson HJ, Krikorian AD (1975) Observations on growth and morphogenesis in cultured cells of carrot (*Daucus carota* L.). Philos Trans R Soc Lond 273:33–53
Tepfer D (1985) Transformation of several species of higher plants by *Agrobacterium rhizogenes*: sexual transmission of the transformed genotype and phenotype. Cell 37:359–367
Thomas JC, Guiltinan MJ, Bustos S, Thomas T, Ressler C (1989a) Carrot (*Daucus carota*) hypocotyl transformation using *Agrobacterium tumefaciens*. Plant Cell Rep 8:354–357
Thomas JC, Nessler C, Katterman F (1989b) Interuption of somatic embryogenesis in *Daucus carota* L. by 5-bromodeoxyuridine. Plant Physiol 90:921–927
Van Sluys MA, Tempé J (1989) Behaviour of the maize transposable element *Activator* in *Daucus carota*. Mol Gen Genet 219:313–319
Van Sluys MA, Tempé J, Federoff N (1987) Studies on the induction and mobility of the maize activator element in *Arabidopsis thaliana* and *Daucus carota*. EMBO J 6:3881–3890
Wilde HD, Nelson WS, Booiji H, de Vries SC, Thomas TL (1988) Gene expression programme in embryogenic and non-embryogenic carrot cultures. Planta 176:205–211

II.6 Transformation in *Digitalis purpurea* L. (Foxglove)

KAZUKI SAITO[1], MAMI YAMAZAKI[1], KOICHIRO SHIMOMURA[2], KAYO YOSHIMATSU[2], and ISAMU MURAKOSHI[1]

1 Introduction

Foxglove (*Digitalis purpurea* L., common foxglove, *D. lanata* Ehrh., Grecian foxglove, and other related species) is one of the most commonly used medicinal plants in the world (Trease and Evans 1983). The dried leaves are used as a crude drug and contain a number of cardioactive glycosides, e.g., digitoxin, gitoxin, etc. A large number of reports have been published on tissue culture, regeneration, production, and biotransformation of cardenolides in *Digitalis* species (for review see Rücker 1988). However, only one paper has appeared on genetic transformation of *D. purpurea* (Saito et al. 1990a). The efficient transfer and the strong expression of foreign genes in foxglove offer promising possibility of genetic manipulation of secondary metabolism of this plant.

The binary vector system based on an *Agrobacterium*-Ri plasmid can be efficiently used to produce transgenic hairy roots integrated with any desired foreign genes. By this method, selection of transformants using growth inhibitors is not necessary, because high frequency of double transformation with T-DNAs of a helper Ri plasmid and a disarmed vector Ti plasmid can be expected (Hamil et al. 1987). This technique provides a short way to obtain transgenic tissues of the plants for which transformation methods by means of a completely disarmed Ti plasmid are not yet established.

In this chapter, we describe the stable integration and expression of chimeric *neo* and *gus* genes in *D. purpurea* using an *Agrobacterium*-Ri plasmid binary vector. The green hairy roots expressing chimeric *neo* and *gus* genes could produce cardioactive glycosides.

2 Transformation

2.1 Infection with *Agrobacterium*

Leaf discs of aseptically growing *D. purpurea* were co-cultured with *A. rhizogenes* harboring both a wild agropine-type Ri plasmid, pRi15834, and a binary vector,

[1] Faculty of Pharmaceutical Sciences, Chiba University, Yayoi-cho 1-33, Inage-ku, Chiba 263, Japan
[2] Tsukuba Medicinal Plant Research Station, National Institute of Hygienic Sciences, 1 Hachimandai, Tsukuba, Ibaraki 305, Japan

Biotechnology in Agriculture and Forestry, Vol. 22
Plant Protoplasts and Genetic Engineering III (ed. by Y.P.S. Bajaj)

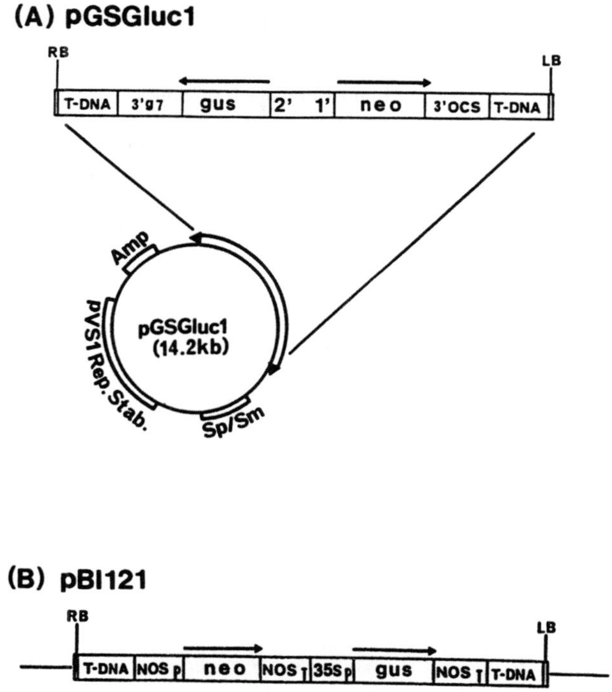

Fig. 1A, B. Binary vectors pGSGluc1 (**A**) and pBI121 (**B**). *1′* and *2′* TR1′ and 2′ promoters, respectively; *3′OCS* 3′ region of octopine synthase gene; *3′g7* 3′ region of gene 7; NOS_P promoter of nopaline synthase gene; NOS_T 3′ region of nopaline synthase gene; $35S_P$ promoter of CaMV 35S RNA; *RB* right border; *LB* left border. (Saito et al. 1990a)

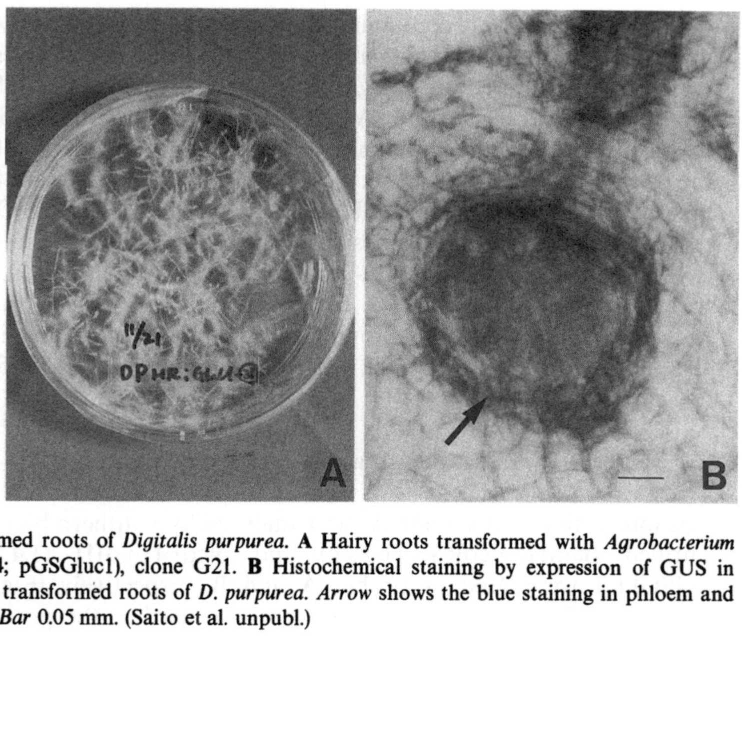

Fig. 2A, B. Transformed roots of *Digitalis purpurea.* **A** Hairy roots transformed with *Agrobacterium rhizogenes* (pRi15834; pGSGluc1), clone G21. **B** Histochemical staining by expression of GUS in transverse section of transformed roots of *D. purpurea. Arrow* shows the blue staining in phloem and surrounding tissues. *Bar* 0.05 mm. (Saito et al. unpubl.)

Table 1. Transformation of *D. purpurea* with *A. rhizogenes* harboring pRi15834 and a mini Ti vector. (Saito et al. 1990a)

Vector	Number of hairy root clones			
	Total	Km^R/total	Opine (+)/ Km^R or total	GUS (+)/ opine (+)
pGSGlucl	38	10/38	6/10	4/6
pBI121	27	–	8/27	3/4

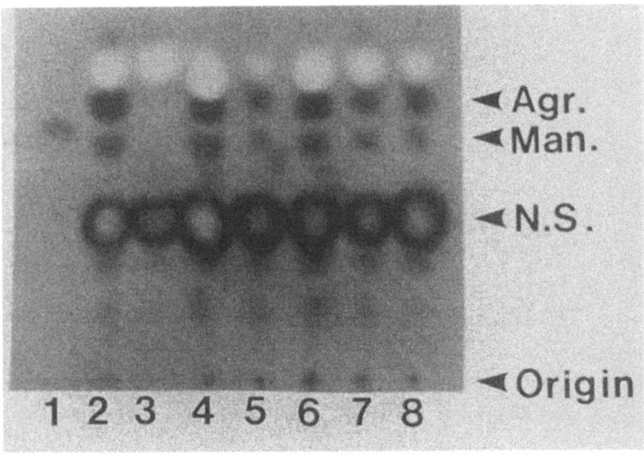

Fig. 3. Agropine and mannopine detection in transformed roots. *Lane 1* standard mannopine; *lane 2* hairy roots with pRi15834; *lane 3* leaf of untransformed plant; *lane 4* clone G1; *lane 5* clone G21; *lane 6* clone B2; *lane 7* clone B4; *lane 8* clone B20; *Agr.* agropine; *Man.* mannopine; *N.S.* neutral sugars. (Saito et al. 1990a)

pGSGlucl or pBI121, containing the chimeric genes (Fig. 1). After 2–3 weeks of infection, very fine hairy roots appeared at the veins of the leaves. These roots were removed and cultured on B5 agar plate supplemented with Claforan (200 mg/l) to remove *Agrobacterium* (Fig. 2A). In the experiments using pGSGlucl, out of 38 clone of excised roots, ten clones grew on kanamycin (Km) (50 mg/l) containing medium; of these ten clones, six were positive for the opines; of these six, four were positive for β-glucuronidase (GUS) enzymatic assay (Table 1). Of these four clones, the two rapidly growing clones (G1, G21) were chosen for further study. In the experiments using pBI121, Km-resistant selection was omitted. For further study, we chose three clones (B2, B4, B20) positive in both opine and GUS assays. All these five clones produced agropine and mannopine as shown in Fig. 3, indicating the integration of TR-DNA of pRi15834 (De Paolis et al. 1985).

2.2 Southern Hybridization Analysis

The integration of the T-DNAs was analyzed by Southern blot hybridization. The purified BamHI-EcoRV (1.3 kb) and NcoI-HindIII (1.4 kb) fragments of pGSGlucl were used as probes A and B, respectively (Fig. 4A). The EcoRV fragments of pLJ1 (Jouanin 1984) were also used as the probe for TL-region of

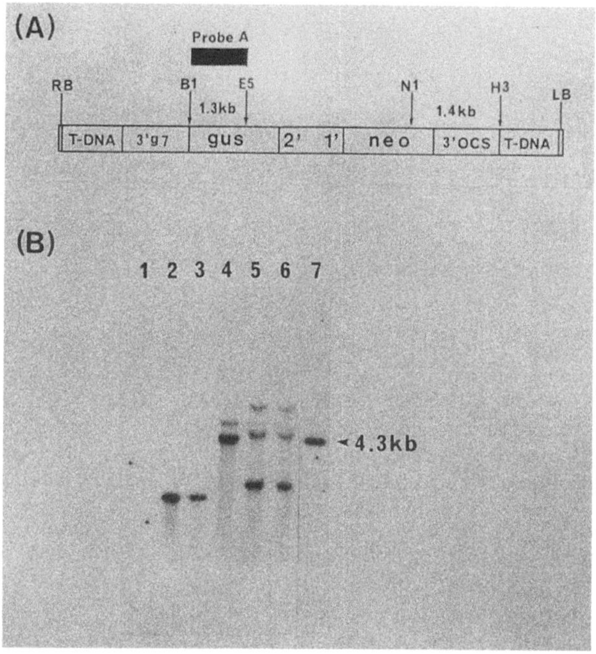

Fig. 4A, B. Southern blot hybridization of transgenic foxglove **A** T-DNA region of pGSGluc1 The chimeric *neo* and *gus* genes are driven by TR 1′ and 2′ promoters, respectively. *Probe A* is the 1 3 kb BamHI-EcoRV fragment *RB* right border, *LB* left border, *B1* Bam HI, *E5* EcoRV, *N1* NcoI, *H3* HindIII, *3′g7* 3′ end of gene 7, *3′OCS* 3′ end of octopine synthase gene. **B** Genomic hybridizations The isolated plant DNA was digested with EcoRV, separated on 0 7% agarose gel, blotted onto Nylon filter, and hybridized with ^{32}P-labeled probe A *Lane* 1 control hairy root transformed with only pRi15834, *lane* 2 clone G1, lane *3* clone G21, *lane 4* clone B2, *lane 5* clone B4, *lane 6* clone B20; *lane 7* the purified 4 3 kb BamHI-HindIII fragment of pGSGluc1 (Saito et al 1990a)

pRi15834. Plant DNA was digested with EcoRV to produce the junction fragments composed of T-DNA of a mini Ti plasmid and plant genomic DNA. The blotted DNA on the filter was hybridized with these probes.

Clones G1 and G21 emerged independently from the same infected site with *A. rhizogenes* (pRi15834; pGSGluc1) and were established separately. However, these clones gave exactly the same hybridization bands with probe A (Fig. 4B) and probe B. This suggested that these clones were derived from the same infection event and contained one T-DNA copy of pGSGluc1. This was also the case in clones B4 and B20 transformed with *A. rhizogenes* (pRi15834; pBI121), while these clones were integrated with three T-DNAs of pBI121. All these five clones (G1, G21, B2, B4, B20) gave the hybridization bands not only with probe A but also probe B and the probe of pLJ1. The control hairy root transformed with only pRi15834 showed no hybridization signals with pGSGluc1 T-DNA.

3 Expression of Chimeric Genes in Transgenic *Digitalis* Roots

In pGSGluc1, the chimeric *neo* and *gus* genes are controlled by TR 1′ and 2′ promoters of the mannopine synthase genes in TR-DNA of pTiAch5 (Velten et al. 1984). In pBI121, *nos* and CaMV35S promoters drive *neo* and *gus* genes, respectively. All the established clones (lane 2–6 in Fig. 5) showed both neomycin phosphotransferase-II (NPT-II) and GUS activities, while the control hairy root

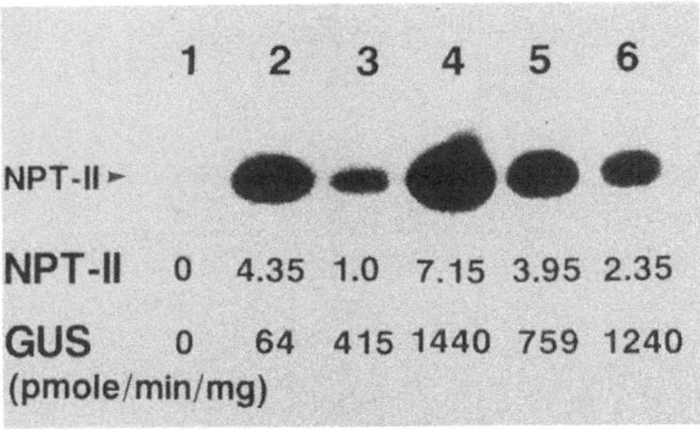

Fig. 5. Expression of chimeric *neo* and *gus* genes in enzymatic assay. *Lane 1* control hairy root of *D. purpurea* transformed with only pRi15834; *lane 2* clone G1; *lane 3* clone G21; *lane 4* clone B2; *lane 5* clone B4; *lane 6* clone B20 The NPT-II assay was carried out using 10 μg of protein. The activities were quantified with a densitometer and normalized to that of clone G21 (*lane 3*). GUS activities were measured by the fluorometric assay using the same protein extracts supplied for the NPT-II assay and expressed as the amount of formation of 4-methylumbelliferone from 4-methylumbelliferone *β*-glucuronide. (Saito et al. 1990a)

transformed with only pRi15834 (land 1) had no detectable NPT-II and GUS activities. All four promoters used in these experiments showed rather high activities in the hairy roots of *D. purpurea* compared with those in tobacco transformants.

The expression of the *gus* gene was histochemically analyzed by using 5-bromo-4-chloro-3-indolyl-*β*-D-glucuronide (X-gluc) (Fig. 2B). The vascular tissues of the transformed roots were strongly stained in all transformants examined. A high level of expression of CaMV35S promoter was observed in tobacco phloem tissues (Jefferson et al. 1987). Recently, we have also shown that TR2′ promoter is phloem-specific in transgenic *Glycyrrhiza uralensis* (licorice) roots (Saito et al. 1990b, 1991) and in regenerated tobacco plants (Saito et al. 1991). The TR2′ promoter is also specifically expressed in phloem and surrounding tissues of *D. purpurea* roots.

4 Production of Cardioactive Glycosides

The production of cardenolides in transformed roots were determined by enzyme-linked immunosorbent assay (ELISA) as shown in Fig. 6. The production of cardioactive glycosides was positively correlated to the concentration of chlorophyll in the cells. This result suggested that the biosynthesis of cardenolides in the transformed roots is induced by greening of the tissues.

5 Summary and Conclusions

Genetically engineered foreign genes have been transferred into 30 species of plants (Gasser and Fraley 1989) by means of *Agrobacterium* transformation and direct DNA transfer techniques. However, these researches are mostly on agriculturally

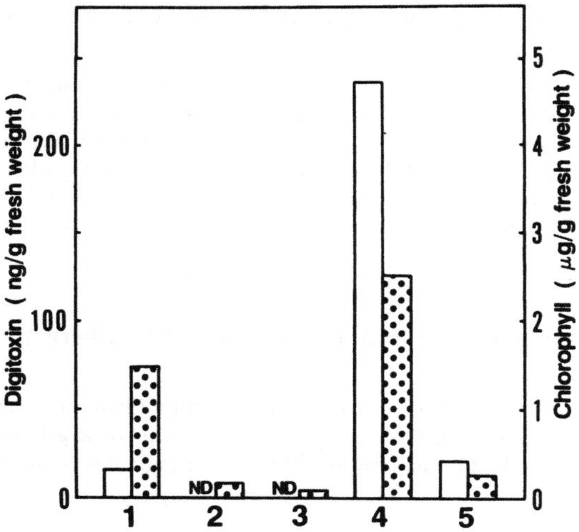

Fig. 6. Production of cardenolides in the transgenic hairy roots. The concentration of cardenolides in the tissue (□) was determined by ELISA using digitoxin as the standard and shown as the equivalent amount of digitoxin. The content of chlorophyll (⊡) was measured as described previously (Saito et al. 1989a). *1* clone G1; *2* clone G21; *3* clone B2; *4* clone B4; *5* clone B20; *ND* not detected. (Saito et al. 1990a)

important crop plants as target species. Little information is available on the stable transfer and expression of foreign genes in pharmaceutically important medicinal plants, which produce a large number of pharmacologically active secondary metabolites. In this chapter, a method is presented to obtain transgenic foxglove roots, in which the foreign genes expressed quite efficiently and the cardioactive glycosides were produced. This method is based on an Ri plasmid binary vector system. The advantages of this technique are as follows: (1) one can easily obtain transgenic roots integrated with any desirable foreign genes without selection of transformants with any growth inhibitors, e.g., antibiotics and herbicides; (2) this technique is suitable for genetic manipulation of secondary metabolism of rapidly growing hairy roots which produce secondary products in high yield; (3) some plant species can regenerate from hairy roots to mature plants and set seeds (Tepfer et al. 1989). In our present experiments, the high production of cardenolides in transformed *D. purpures* is observed in green hairy roots. This greening-induced production of cardenolides has also been the case in the shoot-forming callus of several *Digitalis* species (Rücker 1988) and in hairy roots of *D. lanata* (Yoshimatsu et al. 1990). The production of lupin alkaloids in some Leguminosae plants was also reported to be inducible by greening of the tissues (Saito et al. 1989a, b). These greening-induced productions of the secondary metabolites are presumably related to the fact that these are normally produced in the leaves of these plants. Recently, Moldenhauer et al. (1990) also reported the stable transformation of *D. lanata* with Ti plasmids. This report also presented the production of cardenolides in transformed galls.

6 Protocols

6.1 Plasmids and Bacteria

The plasmids, pGSGluc1 and pBI121, were conjugatively transferred by triparental mating using pRK2013 as a helper plasmid into *A. rhizogenes* Rif^R harboring an agropine-type plasmid pRi15834. Before infection, *Agrobacterium* was cultured in liquid minimal A medium at 28 °C for 2 days.

6.2 Transformation and Tissue Culture of *D. purpurea*

Axenic plants of *D. purpurea* grown from seeds were maintained in medium containing half-strength Murashige and Skoog (1962) salts, 1% sucrose and 0.8% agar. Leaf discs of *D. purpurea* (ca. 1 × 1 cm) were co-cultured with *Agrobacterium* with essentially the same method described for tobacco leaf discs (De Block et al. 1987) without addition of any phytohormones. Hairy roots emerged after 2–3 weeks of inoculation. They were subcultured on B5 medium (Gamborg et al. 1968) supplemented with 200 mg/l Claforan (Hoechst) under illumination (2000 lx, 16 h/day). Claforan could be omitted after three or four transfers.

6.3 Plant DNA Extraction and Southern Blot Hybridization

Plant DNA was isolated from the transformed roots as reported (Saito et al. 1990b). DNA was cut with EcoRV and Electrophoresed in a 0.7% agarose gel, transferred to a Nylon filter, and hybridized with random-prime labeled ^{32}P probes. The final wash of the filter was carried out in 0.1 × SSC at 65 °C.

6.4 NPT-II and GUS Enzymatic Assay and Histochemical Staining

Protein was extracted from plant tissue with extraction buffer containing 50 mM Tris-HCl (pH 6.8) and 2% 2-mercaptoethanol. The same protein extracts were used for both NPT-II and GUS enzymatic assays. NPT-II activity was determined by in situ phosphorylation in nondenaturing polyacrylamide gels, as described elsewhere (Peleman et al. 1989). Both fluorometric and histochemical assays of GUS activity using 4-methylumbelliferone β-glucuronide and X-gluc as the substrates, respectively, were carried out by the methods described by Jefferson et al. (1987) with some modifications. Histochemical staining was carried out as described previously (Saito et al. 1991).

6.5 ELISA of Cardioactive Glycosides

ELISA was carried out as reported elsewhere (Yoshimatsu et al. 1990). The antibody was raised against digoxin and reactable against digitoxin. The fresh plant tissue (100–200 mg) was homogenized with a glass rod in a microtube, and the crude extract after centrifugation at 18 000 *g* for 2 min was immediately subjected to the assay. The quantification was performed using digitoxin as the standard in the linear range of the standard curve from 4 ng/ml to 800 ng/ml.

Acknowledgements. We thank Drs. Jan Leemans and Van Rompaey (Plant Genetic Systems, Gent, Belgium) for generous supply of pGSGluc1. This research was supported by Grants-in-Aid from the Ministry of Education, Science and Culture, Special Cooperation Funds for Promoting Science and Technology Agency, Japan. M.Y. was supported by JSPS Fellowhips for Japanese Junior Scientists.

References

De Block M, Botterman J, Vandewiele M, Dockx J, Thoen C, Gossele V, Rao Movva N, Thompson C, Van Montagu M, Leemans J (1987) Engineering herbicide resistance in plants by expression of a detoxifying enzyme. EMBO J 6:2513–2518

De Paolis A, Mauro ML, Pomponi M, Cardarelli M, Spano L, Constantino P (1985) Localization of agropine-synthesizing functions in the TR region of the root-inducing plasmid of *Agrobacterium rhizogenes*. Plasmid 13:1–7

Gamborg OL, Miller RA, Ojima K (1968) Nutrient requirements of suspension cultures of soybean root cells. Exp Cell Res 50:151–158

Gasser CS, Fraley RT (1989) Genetically engineering plants for crop improvement. Science 244:1293–1299

Hamil JD, Prescott A, Martin C (1987) Assessment of the efficiency of cotransformation of the T-DNA of disarmed binary vectors derived from *Agrobacterium tumefaciens* and the T-DNA of *A. rhizogenes.* Plant Mol Biol 9:573–584

Jafferson RA, Kavanagh TA, Bevan MW (1987) GUS fusions: b-glucuronidase as a sensitive and versatile gene fusion marker in higher plants. EMBO J 6:3901–3907

Jouanin L (1984) Restriction map of an agropine-type Ri plasmid and its homologies with Ti plasmids. Plasmid 12:91–102

Moldenhauer D, Fürst B, Diettrich B, Luckner M (1990) Cardenolides in *Digitalis lanata* cells transformed with Ti plasmids. Planta Med 56:435–438

Murashige T, Skoog F (1962) A revised medium for rapid growth and bioassays with tobacco tissue cultures. Physiol Plant 15:473–497

Peleman J, Saito K, Cottyn B, Engler G, Seurinck J, Van Montagu M, Inze D (1989) Structure and expression analysis of the S-adenosylmethionine synthetase gene family in *Arabidopsis thaliana.* Gene 84:359–369

Rücker W (1988) *Digitalis* spp.: In vitro culture, regeneration, and the production of cardenolides and other secondary products. In: Bajaj YPS (ed) Biotechnology in agriculture and forestry, vol 4. Medicinal and aromatic plants I. Springer, Berlin Heidelberg New York, pp 388–418

Saito K, Yamazaki M, Yamakawa K, Fujisawa S, Takamatsu S, Kawaguchi A, Murakoshi I (1989a) Lupin alkaloids in tissue culture of *Sophora flavescens* var. *angustifolia*: greening-induced production of matrine. Chem Pharm Bull 37:3001–3004

Saito K, Yamazaki M, Takamatsu S, Kawaguchi A, Murakoshi I (1989b) Greening-induced production (+)-lupanine in tissue culture of *Thermopsis lupinoides.* Phytochemistry 28:2341–2344

Saito K, Yamazaki M, Shimomura K, Yoshimatsu K, Murakoshi I (1990a) Genetic transformation of foxglove (*Digitalis purpurea*) by chimeric foreign genes and production of cardioactive glycosides. Plant Cell Rep 9:121–124

Saito K, Kaneko H, Yamazaki M, Yoshida M, Murakoshi I (1990b) Stable transfer and expression of chimeric genes in licorice (*Glycyrrhiza uralensis*) using an Ri plasmid binary vector. Plant Cell Rep 8:718–721

Saito K, Yamazaki M, Kaneko H, Murakoshi I, Fukuda Y, Van Montagu M (1991) Tissue-specific and stress enhancing expression of the TR promoter for mannopine synthase in transgenic medicinal plants. Planta 184:40–46

Tepfer D, Metzger L, Prost R (1989) Use of roots transformed by *Agrobacterium rhizogenes* in rhizosphere research: applications in studies of cadmium assimilation from sewage. Plant Mol Biol 13:295–302

Trease GE, Evans WC (1983) Pharmacognosy, 12th edn. Bailliere Tindall, London, pp 500–509

Velten J, Velten L, Hain R, Schell J (1984) Isolation of a dual plant promoter fragment from the Ti plasmid of *Agrobacterium tumefaciens.* EMBO J 3:2723–2730

Yoshimatsu K, Satake M, Shimomura K, Sawada J, Terao T (1990) Determination of cardenolides in hairy root cultures of *Digitalis lanata* by enzyme-linked immunosorbent assay. J Nat Prod 53:1498–1502

II.7 Transformation in *Duboisia* spp.

YOSHIHIRO MANO[1]

1 Introduction

The genus *Duboisia* consists of three species, *Duboisia leichhardtii* F. Muell, *D. myoporoides* R. Br., and *D. hopwoodii* F. Muell. *Duboisia* species are woody plants and the former two species produce the largest amounts of tropane alkaloids among the Solanaceae, which includes the other alkaloid-producing plants *Atropa, Datura, Hyoscyamus* and *Scopolia.* The major alkaloids present in *D. leichhardtii* and *D. myoporoides* are scopolamine and hyoscyamine, important compounds that have been used as spasmolytics and anesthetics. The third species, *D. hopwoodii,* produces mainly nicotine and nornicotine, both of which are nicotine alkaloids normally found in tobacco plants. Both tropane and nicotine alkaloids are synthesized mainly in the roots of these three species.

The methods for the genetic transformation of plant cells have been developed for use on a variety of plant species. The transformation methods are broadly grouped into three categories: (1) *Agrobacterium*-mediated transformation, (2) direct gene transfer, and (3) protoplast fusion.

This chapter is concerned with the characteristics of hairy root clones transformed by *A. rhizogenes* and the somatic hybrids obtained by protoplast fusion.

2 Genetic Transformation in *Duboisia* by the Ri T-DNA of *Agrobacterium rhizogenes*

Agrobacterium tumefaciens and *Agrobacterium rhizogenes* have the large (200–250 kb) tumor-inducing (Ti) plasmid and root-inducing (Ri) plasmid (Fig. 1), respectively. *Agrobacteria* have the ability to infect most dicotyledonous plants and to transfer the portion (T-DNA) of these plasmids to the plant genome (Koukolikova-Nicola et al. 1987). This natural gene transfer system has been exploited in the development of a variety of plant transformation vectors (Rogers and Klee 1987; Armitage et al. 1988). Another application for this system is the use of the T-DNA as an insertional mutagen to produce the transformed cellular clones. Hairy roots are induced at the *A. rhizogenes*-infection site of the host plant as a result of the

[1] Department of Biological Science and Technology, Tokai University, 317 Nishino, Numazu, Shizuoka 410-03, Japan

Biotechnology in Agriculture and Forestry, Vol. 22
Plant Protoplasts and Genetic Engineerng III (ed. by Y.P.S. Bajaj)

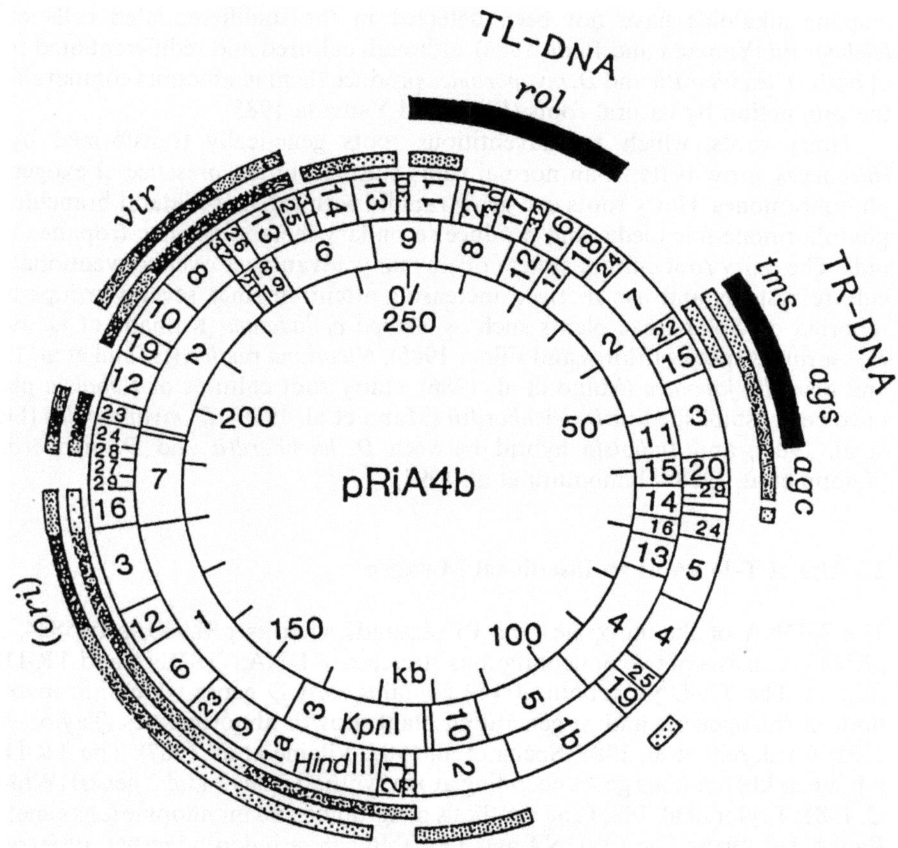

Fig. 1. Restriction map of an agropine type Ri plasmid pRiA4b. *Black arcs* show TL-DNA and TR-DNA regions. *Shaded arcs* represent homology to an octopine type Ti plasmid pTiA6 (*outer*) and a nopaline type Ti plasmid pTiT37 (*inner*). Relative intensities of Southern blot autoradiogram signals are indicated by the *degree of shading*. Symbols: *agc* genes involved in agropine catabolism; *ags* genes involved in agropine synthesis; *ori* origin of replication; *rol* genes involved in the hairy root formation; *tms* genes involved in auxin synthesis; *vir* genes involved in the transfer of T-DNA. (Data from Huffman et al. 1984; Jouanin 1984; and White et al. 1985)

integration and expression of the Ri T-DNA into the plant genome (for a review see White and Sinkar 1987). Here, variation caused by the insertional mutagenesis among hairy root clones is discussed in connection with tropane alkaloid production by *Duboisia* hairy root cultures.

2.1 Hairy Root Cultures

Duboisia leichhardtii F. Muell and *D. myoporoides* R. Br. are cultivated in Australia as commercial sources of scopolamine and hyoscyamine, but their productivities vary considerably depending on the cultivation conditions. Attempts have been made to produce tropane alkaloids by cell and root cultures in these plants; but tropane alkaloids have not been detected in the undifferentiated cells of *D. leichhardtii* (Yamada and Endo 1984), although cultured and redifferentiated roots of both *D. leichhardtii* and *D. myoporoides* produce them in amounts comparable to the production by natural roots (Endo and Yamada 1985).

Hairy roots, which are adventitious roots genetically transformed by *A. rhizogenes*, grow better than normal roots cultured in the presence of exogenous phytohormones. Hairy roots can grow rapidly with extensive lateral branching in phytohormone-free media and produce secondary metabolites like tropane alkaloids. The hairy root culture system offers many advantages over conventional cell culture systems and has received increasing attention since several groups have reported on Solanaceae plants such as *Atropa belladonna* (Kamada et al. 1986), *Hyoscyamus* species (Flores and Filner 1985), *Nicotiana rustica* (Hamill et al. 1986) and *Scopolia japonica* (Mano et al. 1986). Hairy root cultures of *Duboisia* plants have been established in *D. leichhardtii* (Mano et al. 1989), *D. myoporoides* (Deno et al. 1987), and *Duboisia* hybrid between *D. leichhardtii* and *D. myoporoides* (Knopp et al. 1988; Shimomura et al. 1991).

2.2 Use of T-DNA as an Insertional Mutagen

The T-DNA of the agropine type Ri plasmids such as pRiA4b, pRi15834, and pRi1855 consists of two noncontiguous stretches of DNA, TL-DNA and TR-DNA (Fig. 1). The TL-DNA (about 20 kb) contains *rol*A-D genes which are involved both in rhizogenesis and in generating plant growth abnormalities (Taylor et al. 1985; Cardarelli et al. 1987; Spena et al. 1987; Vilaine et al. 1987). The TR-DNA (about 20 kb) contains genes encoding auxin synthesis (*tms* 1 and 2 genes) (White et al. 1985; Taylor et al. 1985) and synthesis of agropine and mannopine (*ags* gene) (De Paolis et al. 1985). The TL-DNA and TR-DNA can be independently transferred to the nuclear genome of infected plant cells (Vilaine and Casse-Delbart 1987).

Hairy root development consists of two processes: (1) primary root meristem initiation and (2) root growth and secondary meristem formation (Shen et al. 1988). The transformed cells harboring the *tms* genes on the TR-DNA produce and diffuse auxin in the inoculated area. The diffusion of auxin from these cells is sufficient to trigger the development of roots from cells that have received the TL-DNA. Roots transformed by the TL-DNA alone as well as those by both TL- and TR-DNAs can grow in the phytohormone-free medium and exhibit the typical hairy-root phenotype, while roots transformed by the TR-DNA alone have the phenotype of nontransformed roots (Vilaine and Casse-Delbart 1987; Vilaine et al. 1987). Hairy roots harboring the TL-DNA alone do not synthesize agropine and mannopine (Mano et al. 1986).

The T-DNA is present in different chromosomal locations in different hairy root lines transformed by *A. rhizogenes* (Ambros et al. 1986). The number of T-DNA inserts is quite variable from one line to the other line of hairy roots and some hairy root lines harbor more than one insert (Byrne et al. 1983; Ambros et al. 1986). The inserts derived from agropine type Ri plasmid TR-DNA region or from mannopine type Ri plasmid T-DNA are also quite variable in length, although the insert lengths are relatively constant for those derived from TL-DNA region (Byrne et al. 1983; Birot et al. 1987; Jouanin et al. 1987; David et al. 1988). Thus the copy number, size, and chromosomal location of the T-DNA region integrated stably into the plant genome are quite variable from clone to clone of hairy roots. The T-DNA is therefore a useful tool for an insertional mutagenesis to produce various types of transformants.

2.3 Establishment of Hairy Root Clones

The term "clone" is commonly used in the following four ways: (1) plants propagated vegetatively, (2) plants regenerated from tissue cultures, (3) the progeny derived by mitosis from a single cell, and (4) the selected host cell containing recombinant DNA molecules (Dougall 1987). It will be used to express the third meaning here.

The primary hairy root, which is induced at the inoculation site, may consist of heterogeneous cells. To obtain the "clone", which consists of homogeneous cells

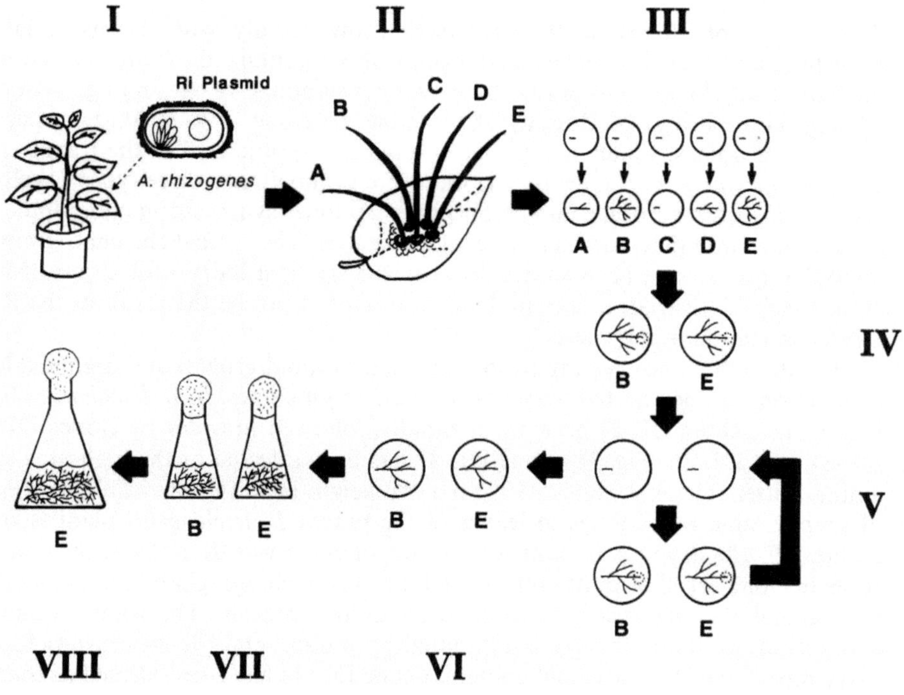

Fig. 2 (I–VIII). Selection method for hairy root clones. **I** Inoculation. This step is illustrated by the whole plant. Leaf discs, stem segments, root segments, callus, and protoplasts also can be used. **II** Induction of hairy roots. **III** Isolation and culture. Tips from adventitious roots generated at the inoculation sites are excised and each cultured separately in phytohormone-free medium containing antibiotics to eliminate bacteria. **IV** Selection of rapidly growing roots in phytohormone-free medium. **V** Cloning. Tips of rapidly growing roots are excised and each cultured separately at least five times in phytohormone-free medium. The parts excised from roots are shown by the *dotted circles*. **VI** Establishment of hairy root clones. **VII** Examination of the useful properties. **VIII** Selection of the desired hairy root clones. The desired clones expressing stably the useful properties are selected from various hairy root clones. (Mano 1989)

derived by mitosis from a transformed single cell, it is essential to reculture the single root tip in phytohormone-free medium more than five times (Fig. 2). Thus the clone can be established by the selection pressure to initiate the root meristem and to form the root tissue in a phytohormone-free medium, and is designated as hairy root clone (Mano et al. 1986; Mano 1989). The cloning and selection are very important to establish the stable transformants.

In *D. leichhardtii* the 45 hairy root clones have been established from tips of about 750 primary hairy roots which had been induced by inoculation with the agropine type of *A. rhizogenes* strain (Mano et al. 1989). About 5–6% of the total primary hairy roots tested could continue to grow in phytohormone-free medium, while the other 94–95% ceased to grow in the medium although some roots grew slowly at the first transfer. These established hairy root clones have been stably subcultured in phytohormone-free medium.

2.4 Variation Among Hairy Root Clones and Selection of the Desired Clones

The hairy root clones of *D. leichhardtii* grow rapidly with extensive lateral branching and their properties have been stably maintained. Characterization of the hairy root clones showed that there was considerable variation in growth rate, alkaloid content and productivity from clone to clone (Fig. 3). With respect to synthesis of opines, which are specific compounds synthesized in the transformed cells by *Agrobacterium*, hairy root clones were grouped into three types: (1) clones synthesizing both agropine and mannopine, (2) clones synthesizing mannopine but no agropine, and (3) clones synthesizing no opines. The typical chromatograms of the opine assay have been shown in *Scopolia japonica* hairy root clones (Mano et al. 1986). The variation among hairy root clones can be the result of the Ri T-DNA insertional mutagenesis.

The desired clones, having properties such as rapid growth and high alkaloid-production, can be selected from various hairy root clones. In *D. leichhardtii* hairy root clones, clone DL-27 grew most rapidly, followed in order by clones DL-34, DL-37, and DL-26 (Fig. 3). Clone DL-47 has been selected as the highest scopolamine-content clone, which is 2.1% on a dry weight basis (% dry wt.). This content of scopolamine is 2.2 times in leaves of the parent *D. leichhardtii* plant (scopolamine, 0.9–1% dry wt.). In contrast, no scopolamine was detected in the cultured roots of clone DL-8 or in its culture medium. These clones will be useful materials to examine the regulation of tropane alkaloid synthesis. The total amount of alkaloid in the batch culture is estimated by multiplying the biomass yield of a hairy root clone by its alkaloid content. Clone DL-34 has been selected as the best scopolamine-producing clone. The alkaloid productivities of these clones have been also stably maintained.

2.5 Production of Tropane Alkaloids by *Duboisia* Hairy Root Cultures

Suitable culture conditions for scopolamine production were established for hairy root clone DL-34 of *D. leichhardtii* after examining the effects of the composition and concentration of the medium and of the inoculum size on growth and alkaloid

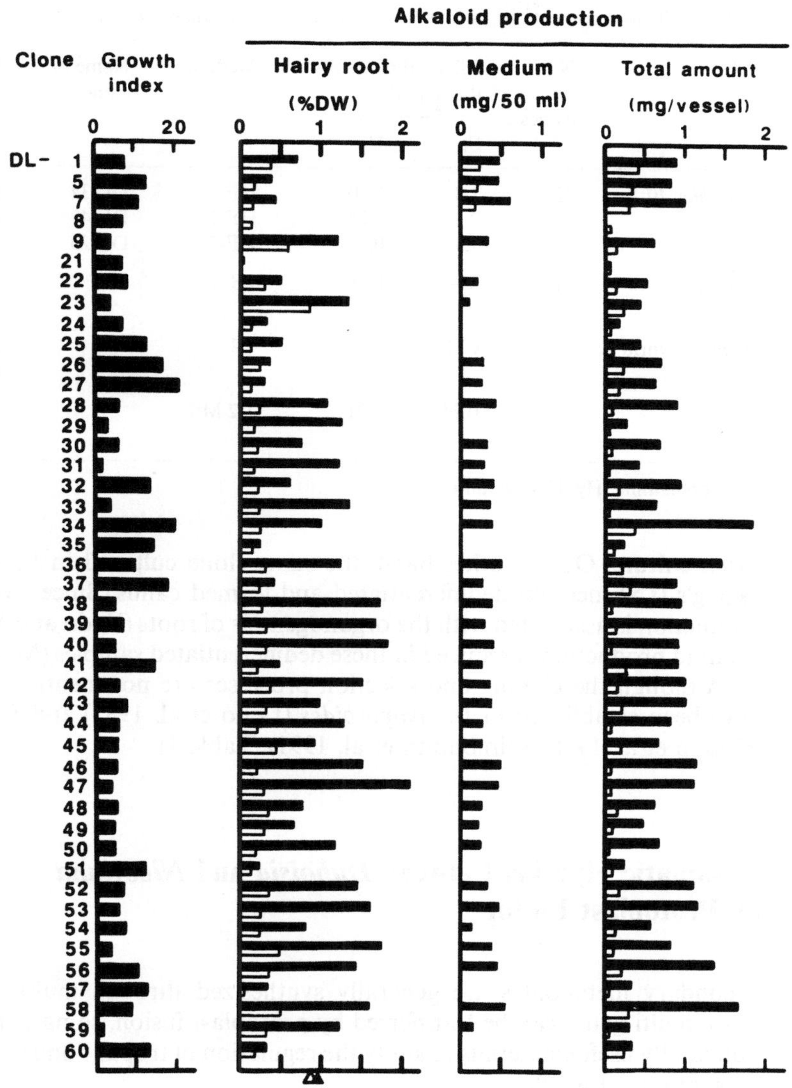

Fig. 3. Variation in growth and tropane alkaloid production among hairy root clones of *Duboisia leichhardtii*. Forty-five hairy root clones were cultured separately in phytohormone-free liquid medium. The growth rates (*Growth index*), alkaloid contents of the root tissues (*Hairy root*), amounts of alkaloid released into the culture medium (*Medium*) and total amounts of alkaloid (*Total amount*) were measured after 4 weeks of culture. *Triangles* indicate the contents of scopolamine (▲) and hyoscyamine (△) in the leaves of the parent plant. ■ growth index; ▬ scopolamine; ▭ hyoscyamine; *DW* dry weight. (Mano et al. 1989)

production (Mano et al. 1989). Hairy root clone DL-34 cultured in HF (modified Heller's) medium grew rapidly with extensive lateral branching and produced large amounts of tropane alkaloids; its scopolamine content was 1.8% dry wt. Clone DL-34 cultured in HF medium produced 78 mg/l of scopolamine after 4 weeks of

Table 1. Alkaloid contents in various hairy root cultures of *Duboisia* plants

Plant	No. of clones	Alkaloid Contents (% dry wt.)		Medium	Name of clone	Reference
		Sc	Hy			
D. leichhardtii	45	1.8	0.40	HF	DL-34	Mano et al. (1989)
		2.1	0.26	HF-2	DL-47	
		1.4	0.82	HF-2	DL-23	
D. myoporoides	1	0.24	0.78	LS		Deno et al. (1987)
Duboisia hybrid	1	0.112	0.023	LS		Knopp et al. (1988)
	1	0.247	0.21	1/2 MS		Shimomura et al. (1991)

Sc; Scopolamine. Hy; Hyocyamine.

batch culture. On the other hand, the same clone cultured in B5 or Linsmaier-Skoog's (LS) medium dedifferentiated and formed callus. Since tropane alkaloid production is associated with the organogenesis of roots (Endo and Yamada 1985), alkaloid production is reduced in these dedifferentiated samples (Mano et al. 1989).

Although the cloning and selection processes are not mentioned, hairy roots have been established in *D. myoporoides* (Deno et al. 1987) and *Duboisia* hybrid (Knopp et al. 1988; Shimomura et al. 1991) (Table 1).

3 Somatic Hybrids Between *Duboisia* and *Nicotiana* by Protoplast Fusion

Secondary metabolites are generally synthesized through multi-step pathways. Since multi-genes can be transferred by protoplast fusion, somatic hybrids will be potentially useful materials to study the regulation of tropane and nicotine alkaloid biosynthesis (Fig. 4).

3.1 Isolation of the Somatic Hybrid Cell Lines Between *D. hopwoodii* and *N. tabacum*

Protoplasts from cell suspension cultures of *D. hopwoodii* were fused with mesophyll protoplasts of *N. tabacum* using polyethylene glycol (Endo et al. 1987). Heterokaryons were identified microscopically, isolated with a micropipet and were cultured individually in small nylon mesh chambers surrounded by tobacco nurse cells. One putative intergeneric hybrid cell line was obtained, and the hybrid nature of this cell line (HT1031) was confirmed by the following biochemical and cytological investigations about 1 year after fusion. By the biochemical studies of the electrophoretic patterns of peroxidase and acid phosphatase, and the thin-layer

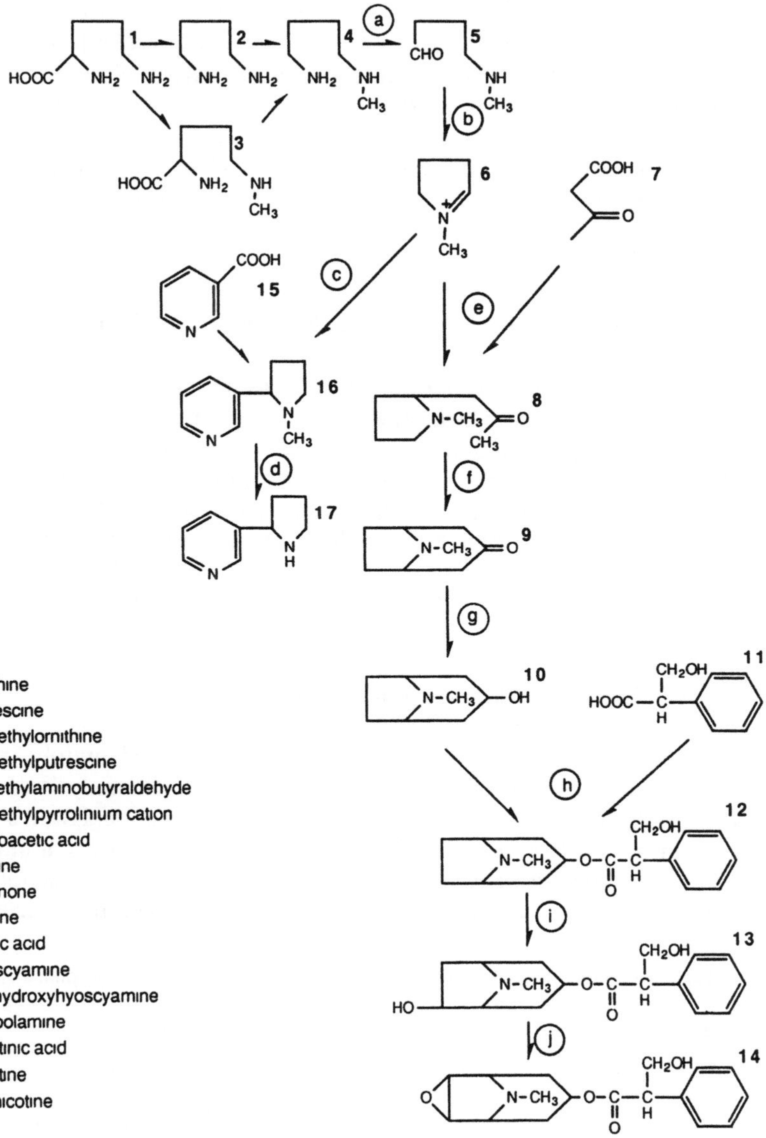

Fig. 4. Biosynthetic pathway of tropane and nicotine alkaloids. (Endo et al. 1991)

chromatography of fluorescent compounds, it was evident that this cell line shows intermediate patterns of both parental species. Observations of metaphase chromosomes indicated that this cell line contained both types of chomosomes.

3.2 Genetic Instability of Somatic Hybrid Cell Lines

The hybrid cell line HT1031 was divided into 500 small cell aggregates about 1 year after fusion. These were cultured individually and ten subclones were used to investigate nuclear DNA content, chromosome constitution, and the electrophoretic patterns of peroxidase about 3 years after fusion (Endo et al. 1988). Nuclear DNA content differed in each subclone and the extent of the elimination of tobacco chromosomes varied among the subclones. Peroxidase isozyme analysis indicated the loss of a tobacco-specific isozyme. Thus these studies show the genetic instability in this somatic hybrid cell line after 3 years of culture. Since the organogenesis was poor in this hybrid cell line, alkaloid biosynthesis could not be investigated.

3.3 Isolation of the Somatic Hybrids Between *D. leichhardtii* and *N. tabacum*

Protoplasts from *D. leichhardtii* cell suspension culture, which was derived from a root culture producing large amounts of scopolamine, hyoscyamine, and nicotine, were fused with mesophyll protoplasts of *N. tabacum* by electrofusion (Endo et al. 1991). Six putative hybrid colonies were obtained after culturing heterokaryons individually and regenerated shoots upon transfer to the regeneration medium. Regenerated shoots could be subcultured on the regeneration medium. The hybrid nature of the cloned cells and regenerated shoots was confirmed by cytological investigation and ribosomal DNA analysis, respectively.

3.4 Alkaloid Biosynthesis in Somatic Hybrids

Since tropane and nicotine alkaloids are synthesized mainly in the roots, it is essential to induce roots from the shoot cultures for analysis of alkaloid biosynthesis. Alkaloid composition in hybrid plantlets was analyzed 2 to 4 weeks after root initiation (Endo et al. 1991). The hybrid plantlets predominantly produced nicotine, while *Duboisia* plantlets produced equal levels of scopolamine, hyoscyamine, and nicotine, and *Nicotiana* plantlets produced only nicotine.

Table 2. Alkaloid biosynthetic activities in shoots of parent and hybrid plants[a]. (Endo et al. 1991)

Reaction step	Conversion (μmol)							
	NT	DL	Hybrids					
			A	B	C	D	E	F
a) Methylputrescine → MABA	2.34	1.28						
c) MABA + nicotinic acid → nicotine	0.19	nd	0.20	0.03	0.03	0.10	0.01	0.04
d) Nicotine → nornicotine	nd	0.30	+	0.05	nd	+	nd	0.05
f) Hygrine → tropinone	nd	nd						
g) Tropinone → tropine	nd	3.30	6.33	5.18	nd	4.51	5.19	5.27
h) Tropine + tropic acid → hyoscyamine	nd	nd						
i) Hyoscyamine → 6β-hydroxyhyoscyamine	nd	0.25	0.40	0.23	nd	0.02	0.18	0.23
j) 6β-Hydroxyhyoscyamine → scopolamine	nd	0.63	0.29	0.07	nd	0.05	0.09	0.10

[a] Precursor(s) (10 μmol) of each reaction step (a-j as in Fig. 4) were fed to the shoots, and the amounts of conversion products were measured. NT, *Nicotiana tabacum*; DL, *Duboisia leichhardtii*; nd, not detected; +, low activity detected. MABA, methylaminobutyraldehyde.

On the other hand, the shoot cultures serve the purpose of the detection and identification of small amounts of alkaloids formed upon precursor feeding, because alkaloid production is completely repressed in the shoot cultures. By the precursor feeding experiments, the hybrid shoots in all except one line showed the presence of the entire set of enzymes of both parental species (Table 2).

4 Summary and Conclusions

Hairy root clones stably expressing the desired properties have been established in *D. leichhardtii* by the transformation by the Ri T-DNA of *A. rhizogenes*, followed by cloning and selection. The hairy root culture system has markedly enhanced the productivity of tropane alkaloids in vitro more than in the conventional cell culture systems. Some problems still remain for the commercial production of tropane alkaloids. A profitable in vitro production of scopolamine would require a product titer of about 2 g/l (Strauss 1989). Therefore the productivity must be improved by more than 25 times higher than the results obtained in hairy root cultures of *D. leichhardtii*.

Hairy root clones of *D. leichhardtii* have released scopolamine as well as hyoscyamine into the culture medium during the logarithmic phase of growth (Mano et al. 1989). These characteristics may be applicable to continuous culture systems for the commercial production of tropane alkaloids.

Hyoscyamine 6β-hydroxylase (H6H), which catalyzes the hydroxylation of hyoscyamine in the biosynthetic pathway leading to scopolamine (Fig. 4), has been purified from the cultured roots of *Hyoscyamus niger* (Hashimoto and Yamada 1986; Yamada et al. 1990) and its cDNA clone has also been isolated (Matsuda et al. 1991). The recombinant DNA technology can be applied to the modification and regulation of tropane alkaloid production.

Duboisia plants have high potential to produce tropane alkaloids. Improving the hairy root clones of *Duboisia* by introducing H6H gene fused to the promoter highly expressing in root tissues will probably enhance the scopolamine production to a higher level by combining with the continuous culture system.

References

Ambros PF, Matzke AJM, Matzke MA (1986) Localization of *Agrobacterium rhizogenes* T-DNA in plant chromosomes by in situ hybridization. EMBO J 5:2073–2077

Armitage P, Walden R, Draper J (1988) Vectors for the transformation of plant cells using *Agrobacterium*. In: Draper J, Scott R, Armitage P, Walden R (eds) Plant genetic transformation and gene expression, a laboratory manual. Blackwell, Oxford, pp 1–67

Birot AM, Bouchez D, Casse-Delbart F, Durant-Tardif M, Jouanin L, Pautot V, Robaglia C, Tepfer D, Tepfer M, Tourneur J, Vilaine F (1987) Studies and uses of the Ri plasmids of *Agrobacterium rhizogenes*. Plant Physiol Biochem 25:323–335

Byrne MC, Koplow J, David C, Tempe J, Chilton MD (1983) Structure of T-DNA in roots transformed by *Agrobacterium rhizogenes*. J Mol Appl Genet 2:201–209

Cardarelli M, Mariotti D, Pomponi M, Spano L, Capone I, Costantino P (1987) *Agrobacterium rhizogenes* T-DNA genes capable of inducing hairy root phenotype. Mol Gen Genet 209:475–480
David C, Petit A, Tempe J (1988) T-DNA length variability in mannopine hairy roots: more than 50 kilobasepairs of pRi T-DNA can integrate in plant cells. Plant Cell Rep 7:92–95
Deno H, Yamagata H, Emoto T, Yoshioka T, Yamada Y, Fujita Y (1987) Scopolamine production by root cultures of *Duboisia myoporoides*. II. Establishment of a hairy root culture by infection with *Agrobacterium rhizogenes*. J Plant Physiol 131:315–323
De Paolis A, Mauro ML, Pomponi M, Cardarelli M, Spano L, Costantino P (1985) Localization of agropine-synthesizing function in the Tr region of the root-inducing plasmid of *Agrobacterium rhizogenes* 1855. Plasmid 13:1–7
Dougall DK (1987) Cell cloning and the selection of high yielding strains. In: Constabel F, Vasil IK (eds) Cell culture and somatic cell genetics of plants vol 4. Academic Press, San Diego, pp 117–124
Endo T, Yamada Y (1985) Alkaloid production in cultured roots of three species of *Duboisia*. Phytochemistry 24:1233–1236
Endo T, Komiya T, Masumitsu Y, Morikawa M, Yamada Y (1987) An intergeneric hybrid cell line of *Duboisia hopwoodii* and *Nicotiana tabacum* by protoplast fusion. J Plant Physiol 129:453–459
Endo T, Komiya T, Mino M, Nakanishi K, Fujita S, Yamada Y (1988) Genetic diversity among sublines originating from a single somatic hybrid cell of *Duboisia hopwoodii* + *Nicotiana tabacum*. Theor Appl Genet 76:641–646
Endo T, Hamaguchi N, Eriksson T, Yamada Y (1991) Alkaloid biosynthesis in somatic hybrids of *Duboisia leichhardtii* F. Muell. and *Nicotiana tabacum* L. Planta 183:505–510
Flores HE, Filner P (1985) Metabolic relationships of putrescine, GABA and alkaloids in cell and root cultures of Solanaceae. In: Neumann KH, Barz W, Reinhard E (eds) Primary and secondary metabolism in plant cell cultures. Springer, Berlin Heidelberg New York, pp 174–185
Hamill JD, Parr AJ, Robins RJ, Rhodes MJC (1986) Secondary product formation by cultures of *Beta vulgaris* and *Nicotiana rustica* transformed with *Agrobacterium rhizogenes*. Plant Cell Rep 5:111–114
Hashimoto T, Yamada Y (1986) Hyoscyamine 6β-hydroxylase, a 2-oxoglutarate-dependent dioxygenase, in alkaloid-producing root cultures. Plant Physiol 81:619–625
Huffman GA, White FF, Gordon MP, Nester EW (1984) Hairy-root-inducing plasmid: physical map and homology to tumour-inducing plasmids. J Bacteriol 157:269–276
Jouanin L (1984) Restriction map of an agropine-type Ri plasmid and its homologies with Ti plasmids. Plasmid 12:91–102
Jouanin L, Guerche D, Pamboukdjian N, Tourneur C, Casse-Delbart F, Tourneur J (1987) Structure of T-DNA in plants regenerated from roots transformed by *Agrobacterium rhizogenes* strain A4. Mol Gen Genet 206:387–392
Kamada H, Okamura N, Satake M, Harada H, Shimomura K (1986) Alkaloid production by hairy root cultures in *Atropa belladonna*. Plant Cell Rep 5:239–242
Knopp E, Strauss A, Wehrli W (1988) Root induction on several *Solanaceae* species by *Agrobacterium rhizogenes* and the determination of root tropane alkaloid content. Plant Cell Rep 7:590–593
Koukolikova-Nicola Z, Albright L, Hohn B (1987) The mechanism of T-DNA transfer from *Agrobacterium tumefaciens* to the plant cell. In: Hohn T, Schell J (eds) Plant DNA infectious agents. Springer, Berlin Heidelberg New York, pp 109–148
Mano Y (1989) Variation among hairy root clones and its application. Plant Tissue Cult. Lett 6:1–9
Mano Y, Nabeshima S, Matsui C, Ohkawa H (1986) Production of tropane alkaloids by hairy root cultures of *Scopolia japonica*. Agric Biol Chem 50:2715–2722
Mano Y, Ohkawa H, Yamada Y (1989) Production of tropane alkaloids by hairy root cultures of *Duboisia leichhardtii* transformed by *Agrobacterium rhizogenes*. Plant Sci 59:191–201
Matsuda J, Okabe S, Hashimoto T, Yamada Y (1991) Molecular cloning of hyoscyamine 6β-hyroxylase, a 2-oxoglutarate-dependent dioxygenase, from cultured roots of *Hyoscyamus niger*. J Biol Chem 266:9460–9464
Rogers SG, Klee H (1987) Pathways to plant genetic manipulation employing *Agrobacterium*. In: Hohn T, Schell J (eds) Plant DNA infectious agents. Springer, Berlin Heidelberg New York, pp 179–203
Shen WH, Petit A, Guern J, Tempe J (1988) Hairy roots are more sensitive to auxin than normal roots. Proc Natl Acad Sci USA 85:3417–3421
Shimomura K, Sauerwein M, Ishimaru K (1991) Tropane alkaloids in the adventitious and hairy root cultures of solanaceous plants. Phytochemistry 30:2275–2278
Spena A, Schmulling T, Koncz C, Schell J (1987) Independent and synergistic activity of rol A, B and C loci in stimulating abnormal growth in plants. EMBO J 6:3891–3899

Strauss A (1989) *Hyoscyamus* spp: In vitro culture and the production of tropane alkaloids. In: Bajaj YPS (ed) Biotechnology in agriculture and forestry vol 7. Medicinal and aromatic plants II Springer, Berling Heidelberg New York, pp 286–314

Taylor BH, White FF, Nester EW, Gordon MP (1985) Transcription of *Agrobacterium rhizogenes* A4 T-DNA. Mol Gen Genet 201:546–553

Vilaine F, Casse-Delbart F (1987) Independent induction of transformed roots by the TL and TR regions of the Ri plasmid of agropine type *Agrobacterium rhizogenes*. Mol Gen Genet 206:17–23

Vilaine F, Charbonnier C, Casse-Delbart F (1987) Further insight concerning the TL-region of the Ri plasmid of *Agrobacterium rhizogenes* strain A4: Transfer of a 1.9 kb fragment is sufficient to induce transformed roots on tabacco leaf fragments. Mol Gen Genet 210:111–115

White FF, Sinkar VP (1987) Molecular analysis of root induction by *Agrobacterium rhizogenes*. In: Hohn T, Schell J (eds) Plant DNA Infectious Agents. Springer, Berlin Heidelberg New York, pp 149–177

White FF, Taylor BH, Huffman GA, Gordon MP, Nester EW (1985) Molecular and genetic analysis of the transferred DNA regions of the root-inducing plasmid of *Agrobacterium rhizogenes*. J Bacteriol 164:33–44

Yamada Y, Endo T (1984) Tropane alkaloid production in cultured cells of *Duboisia leichhardtii*. Plant Cell Rep 3:186–188

Yamada Y, Okabe S, Hashimoto T (1990) Homogeneous hyoscyamine 6β-hydroxylase from cultured roots of *Hyoscyamus niger*. Proc Jpn Acad Ser B Phys Biol Sci 66:73–76

II.8 Transformation in *Eustoma grandiflorum* (Lisianthus)

S.C. Deroles[1], S.E. Ledger[1], R.M. Miller[1], K.M. Davies[1], and N.K. Given[2]

1 Introduction

Lisianthus [*Eustoma grandiflorum* (Griseb) Shinn.] is a gentian native to the central and southern United States (Bailey 1949, Shinners 1957). Due to its large flowers, long stems, and extended vase life, it is an increasingly popular cut flower. The typical wild plant has blue-purple flowers, but intensive breeding in Japan over the last 30 years has produced varieties with white, pink, plum, and mauve flowers. To enhance the popularity of lisianthus, new varieties need to be developed. Desirable traits include; an expansion of the color range to red and yellow, sectoring of the pigmentation to produce stripes and star patterns, and a more compact plant form for the pot plant market.

To date, breeding of lisianthus has been by sexual hybridization. Genetic engineering makes it possible to introduce genes that are not currently available in the lisianthus sexual gene pool, allowing novel phenotypes to be generated. In addition, specific characteristics can be targeted, such as flower color (Mol et al. 1989).

This chapter describes research carried out at the Crop & Food's Levin Research Centre (Levin RC) in New Zealand on developing a genetic transformation system for lisianthus. As a result of its breeding program, Levin RC has established techniques for the in vitro propagation of lisianthus. Attempts are now being made to build upon this work using *Agrobacterium*-mediated transformation to produce novel cut flower varieties of lisianthus.

2 Breeding and In Vitro Propagation

2.1 Selection of New Varieties via Breeding

Breeding of lisianthus at Levin RC during the last 7 years has focused on producing varieties with new flower colors and form. The source material was a lisianthus seed line produced by the Dai-Ichi Seed Company, Japan. These seeds

[1] Levin Research Centre, Crop and Food Research Ltd, Private Bag, Levin, New Zealand

[2] Dr. Nigel Given died in June 1991. The authors would like to record their appreciation for the advice and support given by Dr. Given during the course of the work reported

Biotechnology in Agriculture and Forestry, Vol. 22
Plant Protoplasts and Genetic Engineering III (ed. by Y.P.S. Bajaj)

were grown to flowering and 100 plants (numbered 1–100) selected for further study. Tissue cultured lines from these plants became the source of lisianthus used in the transformation experiments. Breeding of lisianthus is continuing, through gamma irradiation of seeds and somaclonal variation from in vitro culture (Butcher, Levin RC, pers. comm.).

2.2 In Vitro Propagation of Lisianthus

Early work on the in vitro propagation of lisianthus by Semeniuk and Griesbach (1987) described methods for the production of plants from shoot tips, and leaf and internodal stem segments via a callus phase. These techniques provided not only a rapid method for clonal propagation and the production of secondary metabolites (Skrzypczak and Wesolowska 1993), but also a source of somaclonal variants (Griesbach and Semeniuk 1987). For genetic transformation, however, an in vitro regeneration method is required that gives rise to shoots directly from an explant, thus reducing the potential for somaclonal variation by removing the callus phase. Such a method was developed at Levin RC (Seelye and Butcher 1987), using levels of auxin and cytokinins lower than previously published (Semeniuk and Griesbach 1987). The use of this technique, in conjunction with *Agrobacterium*-mediated transformation, to produce genetically modified lisianthus plant is described here.

In vitro-grown lisianthus tissue was used as the source material for transformation experiments, in order to avoid micro-organism contamination encountered with glasshouse-grown plants. A basal medium (BM) consisting of MS salts (Murashige and Skoog 1962), LS growth factors (Linsmaier and Skoog 1965), 30 g/l sucrose, and 0.78 g/l Davis agar was used. Cultures were grown at 25 °C under a 16/8 h light/dark photoperiod, with 35 μmol/m/s cool white fluorescent light. In vitro shoots for stock plants were produced from nodal segments on proliferation media (PM), consisting of BM with 0.05 mg/l indole-3-butyric acid (IBA), 0.3 mg/l benzyladenine (BA) and 0.1 mg/l gibberellic acid (GA_3). The addition of GA_3 increased the average length of the shoots, making further manipulations easier.

For shoot regeneration from leaf pieces, young rapidly expanding leaves were removed from in vitro stock plants and cut into 5-mm square pieces. The leaf pieces were placed abaxial surface down on shooting media (SM), consisting of BM plus 0.05 mg/l IBA, 1 mg/l BA and 0.1 mg/l GA_3.

For root induction individual shoots were placed on BM plus 1 mg/l indole-3-acetic acid for 5 to 7 days and then, to allow both root and shoot growth, transferred to rooting media (RM) consisting of BM with the MS salts reduced by half.

3 *Agrobacterium*-Mediated Transformation

Four cultivars from the original lisianthus lines have now been transformed, all of which have a single whorl of petals: #1 (mauve), #42 (cream-white), #54 (purple), and #65 (pink).

3.1 *Agrobacterium tumefaciens* Strains

Three *Agrobacterium tumefaciens* strains were tested for their ability to produce bacterial galls on lisianthus: C58, a virulent nopaline type: A722, a virulent strain having the chromosomal background of A114 (C58 cured of the Ti plasmid pTiC58) and carrying the octopine Ti plasmid pTiA6NC (Chilton et al. 1976; Garfinkel and Nester 1980): and LBA4404 (Hoekema et al. 1983), which carries the disarmed Ti Plasmid pAL4404 (an octopine type plasmid with no T-DNA region). A722 was the only strain effective in gall production, and was subsequently used as the host for binary vectors.

The binary vector pKIWI110 (Janssen and Gardner 1990) was used to assess the efficiency of DNA transfer by *Agrobacterium*. It contains a kanamycin resistance gene (neomycin phosphotransferase II; NPTII) under the control of a nopaline synthase (*nos*) promoter, and the β-D-glucuronidase (GUS) reporter gene under the control of the Cauliflower Mosaic Virus (CaMV) 35S promoter. For the generation of transformed plants, the binary vector pGA643 was used (An et al. 1988). This vector contains a NPTII gene under the *nos* promoter, and a multiple cloning site downstream from the CaMV 35S promoter, for the insertion of an additional gene. This vector does not carry a GUS gene.

3.2 Transformation Protocol

Leaf pieces were inoculated with *Agrobacterium* using a modification of the technique of Horsch et al. (1985). Leaf pieces were cut on the day of inoculation and placed onto SM. The *Agrobacterium* was grown overnight at 28 °C in YN broth (beef extract 3 g/l, Bacto-peptone 5 g/l, NaCl 8 g/l, yeast extract 10 g/l), containing 25 mg/l kanamycin, at 310 rpm. The cells were then pelleted by centrifugation and resuspended in fresh YN broth. The leaf pieces were submerged in this inoculum, blotted dry between two pieces of sterile filter paper, and replaced onto SM. After 2 days of cocultivation, the leaf pieces were transferred to SM containing 500 mg/l ticarcillin and 70 mg/l kanamycin (SMK), and grown under conditions described previously for shoot regeneration.

Shoots that regenerated in the presence of kanamycin were transferred from the leaf piece and tested further by screening for growth on PM containing 500 mg/l ticarcillin and 70 mg/l kanamycin. Individual shoots from proliferating clumps were rooted as previously described, removed into pumice, and grown under high humidity in a growth chamber. After 3–4 weeks the rooted plants were repotted in soil and grown to flowering at 25 °C under 700 μmol white light/m^2/s.

4 Analysis of Early Transformation Events

Analysis of early transformation events was carried out using the GUS reporter gene (Jefferson 1987). This system allows the identification of single transformed cells, as those expressing the GUS gene are able to convert the colorless substrate X-Gluc (5-bromo-4-chloro-3-indolyl glucuronide) to a blue precipitate.

4.1 GUS Assay and Results

GUS assays of leaf pieces were carried out using a modification of the method of Jefferson (1987). Leaf pieces were incubated overnight at 37 °C in a solution containing 0.3 g/l X-Gluc, 0.003% (v/v) of dimethyl formamide and 50 mM $NaPO_4$ (pH 7.0).

The pattern of GUS expression following *Agrobacterium* inoculation of lisianthus leaf pieces with pKIWI110/A722 was studied using cultivar #42 (Table 1). A simple time course of GUS expression was observed. After 2 days single GUS stained cells were visible, at one loci per ten leaf pieces on average. Eight days after inoculation, four loci per ten leaf discs were present, and these tended to be localized in small groups. By 15 days, 21 independent loci were observed per ten leaf pieces. In subsequent experiments, calli 1–2 mm in diameter had regenerated on the edge of the leaf pieces after 21 days. When assayed for GUS activity, these calli produced an intense blue coloration.

The blue loci were observed only on the cut edge of the leaf piece, parallel to the main vein. No GUS expression was observed from non-inoculated leaf pieces or those inoculated with pKIW110/A2002, which is unable to transfer the T-DNA.

4.2 Discussion

The steady increase observed in the expression of GUS following *Agrobacterium* inoculation of lisianthus is markedly different to the pattern seen with petunia (Jansen and Gardner 1990) and chrysanthemum (Ledger et al. 1991). In both of these plants, large zones of GUS-expressing cells are visible soon after inoculation. These zones then rapidly disappear, leaving behind a few GUS-expressing loci, which subsequently develop into calli.

It is possible that *Agrobacterium* transfers high numbers of GUS-expressing T-DNA copies to the plant cell, but that only a few of these copies integrate into the genome (Jansen and Gardner 1990). Thus, the number of GUS-expressing loci falls sharply with time, leaving only the stable transformation events. This would

Table 1. Expression of GUS in lisianthus, following inoculation with *Agrobacterium.* Lisianthus leaf pieces were inoculated with pKIWI110/A722, and expression of the GUS gene assayed histochemically. Ten leaf pieces were assayed at each time point

Time after inoculation (days)	Average number of loci expressing GUS
2	1
5	2
8	4
12	5
15	21

explain why large numbers of GUS-positive loci are initially observed in petunia and chrysanthemum. In lisianthus, no initial high expression of GUS is observed, suggesting that the T-DNA transfer event may be inefficient. However, a high proportion of the transferred T-DNAs may integrate stably into the plant genome, giving the observed slow increase in GUS expression.

Another unusual feature of GUS expression in lisianthus is that the number of individual loci continues to increase for up to 15 days after inoculation with *Agrobacterium*. In petunia and chrysanthemum, GUS expression peaks after 4 days, and it is thought that this is the period needed to develop maximal expression of the introduced gene (Jansen and Gardner 1990). Thus, in lisianthus, T-DNA transfer may be occurring up to 10 days after inoculation, implying that *Agrobacterium* is active throughout this period. The antibiotic used to control *Agrobacterium* growth, ticarcillin, does not kill the bacterium, and as a result it may still be able to transfer the T-DNA, even though it can no longer replicate. However, this would not explain why the pattern has not been observed with other plants, and a second factor contributing may be the slow regeneration rate of lisianthus. Petunia and chrysanthemum both regenerate quickly, with rapid recovery of the wound sites, giving a relatively short access period for the *Agrobacterium*. Lisianthus regeneration is much slower, and it is possible that *Agrobacterium* is able to access wounded cells over a longer period, leading to the increased phase of GUS expression. The rate of wound recovery in lisianthus, however, has not been determined.

5 Production of Kanamycin-Resistant Shoots

5.1 Results

Leaf pieces on nonselective regeneration media gave rise to shoots mainly from the edges perpendicular to the main vein (i.e., the basal and apical leaf edges) (Fig. 1A). However, when leaf pieces were cocultivated with pKIWI110/A722, and placed under kanamycin selection, the majority of the shoots formed from edges parallel to the main vein, or from corners of the leaf pieces (Fig. 1B).

Selection for kanamycin resistance at the shooting stage is not definitive for lisianthus, whilst selection at the rooting stage is slow due to the poor rate of root initiation. Thus, the primary stage of kanamycin selection used was the proliferation phase of micropagation. Single shoots were excised from the leaf piece and placed onto proliferation media containing kanamycin. Shoots that remained green, formed a callus at the base of the stem, and proliferated, were considered to be kanamycin-resistant. Shoots that became bleached, blackened at the base of the shoot, and failed to proliferate, were considered to be kanamycin-sensitive. Only a low percentage of shoots were kanamycin-resistant (Table 2), with 1–3% of inoculated leaf pieces producing a resistant shoot. These shoots were removed into soil and grown to flowering to test for the inheritance of the NPTII and GUS genes (Fig. 2).

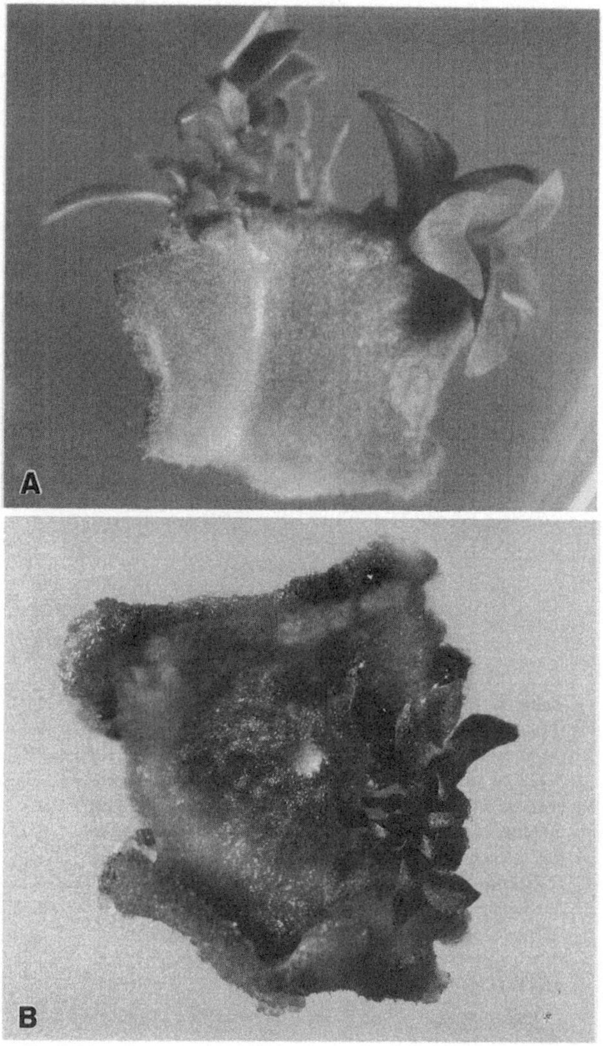

Fig. 1A, B. Shoot regeneration in lisianthus. (**A**) Shoots regenerating from an untreated lisianthus leaf piece in in vitro culture. Sites of shoot regeneration are on the side of the leaf piece perpendicular to the main vein. (**B**) Shoots regenerating from a lisianthus leaf piece cocultivated with *Agrobacterium tumefaciens* and placed on kanamycin selection. Sites of regeneration are on the *side of the leaf piece* parallel to the main vein

5.2 Discussion

A marked difference was observed between the sites of shoot regeneration in lisianthus and those known for petunia and chrysanthemum. Petunia and chrysanthemum follow the pattern of many plants, where the primary regenerative sites are located on the edges perpendicular to the main vein, both in vitro culture and following *Agrobacterium* inoculation (Jansen and Gardner 1990; Ledger et al. 1991). In contrast, in lisianthus, following *Agrobacterium* inoculation the primary area of regeneration changes from the sides perpendicular to the main vein to those parallel. This pattern of shoot development correlates with the localization of GUS expressing cell loci indicating that in lisianthus, the cells highly competent for

Fig. 2. A transgenic lisianthus plant (TL1), regenerated under kanamycin selection, and a wild-type lisianthus plant

Table 2. Production of kanamycin-resistant lisianthus shoots. Leaf pieces of two lisianthus cultivars (#42 and #54) were inoculated with pKIWI110/A722 or pGA643/A722 and placed on shoot regeneration media, containing 70 mg/l kanamycin and 500 mg/l ticarcillin. Shoots that regenerated from the leaf pieces were subsequently tested for kanamycin resistance using proliferation medium

	LP	Shoot	Kan^r
pKIWI110/A722			
#42	574	45	1
#54	70	22	1
pGA643/A722			
#42	342	50	4
#54	92	15	0

LP = number of leaf pieces inoculated; Shoot = number of shoots removed from leaf pieces; Kan^r = number of kanamycin-resistant shoots.

regeneration are different from those receptive to transformation. Such a dichotomy has also been observed for apple (Atkinson, University of Auckland, pers. comm.). The transformation frequency for lisianthus is currently very low. Further studies on the dichotomy between regeneration-competent and transformation-receptive cells could provide information on the control of the processes involved, and could lead to a more efficient transformation system.

6 Southern Hybridization Analysis

Shoots that were able to root in the presence of kanamycin were analyzed by Southern hybridization. This was to establish that the T-DNA was inserted into the plant genome, and to determine the structure and number of T-DNA inserts.

6.1 DNA Isolation and Southern Hybridization

DNA was isolated from wild-type and kanamycin-resistant shoots. Leaf material from proliferating clumps of shoots derived from a single transformant (0.5–1.0 g) were ground in liquid nitrogen, the tissue suspended in extraction buffer [4 M guanidinium isothiocyanate, 26 mM NaOAc, 1.5% (w/v) Na lauryl sarcosine , 2.5% v/v 2-mercaptoethanol], extracted with an equal volume of phenol/chloroform 1:1 (v/v) followed by an equal volume of chloroform, and the DNA precipitated with 0.8 volumes of isopropanol. The DNA was then pelleted by centrifugation and resuspended in 10 mM Tris-HCl, 1 mM EDTA. The DNA was digested with restriction enzymes, separated electrophoretically on a 0.7% agarose gel, and transferred onto Hybond N + membrane following the manufacturer's instructions (Amersham International Pty). The membrane was hybridised against the coding region of the NPTII gene, isolated as a *Hind*III fragment from pKIWI110 and radiolabeled with [$^{\alpha 32}$P]-dCTP. Hybridization was carried out as described by Deroles and Gardner (1988).

6.2 Results

Analysis was performed on three kanamycin-resistant shoots: TL1#42 and TL139#54 inoculated with pKIWI110/A722, and TL91#42 inoculated with pGA643/A722. DNA from TL1 and TL139 was digested with *Eco*RI, while DNA from TL91 was cut with *Bam*HI and *Hind*III. All three samples were hybridized with the 8 kb *Hind*III fragment of pKIWI110 containing the NPTII coding region.

All of the DNA samples from the kanamycin-resistant plants gave the hybridization pattern expected for an intact NPTII gene; a single band at 1.6 kb for TL1 and TL139, and a single band at 2.5 Kb for TL91 (Fig. 3). Additional bands were also observed for TL1 and TL139. The bands at 4.6 kb correspond to an internal region of the pKIWI110 T-DNA which includes the 3′ region of the GUS gene. The remaining additional bands represent junction fragments at the right border of the T-DNA. The size of these bands indicate that they are composed of both T-DNA and plant DNA, thus establishing that the T-DNA is inserted into the plant genome. No positive bands were observed in the noninoculated control plant samples.

6.3 Discussion

Southern hybridization analysis demonstrated the presence of the expected NPTII-containing T-DNA regions in the three putative transformants. The presence of

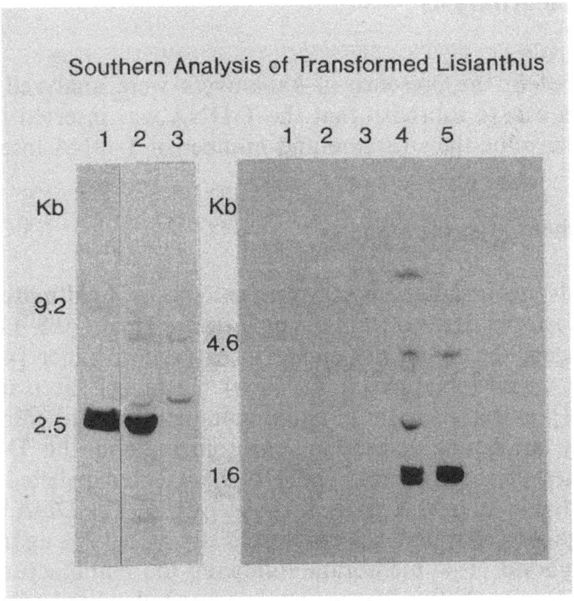

Fig. 3. Southern hybridization analysis of kanamycin-resistant lisianthus plants. DNA was isolated from leaves of kanamycin-resistant lisianthus, digested with restriction enzymes, separated on an agarose gel, and transferred to a nylon membrane as described in the text. The membrane was hybridized to the NPTII coding region of pKIWI110, excised as a *Hind*III fragment and radiolabeled with [^{32}P]-dCTP. Hybridization was visualized by autoradiography. *Left: Lane 1* pGA643 digested with *Bam*HI/*Hind*III showing the expected 2.5 kb fragment containing the NPTII gene; *Lane 2* DNA from transformant TL91 digested with *Bam*HI/*Hind*III; *Lane 3* DNA from an untransformed lisianthus plant digested with *Bam*HI/*Hind*III. *Right: Lane 1* pKIWI110 digested with *Eco*RI, showing the expected 1.6 kb fragment containing the NPTII gene; *Lane 2* DNA from an untransformed lisianthus digested with *Eco*RI; *Lane 4* DNA from transformant TL1 digested with *Eco*RI; *Lane 5* DNA from transformant TL139 digested with *Eco*RI

junction fragments shows that the T-DNA is integrated into the plant genome. In transformant TL1 (Lane 4, Fig. 3), the presence of more than one junction fragment indicates more than one integrated copy of the T-DNA. Inheritance studies on these plants should determine the number of independently expressing T-DNA loci that each plant contains.

7 Future Directions: Manipulation of Color in Lisianthus

The primary research direction of the Plant Molecular Biology Group at Levin RC is the manipulation of color via the genetic alteration of the flavonoid biosynthetic pathway. The *Agrobacterium*-mediated transformation system described in this chapter will enable such experiments to be carried out in lisianthus.

7.1 Alteration of the Flavonoid Glycosylation Pattern

Flavonoids are commonly glycosylated by the enzyme uridine-di-phospho-glucose: flavonoid-glucosyl-transferase (UFGT), which introduces a glucose moiety. Subsequent acylation of this glucose residue increases the stability of the flavonoid, and may also affect its interaction with copigments (Brouillard 1988). In lisianthus, the primary sugar residue attached to the flavonoids is galactose rather than glucose (Asen et al. 1986). The site of acylation of the sugar is the same as the site of stereochemical difference between glucose and galactose (K. Markham, Industrial Research Limited, pers. comm.). Thus, the acylation activity may be specific to the sugar moiety, as distinct substrate specificities have been reported for acylation enzymes (Heller and Forkman 1988). In our laboratory we have the gene coding for UFGT from *Antirrhinum majus* (Martin et al. 1991), which in this species adds a glucose residue to anthocyanins (Martin et al. 1987). We have inserted the *Antirrhinum* UFGT gene into the multiple cloning site of pGA643, and are generating transformed lisianthus using this construct. Our aim is to alter flower color by generating a second class of nonacylated, glycosylated flavonoids in lisianthus, that will interfere with the complex copigmentation effects that we believe occur in the flower.

7.2 Generation of Patterns via Antisense Gene Expression

Antisense RNA has recently emerged as a powerful tool for plant genetic engineering. Antisense genes are genetic constructs that produce a mRNA of complementary sequence to the native sense in mRNA (van der Krol et al. 1988). Current understanding indicates that the two mRNAs form a duplex that is either blocked from translation or degraded. Thus, the expression of an antisense gene results in the specific shut-down of the complementary native gene.

Van der Krol et al. (1988) used this method to down-regulate the level of mRNA for chalcone synthase, a key flavonoid biosynthetic enzyme, and produced a white petunia line from a purple-flowered cultivar. An interesting feature of the study was the generation of patterns in the flower, thought to be caused by shut down of the native gene in discrete areas of the flower petal. Although the mechanism for this is currently unknown, it may provide a convenient mechanism for the generation of novel patterns in commercial flowers.

We have isolated several of the flavonoid biosynthetic genes from lisianthus via the RACE Polymerase Chain Reaction technique (Frohman et al. 1988), and antisense versions of these genes have been inserted into the multiple cloning site of pGA643. With the aim of generating patterned flowers, these vectors are currently being used to generate transgenic lisianthus.

Acknowledgments. We would like to thank Dr. Bart Jansen for allowing us to use pKIWI110, Dr. Gyn An for pGA643, and Dr. Cathie Martin for the UFGT cDNA clone from *Antirrhinum.* Our thanks also to Mr. David Manson for excellent technical help, and to Dr. Eric Walton, Dr. Ellen Podivinsky and Ms. Kathy Slindee for helpful comments on the manuscript. All figures and tables in this manuscript are taken from Ledger et al. (in prep.).

References

An G, Ebert PR, Mitra A, Ha SB (1988) Binary vectors. In: Gelvin SB, Schilperoort RA (eds) Plant molecular biology manual. Kluwer, Dordrecht, ppA3 1–19
Asen S, Griesbach RJ, Norris KH, Leonhardt BA (1986) Flavonoids from *Eustoma grandiflorum* petals. Phytochemistry 25:2509–2513
Bailey L (1949) Manual of cultivated plants. Macmillan, New York
Brouillard R (1988) Flavonoids and flower colour. In: Harborne JB (ed) The flavonoids: advances in research since 1980. Chapman and Hall, London, pp 525–538
Chilton M-D, Farrand SK, Levin R, Nester EW (1976) RP4 promotion of transfer of a large *Agrobacterium* plasmid which confers virulence. Genetics 83:609–618
Deroles SC, Gardner RC (1988) Analysis of the T-DNA structure in a large number of transgenic petunias generated by *Agrobacterium*-mediated transformation. Plant Mol Biol 11:365–377
Frohman MA, Dush MK, Martin GR (1988) Rapid production of full length cDNA's from rare transcripts: amplification using a single gene-specific oligonucleotide primer. Proc Natl Acad Sci USA 85:8998–9002
Garfinkel DJ, Nester EW (1980) *Agrobacterium tumefaciens* mutants affected in crown gall tumorigenesis and octopine catabolism. J Bacteriol 144:732–743
Griesbach RJ, Semeniuk P (1987) Use of somaclonal variation in the improvement of *Eustoma grandiflorum*. J. Hered 78:114–116
Heller W, Forkmann G (1988) Biosynthesis. In: Harborne JB (ed) The flavonoids: advances in research since 1980. Chapman and Hall, London, pp 399–426
Hoekema A, Hirsch PR, Hooykaas PJJ, Schilperoort RA (1983) A binary vector strategy based on the separation of the *vir* and T-DNA regions of the *Agrobacterium tumefaciens* Ti-plasmid. Nature 303:179–181
Horsch RB, Fry JE, Hoffmann NL, Eicholtz D, Rogers SG, Fraley RG (1985) A simple and general method for transferring genes into plants. Science 227:1229–1231
Jansen B, Gardner RC (1990) Localised transient expression of GUS in leaf discs following cocultivation with *Agrobacterium*. Plant Mol Biol 14:61–72
Jefferson RA (1987) Assaying chimeric genes in plants: the GUS gene fusion system. Plant Mol Biol Rep 5:387–405
Ledger SE, Deroles SC, Given NK (1991) Regeneration and *Agrobacterium*-mediated transformation of chrysanthemum. Plant Cell Rep 10:195–199
Ledger SE, Deroles SC, Given NK *Agrobacterium*-mediated transformation of lisianthus (*Eustoma grandiflorum*) (in prep.)
Linsmaier EM, Skoog F (1965) Organic growth factor requirements of tobacco tissue culture. Physiol Plant 18:100–127
Martin C, Carpenter R, Coen ES, Gerats AGM (1987) The control of floral pigmentation in *Antirrhinum majus*. In: Thomas H, Grierson D (eds) Developmental mutants in higher plants. Cambridge University Press, pp 19–52
Martin C, Prescott A, Mackay S, Bartlett J, Vrijlandt E (1991) Control of anthocyanin biosynthesis in flowers of *Antirrhinum majus*. Plant J 1:37–49
Mol JNM, Stuitje AR, van der Krol A (1989) Genetic manipulation of floral pigmentation genes. Plant Mol Biol 13:287–294
Murashige T, Skoog F (1962) A revised medium for rapid growth and bioassays with tabacco tissue cultures. Physiol Plant 15:473–497
Seelye JF, Butcher SM (1987) Micropropagation of lisianthus (*Eustoma grandiflorum*). Selections IAPTC Book of Abstracts August 1987 Maroochydore, Queensland, Australia
Semeniuk P, Griesbach RJ (1987) In vitro propagation of prairie gentian. Plant Cell Tissue Organ Cult 8:249–253
Shinners LH (1957) Synopsis of the genus *Eustoma* (Gentianaceae). Southwest Nat 2:38–43
Skrzypczak L, Wesolowska M (1993) *Eustoma grandiflorum* Shinn (Texas Bluebell): Callus culture, micropropagation, and the production of gentiopicroside and other secondary metabolites. In: Bajaj YPS (ed) Biotechnology in agriculture and forestry, vol 24. Medicinal and aromatic plants V. Springer, Berlin Heidelberg pp 192–201
van der Krol AR, Lenting PE, Veenstra J, van der Meer IM, Koes RE, Gerats AGM, Mol JNM, Stuije AR (1988) An antisense chalcone synthase gene in transgenic plants inhibits flower pigmentation. Nature 333:866–869

II.9 Transformation in *Helianthus annuus* L. (Sunflower)

B. Schrammeijer, A. Hoekema, and P.C. Sijmons[1]

1 Introduction

The successful application of *Agrobacterium* transformation methods to a specific crop has its inherent problems. Firstly, it is dependent on the susceptibility of the species to *Agrobacterium* infection and on its potential to regenerate from transformed cells into fertile plants. Unfortunately, a number of reliable regeneration protocols, developed in the past decade, proved unsuitable for transformation with *Agrobacterium*. A possible explanation is the stress response of the plant tissue during cocultivation with the bacterium, which inhibits efficient regeneration. This is especially true for regeneration methods that need long (more than several months) and complex (several sequential hormone regimes) tissue culture phases. Also, for certain antibiotics that are used to stop bacterial growth, phytohormone activities are known. This may have detrimental effects on regeneration. Furthermore, somaclonal variation as a result of tissue culture poses a problem for plant genetic engineering: it is generally aimed at the introduction of only a specific trait carried by one or several genes into the target crop. All intrinsic qualities of the target cultivars should remain otherwise unchanged. Ideally, the tissue culture phase after transformation should be abandoned completely or kept to a minimal period.

In view of these problems, a procedure for meristem transformation developed by Ulian et al. (1988) is of interest. It was originally demonstrated with *Petunia*, and a more recent research on maize meristems (Gould et al. 1991) indicates that the technique may be more widely applicable. In vitro regeneration from pre-existing meristems is relatively easy for most crops and involves the shortest possible tissue culture phase. A major challenge of this approach, however, is to target exactly those cells present in the meristem that give rise to the reproductive tissue in mature plants.

In this chapter, we describe the potential use of meristem transformation with *Agrobacterium* for a recalcitrant crop, sunflower (*Helianthus annuus*). The improvement of germplasms of characteristics like disease and pest resistances, nutritional quality, and processing characteristics have been restricted to conventional breeding. Sunflower is a suitable host plant for *Agrobacterium* gene transfer. There are several reports on the regeneration of sunflower from several explant types (e.g., Binding et al. 1981; Greco et al. 1984; Bohorova et al. 1986; Power 1987; McCann et al. 1988; Witrzens et al. 1988; Espinasse and Lay 1989; Schmitz and Schnabl

[1] MOGEN Int. nv, Einsteinweg 97, 2333 CB Leiden, The Netherlands

Biotechnology in Agriculture and Forestry, Vol. 22
Plant Protoplasts and Genetic Engineering III (ed. by Y.P.S. Bajaj)

1989). The availability of sunflower lines that have the potential to regenerate from embryogenic callus has led to the production of the first transgenic plants (Everett et al. 1987). All described methods make use of a callus phase to regenerate shoots and most of the regeneration protocols take at least several months.

2 Methodology and Experimental Results

Full experimental details for transformation of apical meristem tissue of sunflower via *Agrobacterium* were published earlier (Schrammeijer et al. 1990); these will only be summarized in this section.

The following tissue culture media were used:

M1: MS medium (Murashige and Skoog 1962) with 10 g/l sucrose and solidified with 8 g/l Daichin agar.
M2: M1 medium supplemented with 0.1 mg/l benzyladenine
M3: M2 medium supplemented with 500 mg/l carbenicillin.
M4: M3 medium supplemented with 25 mg/l kanamycin.
M5: M1 medium containing 0.5 MS salts and supplemented with 125 mg/l carbenicillin.

For transformation we made use of different *Agrobacterium tumefaciens* strains, namely the Ach5 strain LBA4404 (Hoekema et al. 1983), the C58 strain LBA1010 (Koekman et al. 1982), and the hypervirulent helper strain EHA101 (Hood et al. 1986). In these strains the binary vectors pMOG121, pMOG131, pGUS-INT were introduced, carrying chimaeric NOS-NPTII and CaMV 35S-GUS genes for selection and detection of transformed tissue (described in Schrammeijer et al. 1990). Cocultivations were generally done with exponentially growing cells, resuspended in MS medium to a final density of about 10^8 cells/ml.

2.1 Dissection Method

Preliminary screening experiments with diverse tissue explants inoculated with *Agrobacterium* indicated that the most promising tissue for meristem transformation was the apical meristem from dissected embryos. Shoot development occurred rapidly and at high frequencies.

For excision of the shoot apical meristem, a dissection method according to Power (1987) was used with some modifications. After sterilization of the seed (*Helianthus annuus* L. cv. Zebulon), the seedcoat was removed. The seed contained a papery translucent seedcoat and a thin, transparent endosperm layer within. Both layers were carefully peeled away. To dissect the shoot meristem the first cut was made through the cotyledons, parallel to the line where the two cotyledons are attached. If present, the primordial leaves from the tip of the shoot apical meristem became visible, and were carefully cut away. Then tissue containing the root meristem was removed. The last cut was made directly through the center of the shoot apical meristem. This procedure is shown in Fig. 1. With this method shoot

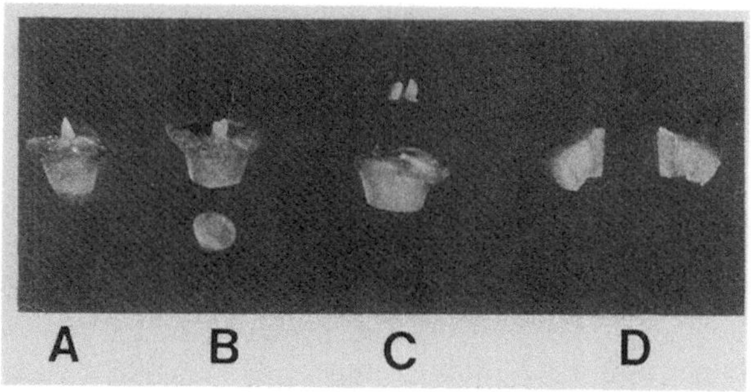

Fig. 1A–D. Dissection of a sunflower shoot apical meristem. **A** The two cotyledons are removed. **B** The root meristem is cut off. **C** The two primordial leaves are cut off. **D** Remaining tissue is sliced in halves through the apical meristem. (Schrammeijer et al. 1990)

development from primary and secondary meristems occurred rapidly and at a high frequency. Cotyledonary tissue around the shoot apical meristem appeared necessary for shoot development. For shoot formation the two explants thus obtained, consisting of cotyledonary tissue and meristematic cells, were cultured on M2 medium. For transformation experiments explants were inoculated with *Agrobacterium*, cocultivated on M2 medium and put on M4 medium for selection.

2.2 Parameters Affecting Transformation and Shoot Development

The influence of various factors on the frequency of meristem transformation and shoot development was analyzed systematically.

The kanamycin sensitivity of the explants was determined after cocultivation with *Agrobacterium* strain EHA101 (pMOG131), and growth on M3 medium supplemented with kanamycin at concentrations between 5 and 100 mg/l. For selection of transformed tissue a kanamycin concentration of 25 mg/l was shown to be optimal. Higher concentrations started to show inhibition of plant growth and bleaching of plant tissue for all the explants, while few inhibitory effects were observed at lower kanamycin concentrations.

The effect of the balance and concentrations of various phytohormones on meristem tissue was optimized for shoot development. Meristem tissue was cocultivated with *Agrobacterium* strain EHA101 (pMOG131). Phytohormones were added to M4 medium as summarized in Table 1. The presence of gibberellic acid (GA_3) in the medium resulted in a strong elongation of cotyledonary tissue. GA_3 induced the development of secondary shoots, but the shoots were thin, pale

Table 1. Hormone additions to MS medium for sunflower

	Phytohormones (mg/l)			
Medium	ZR	IAA	GA_3	BA
A	1.0	0.1	0.1	–
B	1.0	0.1	–	–
C	–	–	–	0.1

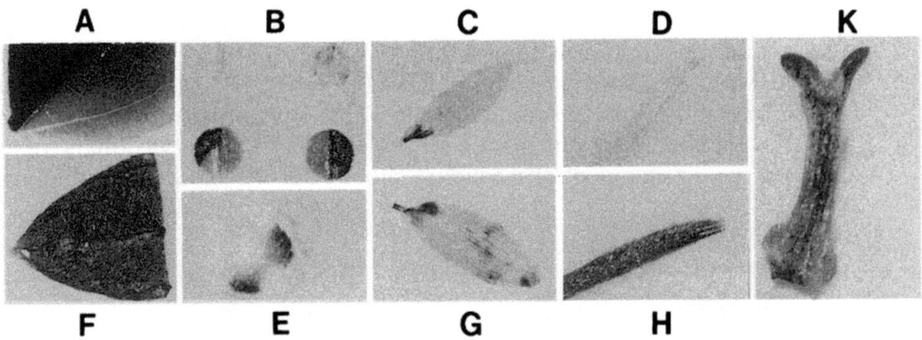
A
B
C
D
K
F
E
G
H

(131-A and 131-B) that were able to grow on selective medium. Shoot 131-A exhibited a normal growth pattern while shoot 131-B had an abnormal growth pattern. It consisted of thick and translucent stem and leaf tissue. Both shoots were propagated on M5 medium in vitro. Two propagated shoots (131-A1 and 131-A2) were obtained from the normally growing shoot and ten propagated shoots (131-B1 to 131-B10) from the deviant one. All shoots were placed on rockwool for root production. Three of these shoots (131-A1, 131-B1 and 131-B2) rooted successfully and were transferred to soil. In this stage the translucent property of shoots 131-B1 and 131-B2 was lost. Only the plant originating from shoot 131-A1 set seed, while the plants originating from shoot 131-B1 and 131-B2 showed premature flowering and did not set seed.

2.3 Analysis of Transformed Plants

Plants 131-A1 and 131-B1 were analyzed for GUS- and NPTII-activity. Plant 131-B2 was too small to be analyzed without the risk of losing this plant. The histochemical GUS assay was used to localize the enzyme activity in the analyzed tissues. From both plants samples were taken from different organs. For plant 131-A1 chimeric expression of the GUS gene was observed in leaves, petals, and stem. A total of nine leaf discs taken across the central vein and from different positions of the plant were analyzed. Three of them showed chimeric gene expression, the other six leaf discs did not express the GUS gene at all. Through the entire stem, gene expression was observed only in vascular tissue. The expression was chimeric and highest in the cambium cells. The GUS gene was not expressed in roots. Plant 131-A1 produced 310 seeds, of which 47 fully matured. In the plants developed from these 47 seeds no GUS-activity could be shown. For plant 131-B1 leaves, petals and roots were also analysed for GUS-activity. The leaves and roots showed nonchimeric expression of the GUS gene while for the petals chimeric gene expression was observed. The results are summarized in Fig. 2. With dot blot analysis, NPTII-activity was shown in the GUS-positive areas of leafdiscs of the tested plants. Primer-directed enzymatic amplification was chosen to confirm the presence of the NPTII gene in the plant cells because of the small amounts of genomic DNA available. To prove that the NPTII enzyme activity in plant 131-B1 was not a result of the presence of *Agrobacterium* strain EHA101 (pMOG131) in the plant, two different sets of oligonucleotide primers were synthesized (Fig. 3). The PCR results indicated that the NPTII activity in plant 131-B1 is not a result of the *Agrobacterium* binary vector but of the T-DNA integrated in the plant genome (Schrammeijer et al. 1990).

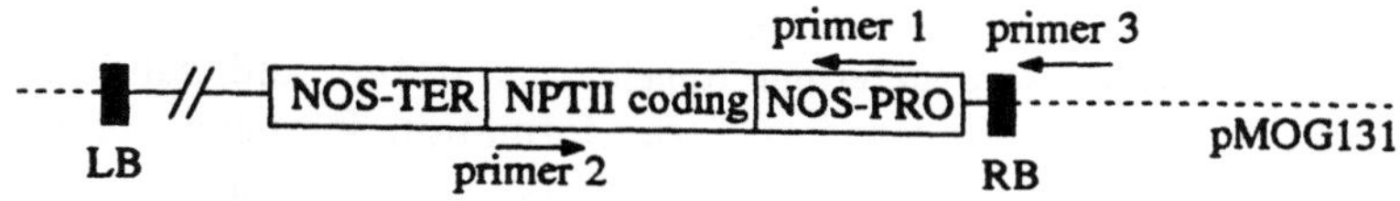

Fig. 3. Segment of the linear map of plasmid pMOG131, in which the position of the PCR-primers *1,2*, and *3* are indicated. (Schrammeijer et al. 1990)

3 Summary and Conclusions

The potential of a transformation system for sunflower via isolated shoot meristems cocultivated with *Agrobacterium* was evaluated. Problems with regeneration from various explants types of sunflower are well documented. Transformation of shoot meristems has the advantage that shoots develop directly from the primary and secondary meristems without an intervening callus phase. The frequency of primary and secondary shoot development using this technique is high and occurs rapidly. However, the large numbers of explants that were used in these studies and the low number of transformed shoots that regenerated from them indicate that meristems are a poor host tissue for *Agrobacterium* infection. Only one regenerated plant was found where we could demonstrate stable integration. In line with the results of Jegla and Sussex (1989) the chimeric expression patter obtained for the GUS gene in the plants 131-A1 and 131-B1 most likely is a result of transformation of only a few cells in the shoot meristem. For plant 131-B1 it concerns here only the petals which showed chimeric expression of the GUS gene. The other analyzed organs of plant 131-B1 showed nonchimeric expression of the GUS gene. No GUS activity could, however, be found in the offspring. The results from Chee et al. (1989) also show chimeric expression after soybean meristem transformation.

A very similar approach was used in attempts to transform pea meristems from imbibed seeds. Over 700 meristem explants were isolated from pea seeds and used for *Agrobacterium* infection. GUS staining at several time points (up to 10 weeks) after cocultivation only yielded one GUS-positive regenerating shoot but no data was obtained for stable integration at DNA level.

It is interesting to note the work of McCabe et al. (1988), who reported the stable transformation of soybean by particle bombardment. With this technique, chimeric shoots were obtained from meristems of immature soybean seeds at a frequency of 2%. However, only for one progeny plant was expression of the NPTII gene observed (McCabe et al. 1988). Also, a very recent and preliminary publication from the laboratories of Pioneer Hi-Bred (Monsan 1991) mentioned an elegant approach that combined the above described techniques; sunflower apical meristems in germinating seeds were used for particle bombardment followed immediately by suspension in an *Agrobacterium* culture. The microprojectiles may be targeted specifically at desired cell layers of the meristem. Subsequently, the efficiency of *Agrobacterium*-mediated transformation is used for gene transfer. As also indicated in this chapter, the subsequent regeneration of the treated meristematic tissue poses no specific problems.

Although at this point the low transformation frequency of the meristematic cells limits the possibilities and general applicability of this procedure, the potent regeneration of meristematic tissue and the recent literature data merits further research. Future improvements on the particle gun technique and a detailed screen with different *Agrobacterium tumefaciens* strains and sunflower cultivars may provide ways to improve the transformation frequency of meristem cells. In conclusion, the usefulness of direct meristem transformation as a routine technique has still to be demonstrated more convincingly.

References

Binding H, Nehls R, Kock R, Finger J, Mordhorst G (1981) Comparative studies on protoplast regeneration in herbaceous species of the Dicotyledoneae class Z Pflanzenphysiol 101:119–130
Bohorova NE, Cocking EC, Power JB (1986) Isolation, culture and callus regeneration of protoplasts of wild and cultivated *Helianthus* species. Plant Cell Rep 5·256–258
Chee PP, Fober KA, Slightom JL (1989) Transformation of soybean (*Glycine max*) by infecting germinating seeds with *Agrobacterium tumefaciens* Plant Physiol 91 1212–1218
Espinasse A, Lay C (1989) Shoot regeneration of callus derived from globular to torpedo embryos from 59 sunflower genotypes Crop Sci 29 201–205
Everett NP, Robinson KEP, Mascarenhas D (1987) Genetic engineering of sunflower (*Helianthus annuus* L) Bio/Technol 5 1201–1204
Gould J, Devey M, Hasegawa O, Ulian UC, Peterson G, Smith RH (1991) Transformation of *Zea mays* L using *Agrobacterium tumefaciens* and the shoot apex Plant Physiol 95 426–434
Greco B, Tanzarella OA, Carrozzo G, Blanco A (1984) Callus induction and shoot regeneration in sunflower (*Helianthus annuus* L) Plant Sci Lett 36.73–77
Hoekema A, Hirsch PR, Hooykaas PJJ, Schilperoort RA (1983) A binary plasmid vector strategy based on separation of vir- and T-regions of the *Agrobacterium tumefaciens* Ti-plasmid. Nature 303 179–180
Hood EE, Helmer GL, Fraley RT, Chilton MD (1986) The hypervirulence of *Agrobacterium tumefaciens* A282 is encoded in a region of pTiBo542 outside of T-DNA J Bacteriol 168 1291–1301
Jefferson RA, Kavanagh TA, Bevan MW (1987) GUS fusions β-glucuronidase as a sensitive and versatile gene fusion marker in higher plants EMBO J 6.3901–3907
Jegla DE, Sussex IM (1989) Cell lineage patterns in the shoot meristem of the sunflower embryo in the dry seed Dev Biol 131 215–225
Koekman BP, Hooijkaas PJJ, Schilperoort RA (1982) A functional map of the replication region of the octopine Ti plasmid. Plasmid 7.119–132
McCabe DE, Swain FS, Martinell BJ, Christou P (1988) Stable transformation of soybean (*Glycine max*) by particle acceleration Bio/Technol 6 923–926
McCann AW, Cooley G, Van Dreser J (1988) A system for routine plantlet regeneration of sunflower (*Helianthus annuus* L) from immature embryo-derived callus Plant Cell Tissue Organ Cult 14 103–110
Monsan P (1991) Pioneer Hi-Bred halves the time for achieving stable sunflower transformation Biotechnol News 11(6).3
Murashige T, Skoog F (1962) A revised medium for rapid growth and bioassays with tobacco tissue cultures Physiol Plant 15 473–497
Power CJ (1987) Organogenesis from *Helianthus annuus* inbreds and hybrids from the cotyledons of zygotic embryos Am J Bot 74 497–503
Schmitz P, Schnabl H (1989) Regeneration and evacuolation of protoplasts from mesophyll, hypocotyl and petioles from *Helianthus annuus* L J Plant Physiol 135 223–227
Schrammeijer B, Sijmons PC, van den Elzen PJM, Hoekema A (1990) Meristem transformation of sunflower via *Agrobacterium* Plant Cell Rep 9 55–60
Ulian E, Smith R, Gould J, McKnight T (1988) Transformation of plants via the shoot apex In Vitro Cell Dev Biol 24 951–954
Vancanneyt G, Schmidt R, O'Connor-Sanchez A, Willmitzer L, Rocha-Sosa M (1990) Construction of an intron-containing marker gene splicing of the intron in transgenic plants and its use in monitoring early events in *Agrobacterium*-mediated plant transformation Mol Gen Genet 220.245–250
Witrzens B, Scowcroft WR, Downes RW, Larkin PJ (1988) Tissue culture and plant regeneration from sunflower (*Helianthus annuus*) and interspecific hybrids (*H. tuberosus* × *H annuus*). Plant Cell Tissue Organ Cult 13 61–76

II.10 Transformation in *Hordeum vulgare* L. (Barley)

Hideyoshi Toyoda[1]

1 Introduction

Genetic transformation of graminaceous plants has been mainly conducted by directly introducing foreign genes into their protoplasts by means of electroporation of polyethylene glycol treatment. In these methods, plant regeneration from protoplasts is a prerequisite for producing transgenic plants. However, the inability of protoplasts of many monocotyledonous plants to regenerate into plants makes it difficult to generalize these methods in a broad range of monocotyledonous species. Therefore, the substitutive method such as a microprojectile bombardment has been developed and applied to direct delivery of foreign genes into monocotyledonous plant materials. Recently the author reported the transient expression of foreign gene introduced into barley coleoptile cells by microinjection and discussed the possible application of coleoptile tissues to a production of transgenic barley (Toyoda et al. 1990; 1992). From this point of view, this chapter describes (1) callus induction and plant regeneration from coleoptile tissues in order to complete a coleoptile system of barley, (2) verification of expression of foreign genes introduced into coleoptile cells and (3) possibility of production of disease-resistant barley. The progress of these investigations would make it possible to produce transgenic plants in a broad range of monocotyledonous species in which protoplast regeneration is difficult.

2 Culture of Barley Coleoptile Tissues

Seeds of barley (*Hordeum vulgare* L. cv. Kobinkatagi) were surface sterilized and germinated at 20 °C on water-moistened filter paper in a Petri dish. Germinating seedlings were transferred to agar-solidified plates supplied with sterile chemical fertilizer and incubated at 20 °C under a continuous illumination of 5000 lx. After 10 days of incubation, coleoptiles were excised from primary leaves and transferred to Murashige-Skoog (1962) (MS) medium supplemented with various concentrations of growth regulators. Callus induction from coleoptile tissue was observed in a broad range of 2,4-D concentrations (1–10 μg/ml), with no relation to concentrations of kinetin or benzylaminopurin combined, and even in the absence of

[1] Faculty of Agriculture, Kinki University, Nakamachi 3327-204, Nara 631, Japan

Biotechnology in Agriculture and Forestry, Vol. 22
Plant Protoplasts and Genetic Engineering III (ed. by Y.P.S. Bajaj)

cytokinin. However, callus tissues were not induced when coleoptile explants were cultured with the medium in which 2,4-D was replaced with NAA. Jelaska et al. (1984) and Rengel and Jelaska (1986) examined regeneration of callus tissues derived from mesocotyl and apical meristem of barley and reported that a transfer of 2,4-D-induced callus tissues to a regeneration medium containing 2,3,5-triiodobenzoic acid was necessary to initiate the shoot formation. In the present study, however, the transfer of callus tissues to a regeneration medium was not an essential step. Induced callus tissues were cut into small pieces, transferred to the same fresh medium, and subcultured till some morphological changes were observed. Greening spots were frequently formed when callus tissues were subcultured for three or four passages with 3 μg/ml 2,4-D and 0.1 μg/ml kinetin. Under this condition, two or three greening spots were formed in 80–85% of callus tissues excised. Greening spots differentiated shoots after incubation of further 10–15 days. A regenerative capability of coleoptile-derived callus tissues was stably maintained during successive subcultures for at least 3 months. Root initiation was observed by transferring leaf-developing shoots to growth regulator-free MS medium. Regenerated plants were transplanted to soil and acclimated in a moist chamber for a few days. These regenerants formed ears under the field condition, and produced viable seeds 2 months after planting.

The present study demonstrated that a combination of 2,4-D and kinetin is effective for plant regeneration of coleoptile-derived callus tissue of barley. In addition, the author evaluated the effect of these growth regulators on callus induction and plant regeneration from explants obtained from different tissues. With the present combination of both growth regulators, callus tissues were induced similarly from both mature embryo and germinating seeds, and regenerated into shoots (Toyoda et al. 1990b).

3 Transient Expression of Foreign Genes Introduced into Barley Coleoptile Cells by Microinjection

In a previous study, the author demonstrated that a microinjection was applicable to various plant materials (Toyoda et al. 1985, 1986, 1988, 1989) including plant pathogenic fungi (Toyoda et al. 1987; Matsuda et al. 1989a). In barley seedlings, basal parts of primary leaves are covered with coleoptile tissues which consist of two separable cell layers and can be observed microscopically without injury (Bushnell et al. 1967). In this study, barley coleoptile cells were used as a material for microinjection of foreign genes. The transient expression of introduced genes is described as a primary verification of transgenic barley.

3.1 System and Optimal Condition for Microinjection

Sterile seedlings of barley (cv. Kobinkatagi) were obtained by the method mentioned above. The inner epidermis was separated from the outer epidermis of coleoptile and laid on the injection plate (Fig. 1). Hormone-free MS medium

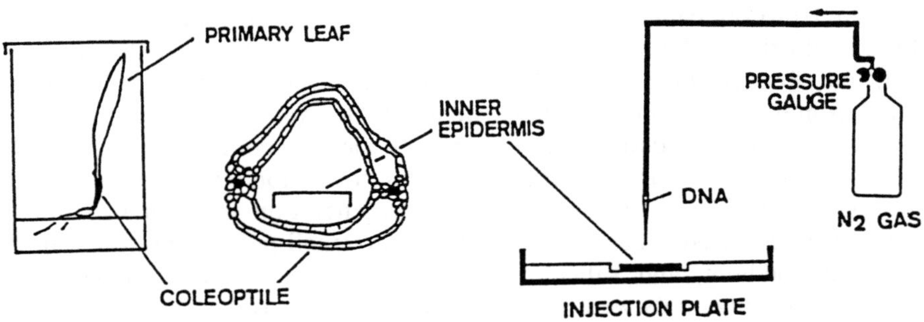

Fig. 1. Microinjection of barley coleoptile cells. Germinating seedling was grown on an agar-solidified plate supplied with fertilizer. Coleoptile tissue was excised and the inner epidermis was transferred to the agar injection plate. Foreign DNA (mRNA) was injected into the cytoplasm of coleoptile cells by constant pressure on silicon oil in a tube linked with a microneedle. The procedures were carried out under a stream of sterile air. (Toyoda et al. 1990a)

(solidified with 2% agar) was used for the injection plate. Microinjection of coleoptile cells was carried out using an Olympus injectoscope (Toyoda et al. 1988).

The optimal conditions for microinjection were defined using the silicon oil injection method described previously (Toyoda et al. 1988). Major factors for microinjection were (1) tip diameter of needles, (2) insertion time, and (3) injection pressure (Toyoda et al. 1985, 1988). In the present study, duration of needle insertion was examined under fixed conditions of tip diameter (0.2 μm) and pressure (2 kg/cm^2). Silicon oil was useful for this purpose because the injected oil formed spherical droplets within the cells and allowed visualization of a successful microinjection. The droplet diameters enlarged with an increase in insertion times. In the present injection, a 5-s insertion formed a 20-μm diameter droplet and allowed coleoptile cells to survive. Longer insertion times caused an immediate cessation of cytoplasmic streaming. These results indicate that cell survival can be determined easily by observing cytoplasmic streaming after injection. Failure of needle insertion resulted in leakage of cell contents and cessation of cytoplasmic streaming. All of the successfully injected cells maintained cytoplasmic streaming stably during the culture (at least 1 week) when diameters of droplets were less than 20 μm. In the following experiments, therefore, a 5-s insertion was used to introduce plasmid DNA into barley coleoptile cells. According to the method described in previous papers (Toyoda et al. 1987, 1988), the amounts of introduced DNA were calculated from sphere volumes of silicon oil injected under the same condition. The volume of the DNA solution (1000 μg/ml) was approximately 4×10^6 μl and about 4 pg of DNA could be injected into the cells.

3.2 Gene Introduction and Detection of Gene Expression

The strategy for introduction of foreign genes (or sense mRNAs) and detection of transcription and translation products was illustrated in Fig. 2. In the present strategy, the genes were directly injected into coleoptile epidermal cells and a RNA

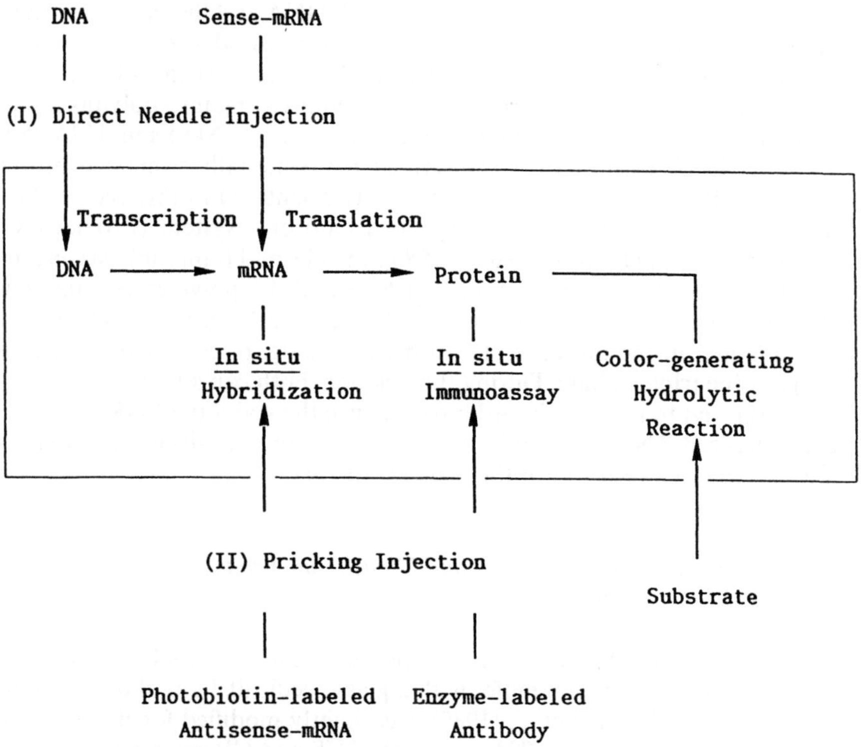

Fig. 2. Strategy for gene transfer and detection of transcription and translation of the introduced gene by dual microinjection. (unpubl.)

probe or labeled antibody was introduced into gene-injected cells by a pricking injection method as previously described by Yamamoto et al. (1982).

3.2.1 Direct Needle Injection

An autoclaved glass needle including injection material was inserted into the cytoplasm of coleoptile cells and the material was forced out of the needle by pressure on silicon oil in a tube connected to the needle. Injection pressure was generated by nitrogen gas and controlled by a pressure gauge (Fig. 1). Under defined conditions, coleoptiles were injected with plasmid DNA or sense mRNA (1000 µg/ml). Injected coleoptiles were incubated at 20 °C, and cell survival was examined soon after injection and before fixation or homogenization. Survivability of cells was determined by examining the maintenance of cytoplasmic streaming in the injected cells.

The plasmids pBI221 and pBI221-CP used for injection contain β-glucuronidase (GUS) gene (Jefferson 1987) and coat protein (CP) gene of tobacco mosaic virus (Toyoda et al. 1992), which are linked with the Cauliflower Mosaic Virus 35S promoter and the nopaline synthase polyadenylation region. The plasmid vector, pB1221, does not contain border regions of T-DNA and therefore has been used to examine transient expression of the gene introduced directly into plant cells (Jefferson 1987). Sense and antisense mRNAs were artificially synthesized by excising the genes from the original vector and inserting into polylinker sites of the plasmid vector pSP64-poly(A) (Promega Corporation, Madison, USA). Sense and antisense RNA strands were transcribed bidirectionally from two promoter sequences (SP6 and T7 promoters) oppositely orientated in this vector. An in vitro transcription was conducted using the Riboprobe System (Promega Corporationson) by the method of Melton (1984); the plasmid (1 mg/ml) was enzymatically linealized and mixed with 0.8 unit/μl SP6 and T7 polymerases and nucleotide mixture (5 mM of ATP, CTP, and UTP, and 0.5 mM of GTP). The transcription was initiated in the presence of 5 mM cap analog, p^1-5′(7-methyl)-guanosine-p^3-5′guanosine triphosphate. During the reaction, both cap structure and poly(A)-tail were attached to mRNAs in order to enhance the stability of mRNA. The validity of synthesized RNAs was confirmed by a northern hybridization analysis with the labeled genes clipped out of the original vectors.

3.2.2 Detection of Transcription and Translation Products by the Pricking Injection Method

For in situ hybridization of transcripts, gene-injected epidermis was fixed and pricked in a hybridization buffer with a glass needle (0.7 μm of outer tip diameter). The method of Raikhel et al. (1989) was slightly modified for fixation and in situ hybridization. Antisense mRNAs of both GUS and CP genes were used as probes for detection of transcription of the corresponding genes, since it has been well recognized than an in situ hybridization of transcripts could be successfully achieved by the use of antisense mRNA strand (Hoefler et al. 1986; Childs et al. 1987). The RNA strands were conjugated with photobiotin according to the method of Forster et al. (1985) and labeled in vivo with avidin-mediated biotinylated enzyme (alkalin phosphate). The hybrid formation was visualized at the site of hybridization by color-generating reaction due to hydrolysis of the sub-

Table 1. Detection of transcription and translation of foreign genes microinjected into barley coleoptile epidermal cells

Materials injected	Cells with Positive reactions[a]/successfully injected cells[b]		
	In situ hybridization with antisense mRNAS	In situ immunoassay with antibodies	Hydrolysis of substrate (X-gluc)
P_{35s}/GUS/T_{nos}	98/132 (74.2)[c]	69/106 (65.1)	98/144 (68.1)
P_{35s}/TMV-CP/T_{nos}	74/105 (70.5)	73/121 (60.3)	–
Cap/GUS-mRNA/PolyA	–	64/111 (57.7)	91/136 (66.9)
Cap/CP-mRNA/PolyA	–	86/129 (66.7)	–

[a] The reactions were examined 24 h after injection.
[b] Cells maintaining cytoplasmic streaming after injection.
[c] Percentage. (unpubl.)

strates (nitro blue tetrazolium and 5-bromo-4-chloro-3-indorylphosphate) by the labeling enzyme. An introduction of antibodies was conducted by the method described previously (Toyoda et al. 1985); the fixed epidermis was rinsed repeatedly with 0.1 M phosphate buffer (pH 7.0) and pricked in a phosphate buffer containing a primary rabbit anti-GUS-antibody or anti-TMV-antibody. After incubation and washing, the specimen was treated with anti-rabbit IgG-antibody labeled with alkaline phosphatase. After 1 h of incubation, the sample was completely washed with a buffer in order to eliminate excess antibody and reacted with the substrates of this enzyme. The results obtained are summarized in Table 1. The data demonstrated that the introduced foreign genes are functionally expressed in barley coleoptile cells and the rates of expression are reproducible with no relation to the detection methods employed, or to the genes introduced.

3.2.3 Histochemical Assay for β-Glucuronidase Activity

In addition to in situ hybridization and in situ immunoassay, transient expression of the reporter gene (GUS gene) was assayed histochemically or fluorogenically. The enzymatic assay was carried out according to the methods described by Jefferson (1987). For histochemical assay of β-glucuronidase production, injected coleoptiles were collected 24 h after incubation, fixed with formamide, and reacted with 5-bromo-4-chloro-3-indolyl-glucuronide (X-Gluc), the substrate of β-glucuronidase. The data indicated that the successfully injected cells gave a positive reaction in the enzymatic assay and the detection rate was comparable to those of in situ hybridization and in situ immunoassay.

Changes in the level of gene expression were examined fluorogenically at the various periods (6–48 h) after injection. Coleoptiles were homogenized in an extraction buffer and the clarified homogenates were mixed with a buffer containing 4-methyl unbelliferyl glucuronide. Enzymatic activities were determined relative to the intensity of authentic 4-methylumbelliferone. As a result, the activity of β-glucuronidase first appeared after 6 h and reached the maximum level 24 h after injection.

4 Assessment of Transformation Strategy with Respect to Disease Resistance

The establishment of infection by the powdery mildew fungus in host plant is dependent on the successful formation of the primary haustorium, the organ taking nutrients from host plants, in epidermal cells. Subsequently, the pathogen produces secondary hyphae to attempt secondary infection in neighboring epidermal cells (Fig. 3). Effective protection of plants from the powdery mildew disease, therefore, would be to inhibit the formation of primary haustoria or to destroy the haustoria before the pathogen develops secondary hyphae. Verifying this assumption will provide support for genetically manipulating crop plants to impart disease resistance.

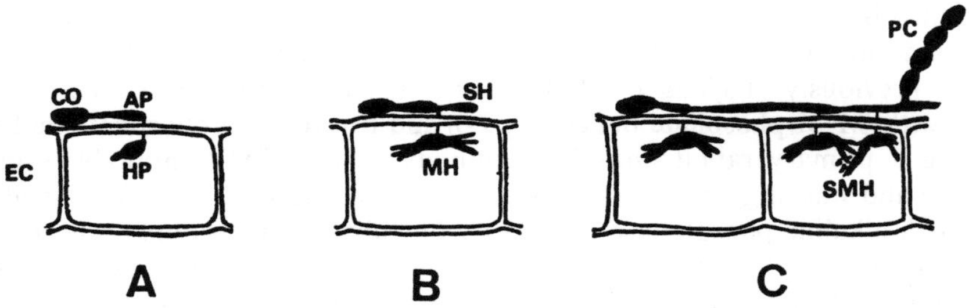

Fig. 3A–C. Life cycle of powdery mildew fungus (*Erysiphe graminis* f. sp. *hordei*) on barley coleoptile tissue. **A** Conidium (*CO*) of this pathogen produces appressorium (*AP*) and develops haustorial primordium (*HP*) in coleoptile epidermal cells (*EC*) 16 h after inoculation. **B** After haustorium (*MH*) matures (20 h after inoculation), the pathogen initiates elongation of secondary hypha (*SH*) and attempts secondary infection at neighboring epidermal cells. **C** After maturation of secondary haustoria (*SMH*), numerous progeny conidia (*PC*) are produced 4–6 days after inoculation. (Toyoda et al. 1991a)

It has been known that mycelial cell walls of some fungi consist of chitin or chitin-related compounds such as chitosan (Davis and Bartnicki-Gracia 1984; Fukamizo et al. 1990), and that the powdery mildew pathogen of barley also contains chitin in infection structures such as appressoria and haustoria (Shiraishi et al. 1976). These results suggest that the enzymatic digestion of fungal haustorium would be a promising strategy for protection from powdery mildew disease without causing any damage in host plants, because it is known that higher plants contain no chitin (Boller 1987).

Powdery mildew of barley invades both the leaf and coleoptile of the host plant and produces numerous progeny conidiospores. Infection can also be achieved in detached coleoptile tissue. This has made it possible to continuously analyze the infection behaviour of the pathogen and host cell response, because the tissue can be observed microscopically without injury to either host or parasite (Bushnell et al. 1967). As discussed earlier, these characteristics of coleoptile tissue are also profitable for microinjection and allowed us to directly introduce foreign materials into barley coleoptile cells (Toyoda et al. 1990a), or into the pathogen inoculated into coleoptile tissue (Toyoda et al. 1987). The direct needle injection technique is a reliable tool for directly introducing foreign materials into specified target cells without causing detrimental effects (Toyoda et al. 1985, 1987, 1988; Potrykus 1990) and therefore would be applicable to an analysis of the effects of materials introduced into epidermal cells on the haustorial growth of this pathogen. In this section, therefore, the direct microinjection of chitinase into barley coleoptile epidermal cells in which haustoria of the pathogen had been formed, in order to assess the suppressive effect of the enzyme on the pathogen development (Toyoda et al. 1991) is described.

4.1 Inoculation with the Pathogen, and Microinjection with Exogenous Chitinase

The inner coleoptile epidermis was separated from primary leaves of non--inoculated seedlings of barley (cv. Kobinkatagi, susceptible to race I), put on a solidified agar of the injection plate, and inoculated with conidiospores of the powdery mildew pathogen, *Erysiphe graminis* f. sp. *hordei* race I. After 16–24 h of incubation (20 °C, 4000 lx), the coleoptile was used for injection.

Microinjection of inner epidermis inoculated with conidiospores was conducted by the direct injection method mentioned above. Epidermal cells in which haustorial primordia had been formed were selected as target cells for injection, and the material [silicon oil, 0.05 M phosphate buffer (pH 6.0), and 0.1% chitinase or fraction V albumin bovine (Sigma)] was introduced. Injected coleoptile tissues were incubated at 20 °C and the growth of haustoria and the elongation of secondary hyphae were examined at various periods after injection.

4.2 Evaluation of Inhibitory Effect of Exogenous Chitinase

Prior to actual injection of chitinase, the effectiveness of the present microinjection protocol was evaluated by injecting silicon oil or buffer solution into haustorium-harboring epidermal cells. As pointed out in previous papers (Toyoda et al. 1987, 1988, 1990a), silicon oil forms a spherical droplet in injected cells. The droplet formation of injected silicon oil was useful not only to visualize the successful microinjection, but also to analyze the effect of injected material on the fungal development. Figure 4A shows a silicon oil-injected epidermal cell harboring a

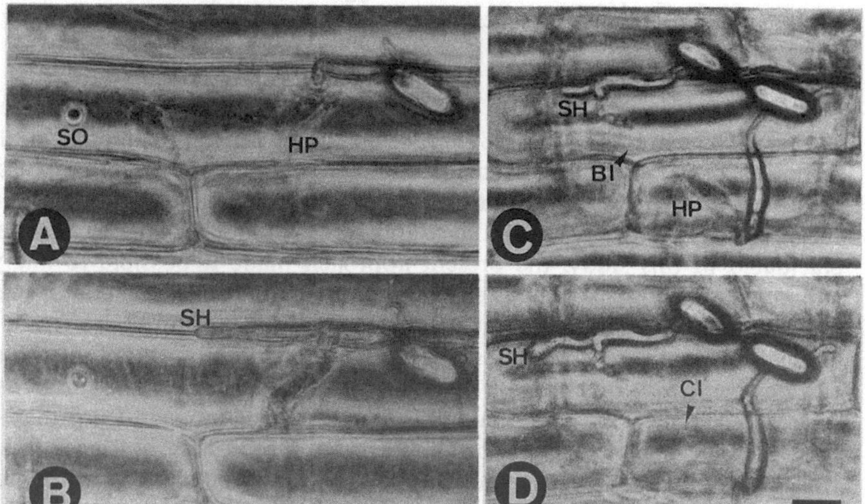

Fig. 4A–D. Successful elongation of secondary hypha of powdery mildew fungus after silicon oil-injection into a barley coleoptile epidermal cell harboring haustorial primordium (**A–B**) and suppressive effect of chitinase injected into a barley coleoptile epidermal cell on the development of the powdery mildew fungus (**C–D**). **A** Silicon oil (*SO*) was injected into the epidermal cell in which haustorial primordium (*HP*) had been formed (16h after inoculation). **B** The injection did not disturb either the maturation of haustorium or the subsequent elongation of secondary hypha (*SH*) (10 h after injection). **C–D** A buffer (*BI*) or 0.1% chitinase (*CI*) was injected into epidermal cell in which haustorial primordium had been formed (16 h after inoculation with the pathogen), and the effect of the injection was examined 6 h (**C**) and 8 h (**D**) after injection. Note that the chitinase injection completely digested the haustorial primordium and suppressed the elongation of secondary hypha, in spite of successful elongation of SH in buffer-injected cell, 8 h after injection. *Bar* 20 µm. (Toyoda et al. 1991a)

haustorial primordium. As clearly shown in Fig. 4B, the haustorium successfully matured and elongated secondary hyphae after silicon oil was injected. There was no difference in the length of elongated secondary hyphae between silicon oil-injected and non-injected epidermal cells. Similar results were also obtained when a buffer solution or albumin were injected into epidermal cells.

Under the present condition for microinjection, more than 90% of chitinase-injected epidermal cells continued to survive throughout the experimental period of 3 days. Thus, the present study revealed that microinjection did not affect fungal development or survival of the host cell. Therefore, the suppressive effect caused by chitinase injection could be considered to be due to the enzymatic action of introduced chitinase. The positive result was demonstrated in Fig. 4C and D. In this experiment, chitinase was injected into epidermal cells when the primordia were formed (16–17 h after inoculation). In epidermal cells injected with chitinase, approximately 70% of the primordia became obscure 4 h after injection and the subsequent formation of secondary hyphae was completely inhibited. On the contrary, a buffer or albumin injection did not cause any suppressive effect on either the haustorial growth or the hyphal elongation. Chitinase was also injected into epidermal cells at the stage when the haustoria had matured (24 h after inoculation). In this microinjection, the introduction of chitinase did not cause the detectable morphological change in haustoria, but notably suppressed the elongation of secondary hyphae. These results suggest that the sensitivity of haustoria to chitinase varied among the stages of haustorial growth, probably depending on the cell wall constitution.

In several plants, chitinase activity increases after inoculation with fungal, bacterial, and viral pathogens, and after treatment with ethylene and elicitors (Boller 1987). Swegle et al. (1989) reported the presence of chitinase in aleurone layers of barley during seed maturation and germination, and discussed the positive role of these enzymes on the defense reaction against soil fungi. Although the production of intrinsic chitinase in powdery-mildewed barley has not been elucidated in this study, the present approach made it possible to analyze the suppressive effect of exogenous chitinase on the pathogen at the real infection site of host plant cells. The data clearly indicated that the haustorial growth or hyphal elongation could be suppressed by the introduction of chitinase, though the suppressive effect of chitinase varied with the growth stages of haustorium attacked. The most effective timing for chitinase injection is at the initial stage of haustorial development, and this implies the effectiveness of transformation of barley with a chitinase gene whose expression could be induced at this infection stage.

5 Promoter Sequences of Barley Responsible for Infection with Powdery Mildew Pathogen

5.1 Isolation of Promoter Sequences by Differential Screening

Chromosomal DNA was extracted from primary leaves of barley and digested with restriction endonuclease *Sau* 3AI. Digested DNA was fractionated by sucrose density-gradient centrifugation, and the fragments (9–12 kb) obtained were linked with *Bam* HI-arms of λ EMBL3 phage vector and in vitro-pakaged into phage particles. The genomic phage clones were transferred to nitrocellulose membrane for plaque hybridization. Total mRNAs were extracted from both inoculated (16 h after inoculation) and noninoculated leaves, purified by oligo-dT membrane, and used as templates for cDNA synthesis. The synthesis of cDNA was carried out by the standard method of Gubler and Hoffman (1983) and labeled with horseradish peroxidase (ECL gene detection system; Amersham International, Buckinghamshire, England) by the method described previously (Toyoda et al. 1991b). By a differential plaque hybridization with cDNAs obtained from inoculated and noninoculated leaves, several phage plaques were isolated as candidates for the genomic clone carrying a certain gene whose transcription could be specifically induced by infection with the powdery mildew pathogen. To further elucidate the timing of transcription of these genes, a Northern dot blot analysis was conducted using total mRNAs obtained at various stages after inoculation (Figs. 5, 6). Consequently, the author selected the clone (BGI-08) showing the highest transcription at the stage of primordium formation (16 h after inoculation). These results imply that transcription of this gene is initiated as a result of function of the regulatory sequences (including a promoter) responsible to the infection. In order to isolate a promoter sequence from the BGI-08 fragment, the insert DNA (approximately 12 kb in length) was excised from a EMBL3 vector and subdivided into 1–2-kb fragments by which a 35S promoter of pB221 was replaced. Barley protoplasts were prepared from primary leaves and electroporated with fragment-containing

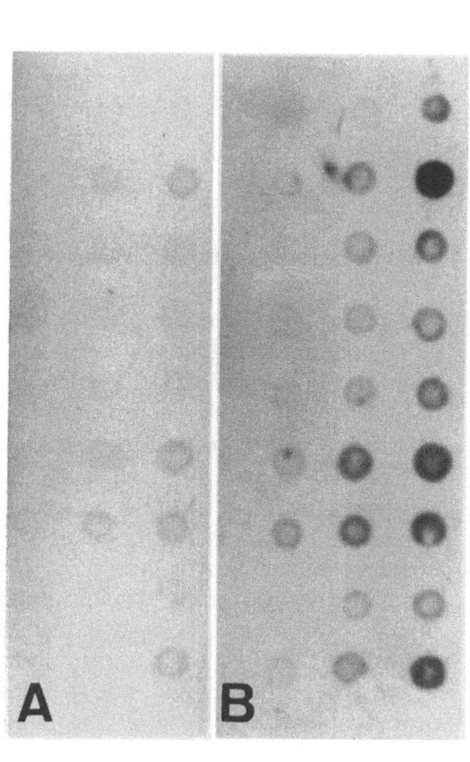

Fig. 5A, B. Dot blot analysis of differentially screened genomic clones of barley. Definite DNAs (0.25, 0.5. and 1 μg, *from left to right*) of each clone were dotted horizontally and hybridized with labeled total mRNAs extracted from non-inoculated (**A**) and inoculated leaves (**B**) of barley. Total mRNAs of inoculated leaves were extracted 16 h after inoculation with the powdery mildew pathogen (the stage of primordium formation). The second horizontal column from top represents the clone (BGI-08) showing the highest transcription at this stage of infection. (unpubl.)

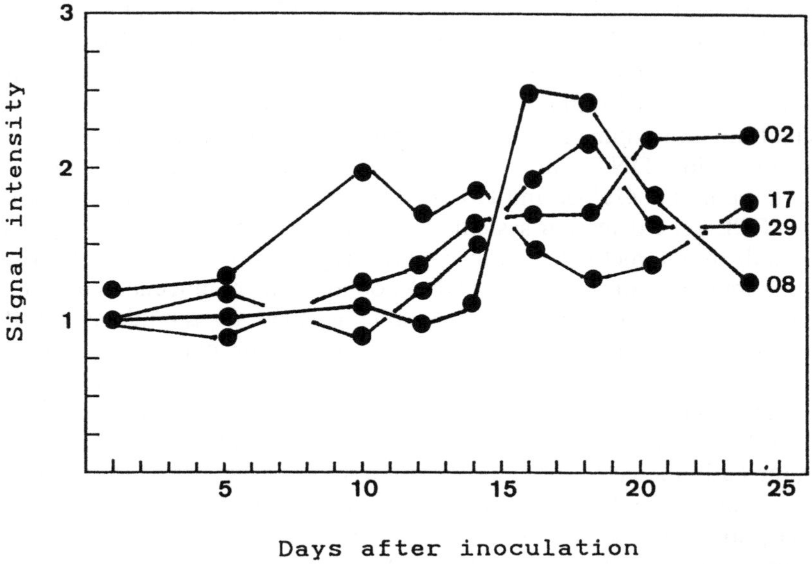

Fig. 6. Changes of signal intensity of Northern dot blot hybridization. Total mRNAs were extracted from primary leaves of barley at the various periods after inoculation and hybridized with labeled-DNAs of genomic clones. The signal intensity of hybridization was densitometrically determined relative to the intensity in mRNAs of non-inoculated leaves. *Numbers* in the figure represent the genomic clones isolated. (unpubl.)

pBI221 plasmids using and electroporation apparatus as described in our previous paper (Matsuda et al. 1989b). A promoter activity of the fragment was determined by fluorogenically examining transient expression of GUS gene in electroporated protoplasts. Finally, the positive fragment (1.2 kb) was obtained and designated BGI-0812.

5.2 Infection-Responsible Promoter Sequence: Assessment by Microinjection

The present protocols for a molecular cloning enabled us to isolate the genomic clone containing a gene expressed at the early stage of infection and an upstream promoter sequence (BIG-0812). However, it was not conclusive whether the BIG-0812 fragment could be responsible to the fungal infection. For verifying this problem, a microinjection technique was applied to detecting transcription initiation of GUS gene at the infected cells. In this experiment, all of coleoptile epidermal cells were injected with the plasmid pBI221 containing a chimeric GUS gene flanked to the BIG-0812 fragment and then inoculated with conidiospores of the powdery mildew fungus. After 16 h of inoculation, the GUS-mRNA was detected by in situ hybridization with photobiotin-labeled antisense mRNA probe mentioned above. Figure 7 clearly indicated that the GUS gene was transcribed only in the cells invaded by the pathogen, but neither in noninfected cells nor in the cells where the pathogen had not invaded. These results strongly suggest that the promoter sequence (BIG-0812) isolated in this study could respond to the powdery

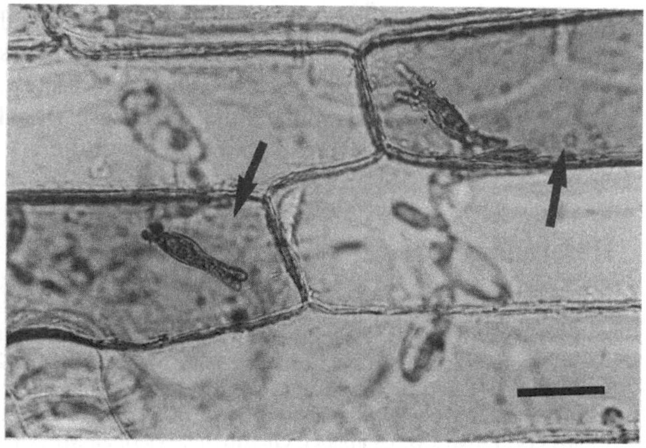

Fig. 7. In situ hybridization detection of transcription of a chimeric GUS gene flanked to the promoter sequence (BGI-0812) microinjected into barley coleoptile epidermal cells. All of coleoptile epidermal cells were injected with the gene and inoculated with the powdery mildew fungus. After 16 h inoculation, the epidermis was fixed and hybridized with photobiotin-labeled antisense GUS-mRNA. Note the positive transcription was induced only in the epidermal cells in which haustorial primordia were formed (*arrows*), but not in non-infected cells or in the cells where the pathogens had not invaded. *Bar* 10 μm. (unpubl.).

mildew infection to initiate transcription of a gene located downstream. An analysis of construction of the promoter is under way.

6 Conclusion and Prospects

In barley plants, various diseases caused by viral, bacterial, and fungal pathogens have been reported and have caused serious problems of yield depression. A powdery mildew disease caused by the fungal pathogen, *E. graminis* f. sp. *hordei* is one of the important diseases in this plant, and the biochemical and morphological aspects of host disease responses have been widely investigated (Goodman et al. 1986). Nevertheless, the molecular basis for the disease is not necessarily efficient enough to analyze genes involved in host resistance. Limited progress of these molecular analyses has been partly due to the technical limitation of gene manipulation in barley plants.

Recently, some workers have applied the direct gene-transfer methods such as microprojectile bombardment and microinjection to delivery of foreign genes to barley, and demonstrated the successful transient expression of introduced genes. Especially, the use of coleoptile tissues enabled us to substantially evaluate the inhibitory effects of the injected material (exogenous chitinase) on growth of the powdery mildew pathogen and to provide an experimental basis for transformation of barley with a chitinase gene. Moreover, the infection-responsible promoter sequence isolated in this study makes it more effective to suppress the pathogen at the infection site and at the early stage of infection. More recently, the author succeeded in isolating a gene encoding endochitinase from *Streptomyces* sp. and determined total DNA sequences of the gene. A success of callus induction and plant regeneration from barley coleoptile tissues will realize a production of disease-resistant transgenic plants, which are transformed with a chitinase gene controlled by the infection-responsible promoter.

References

Boller T (1987) Hydrolytic enzymes in plant disease resistance. In: Kosuge T, Nester EW (eds) Plant-microbe interactions: Molecular and genetic perspectives, vol. 2. Macmillan, New York, pp 385–413

Bushnell WR, Dueck J. Rowell JB (1967) Living haustoria and hyphae of *Erysiphe graminis* f. sp. *hordei* with intact and partly dissected host cells of *Hordeum vulgare*. Can J Bot 45:1179–1732

Childs GV, Lloyd JM, Unabia G, Gharib SD, Wierman ME, Chin WW (1987) Detection of leutinizing hormone β messenger ribonucleic acid (RNA) in individual gonadotropes after castration: use of a new in situ hybridization method with a photobiotinylated complementary RNA probe. Mol Endocrinol 1:926–932

Davis LL, Bartnicki-Garcia S (1984) Chitosan synthesis by the tandem action of chitin synthetase and chitin deacetylase from *Mucor rouxii*. Biochemistry 23:1065–1073

Forster AC, McInnes JL, Skingle DC, Symons RH (1985) Non-radioactive hybridization probes prepared by the chemical labelling of DNA and RNA with a novel reagent, photobiotin. Nucelic Acid Res 13:745–761

Fukamizo T, Sonoda K, Toyoda H, Ouchi S, Goto S (1990) Solid state ^{13}C-NMR analysis of cell wall components of *Fusarium oxysporum*. Agric Biol Chem 54:2761–2762

Goodman RN, Kiraly Z, Wood KR (1986) The biochemistry and physiology of plant disease. Univ Missouri Press, Columbia, pp 433

Gubler U, Hoffman BJ (1983) A simple and very efficient method for generating cDNA libraries. Gene 25:263–269

Hoefler H, Childers H, Montminy MR, Lechan RM, Goodman RH, Wolfe HJ (1986) In situ hybridization methods for the detection of somatostatin mRNA in tissue sections using antisense RNA probes. Histochemical J 18:597–604

Jefferson RA (1987) Assaying chimeric genes in plants: the GUS gene fusion system. Plant Mol Biol Rep 5:387–405

Jelaska S, Rengel Z, Cesar V (1984) Plant regeneration from mesocotyl callus of *Hordeum vulgare* L. Plant Cell Rep 3:125–129

Matsuda Y, Toyoda H, Ouchi S (1989a) Application of microinjection to appressoria and haustoria of *Erysiphe graminis* f sp. *hordei*. Ann Phytopathol Soc Jpn 55:67–68

Matsuda Y, Toyoda H, Nishiguchi T, Ouchi S (1989b) Direct introduction of fluorescein isothiocyanate-conjugated albumin into intact macroconidia of *Fusarium oxysporum* f. sp. *lycopersici* by electroporation. J Phytopathol 125:89–96

Melton DA (1984) Injected anti-sense RNAs specifically block messenger RNA translation in vivo. Proc Natl Acad Sci USA 82:144–148

Murashige T, Skoog F (1962) A revised medium for rapid growth and bioassays with tobacco tissue cultures. Physiol Plant 15:473–497

Potrykus I (1990) Gene transfer to cereals: an assessment. Bio/Technol 8:535–542

Raikhel NV, Bendnarek SY, Lerner DR (1989) In situ RNA hybridization in plant tissue. In: Gelvin SB, Schilperort RA, Verma DPS (eds) Plant moleucular biology manual. Kluwer, Dordrecht, pp B9:1–32

Rengel Z, Jelaska S (1986) Somatic embryogenesis and plant regeneration from seedling tissues of *Hordeum vulgare* L. J Plant Physiol 124:385–392

Shiraishi T, Ouchi S, Oku H (1976) Chitin component in haustorial wall of powdery mildew fungus of barley. Sci Rep Fac Agric Okayama Univ 47:21–24

Swegle M, Hunga J-K, Lee G, Muthukrishnan S (1989) Identification of an endochitinase cDNA clone from barley aleurone cells. Plant Mol Biol 12:403–412

Toyoda H, Matsuda Y, Hirai T (1985) Resistance mechanism of cultured plant cells to tobacco mosaic virus (III) Efficient microinjection of tobacco mosaic virus into tomato callus cells. Ann Phytopathol Soc Jpn 51:32–38

Toyoda H, Matsuda Y, Hirai T (1986) Multiplication and translocation of tobacco mosaic virus microinjected into cell-aggregates of tomato callus. Plant Tissue Cult Lett 3:22–27

Toyoda H, Matsuda Y, Shoji R, Ouchi S (1987) A microinjection technique for conidia of *Erysiphe graminis* f. sp. *hordei*. Phytopathology 77:815–818

Toyoda H, Matsuda Y, Utsumi R, Ouchi S (1988) Intranuclear microinjection for transformation of tomato callus cells. Plant Cell Rep 7:293–296

Toyoda H, Oki T, Matsuda Y, Katsuragi K, Nishiguchi T, Ouchi S (1989) Transformation of constituent cells of tomato callus aggregates by intranuclear microinjection. Plant Tissue Cult Lett 6:95–97

Toyoda H, Yamaga T, Matsuda Y, Ouchi S (1990a) Transient expression of the β-glucuronidase gene introduced into barley coleoptile cells by microinjection. Plant Cell Rep 9:299–302

Toyoda H, Koura Y, Matsuda Y, Horikoshi Y, Tamai T, Yamaga T, Ikeda S, Shimabayashi M, Morita M, Ouchi S (1990b) Callus induction and plant regeneration from barley coleoptile tissues. Plant Tissue Cult Lett 7:202–204

Toyoda H, Matsuda Y, Yamaga T, Ikeda S, Morita M, Tamai T, Ouchi S (1991) Suppression of the powdery mildew pathogen by chitinase microinjected into barley coleoptile epidermal cells. Plant Cell Rep 10:217–220

Toyoda H, Katsuragi K, Tamai R, Ouchi S (1991b) DNA sequence of genes for detoxification of fusaric acid, a wilt-inducing agent produced by *Fusarium* species. J. Phytopathol 133:265–277

Toyoda H, Matsuda Y, Nogi Y, Tamai T, Ouchi S (1992) An improved system for foreign gene expression in barley coleoptile epidermis: gene transfer and in situ detection of gene products by dual microinjection. Plant Tissue Cult Lett 9:154–163

Yamamoto F, Furusawa M, Furusawa I, Obinata M (1982) The "pricking" method. A new efficient technique for mechanically introducing foreign DNA into the nuclei of culture cells. Exp Cell Res 142:79–84

II.11 Transformation in *Kalanchoe*

Shi-Rong Jia[1]

1 Introduction

The genus *Kalanchoe* (Ka-lan-cho-e, pronounced in four syllables with a "k" sound given to the "ch", which appears to be the Latin version of a Chinese plant name) belongs to the Crassulaceae family. It contains more than 200 species; the majority are African or Madagascan, but some are Asian and one of them is native to South America (Everett 1981; Herwig 1984).

Kalanchoes are succulent, herbaceous, or sometimes slightly woody-stemmed herbaceous perennials. The best-known species is *K. blossfeldiana*, of which numerous brightly colored hybrids are marketed in hundreds of thousands. The flowers of an inflorescence after full development remain fresh for a period of 7–8 weeks. In Oklahoma and Texas in the USA, it is considered as the second most important flowering pot plant (Love 1980). In South-East Asia, the plants, with beautiful flowers of different colors (red, pink, yellow, etc.), are mostly sold as "flower of longivity" for Spring festival. In addition, the crushed leaves and leaf juice of *K. laciniata* or *K. pinnata* have been used for their medicinal value to cure headaches, wounds, scratches, ulcers, and insect bites (Morton 1981). *K. blossfeldiana* is a short-day plant which requires at least 2 short days for flower induction. With increasing number of short-day cycles, the number of flowers increases exponentially. The logarithm of the flower numbers is linearly related to the number of inducing short-day cycles (Schwabe 1969). Therefore, this species has been commonly used as a model plant for investigations of flower induction in relation to photoperiod (Schwabe 1969, 1985).

Kalanchoe is highly susceptible to *Agrobacterium*; infection by *A. tumefaciens* or by *A. rhizogenes* usually results in tumor or hairy root formation at the site of infection. Among kalanchoe species, *K. daigremontiana* is most commonly used for the study of Ti-plasmid mutations that affect crown gall tumorigenesis (Garfinkel and Nester 1980; Klee et al. 1982) and for defining the genes in TL-DNA of the Ri-plasmid responsible for hairy root induction (White et al. 1985; Spena et al. 1987). For example, four root loci (*rol*) A, B, C, and D involved in the hairy root formation have been identified through insertional mutagenesis in the TL-DNA. The genetic analysis showed that *A. rhizogenes* with insertions in the *rol* B locus was completely avirulent on kalanchoe leaves, while insertions in the *rol* A, C, and D loci gave an attenuated response on the inoculated tissues. These results consequently show the

[1] Biotechnology Research Center, Chinese Academy of Agricultural Sciences, 30 Baishiqiao Road, Beijing 100081, China

Biotechnology in Agriculture and Forestry, Vol. 22
Plant Protoplasts and Genetic Engineering III (ed. by Y.P.S. Bajaj)

Fig. 1. Plants of *K. laciniata* in blossom

pivotal role of the *rol* B locus in root induction on kalanchoe leaves (White et al. 1985). By analyzing transcripts of the *rol* A, B, and C on TL-DNA, Spena et al. (1987) further confirmed that the DNA fragments containing the *rol* B locus alone are able to induce root formation both in tobacco and kalanchoe tissues. Moreover, they found the *rol* A, B, and C loci had independent and synergistic action in stimulating hairy root formation.

Although tumors and hairy roots can be readily induced on kalanchoe plants by agrobacterial infection, to the best of my knowledge, kalanchoe transformation leading to transgenic plant production has only one report on *K. laciniata* to date (Jia et al. 1989).

K. laciniata is an ornamental plant (Fig. 1) commercially propagated by cuttings or seeds. It usually takes 4–5 months from planting to flowering, if the plants are subjected to short-day length (14-h dark) treatment for 1 month. Based on the high susceptibility of this species to *A. tumefaciens* and the tissue culture system established by Ji-Fang Wang and her colleagues (pers. comm.), we have developed a reproducible, efficient transformation system via *A. tumefaciens*-mediated gene transfer for *K. laciniata*. The transformation frequency reached 75–80% and on the average as high as ten individual transgenic plants were produced per explant (Jia et al. 1989).

2 Transient Gene Expression

In our study, a binary vector system was used in which a plasmid pROA93 in *E. coli* LE392 was mobilized by triparental mating into three strains of *A. tumefaciens*, nopaline strain A208SE, octopine strain GV3111SE, and agropine strain EHA101, that carried the disarmed Ti-plasmid pTiT37-SE, pTiB6S3-SE, and pTiBo542, respectively. The plasmid pROA93 carries the coding sequence of neomycin phosphotransferase II (NPT II) gene and the *E. coli* β-glucuronidase (GUS) gene, both of which are fused to a Cauliflower Mosaic Virus (CMV) 35S promoter. The NPT II gene conferring kanamycin resistance is used for selection of transformed cells and the GUS gene is used as a reporter.

For determination of the GUS gene transient expression and optimization of the transformation system, the wound sites of the inoculated explants were sectioned individually by hand after different durations of cocultivation with the above-mentioned three different bacterial strains. Sections were then histochemically assayed for GUS activity as described by Jefferson (1987). Transient transformation frequency (TTF) was calculated based on the number of explants with transformed cells as a percentage of the total number of explants inoculated. The percent area of transformed tissue (PATT) was calculated based on the mean of the transformed tissue area as a percentage of the total tissue area on the section. By assaying transient expression of GUS gene in the inoculated explants, we found that two major critical factors are:

1. Prolonged coculture dramatically increased transient expression. The TTF was only 10% after 2 days of coculture with all three bacterial strains tested. However, the frequency rapidly increased to 40–70% after 4 days and reached 70–100% after 6 days (Fig. 2). Sections also showed low PATT (0.1–4%) at day 2, while PATT increased considerably when cocultivation time prolonged to 6–8 days (16–57%) (Fig. 3).
2. Agrobacterial strain A208SE was more effective than GV3111SE and EHA101. Of the three strains tested, the PATT at day 6 for the A208SE was 57% while the values for GV3111SE and EHA101 were 18% and 24.5%, respectively (Fig. 3).

When the GUS gene transient expression in explants of *K. laciniata* was compared with that of *Nicotiana tabacum* cv. SR1, both cocultivated with the nopaline strain A208SE carrying a binary vector pROA93, the tobacco seemed to have a better initial response to *A. tumefaciens*. While TTF at day 2 and 4 for kalanchoe was 10 and 70% respectively, the respective TTF for SR1 was 40 and 100%. The PATT at day 4 was also higher in SR1 (25%) than kalanchoe (11%).

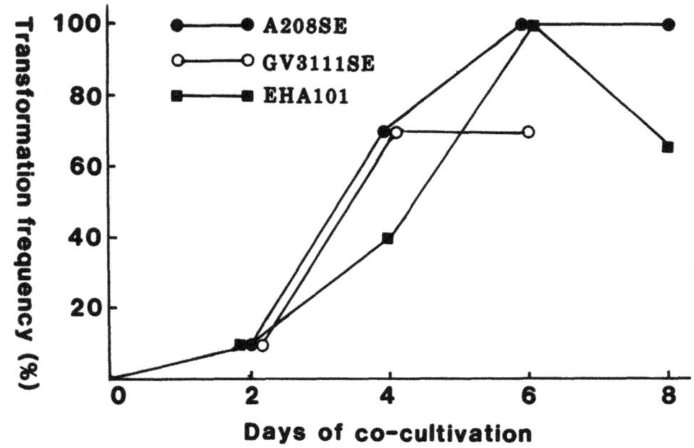

Fig. 2. Effects of cocultivation time on transient transformation frequency of *K. laciniata* leaf explants inoculated with different strains of *A. tumefaciens*

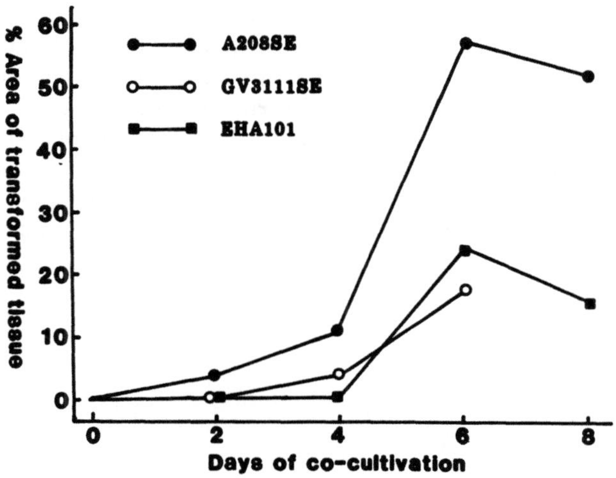

Fig. 3. Effects of cocultivation time on percent area of transformed tissue on transverse leaf section. Leaf explants were cocultured with different strains of *A. tumefaciens*

However, when the coculture time was prolonged to 6 and 8 days, the PATT in kalanchoe was higher (57 and 52%) than that in SR1 (32 and 23%). This indicated that *K. laciniata* was at least as susceptible to *A. tumefaciens* as *N. tabacum*. Explants cocultivated with agrobacteria for 2–4 days have been commonly used in genetic transformation of plants such as tobacco. Apparently, this is not the optimal coculture time for species like *K. laciniata*. Therefore, the optimal coculture time for a given species should be determined case by case.

3 Stable Transformation

Preliminary experiments indicated that 50 mg/l of kanamycin totally inhibited shoot regeneration from both control and inoculated explants. Therefore, a two-step selection scheme was applied in which the concentration of kanamycin was increased stepwise. The media used for selection of transformants are listed in Table 1.

Table 1. Media used for *Kalanchoe laciniata* transformation

Medium	Culture response	Basal medium	Hormones (mg/l)			Antibiotics (mg/l)	
			BA[b]	KT[b]	IAA[b]	Kan[c]	Cb[c]
C1	Shoot induction	MS[a]	1.0		0.2	40	500
C2	Shoot elongation	MS		0.5	0.1	100	500
C3	Root induction	1/2MS			0.2	100	500

[a] MS Murashige and Skoog (1962).
[b] BA, KT, and IAA refer to 6-benzylaminopurine, kinetin, and indol-3-acetic acid, respectively.
[c] Kan and Cb refer to kanamycin and carbenecillin respectively.

Explants were first transferred onto fresh C1 agar medium containing 40 mg/l kanamycin and 500 mg/l carbenecillin for shoot induction and selection of kanamycin resistance. After 2 weeks of culture, shoot bud differentiation was visible. The explants were then subcultured once for another 2 weeks on the same medium. Upon transferring the clumps of multiple shoots onto C2 medium supplemented with 100 mg/l kanamycin, putative dark green transformants could be distinguished from yellowish nontransformed shoots (Fig. 4). When the dark green shoots grew to the two to three leaf pair stage, they were separated individually and

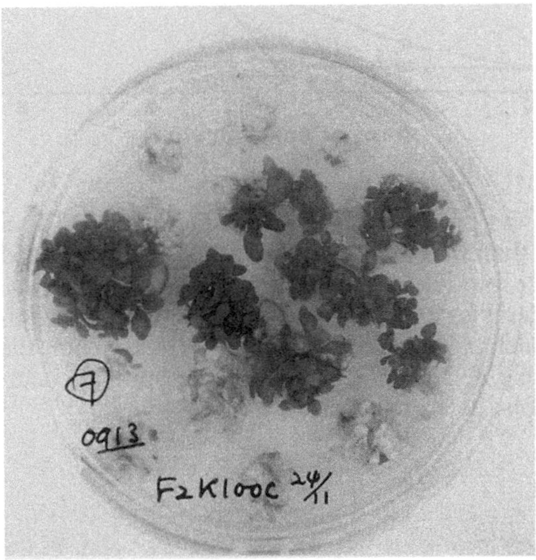

Fig. 4. Putative dark green transformed shoots grown on C2 selective medium supplemented with 100 mg/l kanamycin

Fig. 5. Transgenic plantlets were dark green with vigorous root system on C3 rooting medium containing 100 mg/l kanamycin, while nontransformed shoots were smaller in size and became chlorotic on the same medium

transferred onto C3 medium containing 100 mg/l kanamycin for rooting. Rooted plantlets with GUS activity were considered as stable transformants. Under this selective pressure, approximately 50% of the shoots rooted and showed GUS activity. The transgenic plantlets had dark green leaves and grew vigorously while nontransformed ones grew poorly and turned yellow in color (Fig. 5). Therefore, by using this two-step selection procedure, the transformed and nontransformed plants could be easily differentiated by their morphology.

Experimental data (Table 2) showed that prolonged coculture resulted in a higher frequency of stable transformation. While 4 days of coculture with strain A208SE resulted in 36% of explants producing transgenic plants, the value for 6 and 7 days of coculture was doubled (75 and 82% respectively). The number of transgenic plants produced per explant for 6–7 days coculture (3.4–9.8) was two to eight fold greater than that of 4 days coculture (1.2). Therefore, it is evident that prolonged coculture dramatically increased the transformation frequency as well as the number of transgenic plants produced.

Obviously, the number of explants producing transgenic plants was associated with the number of explants having transformed cells in the transient expression, while the number of transgenic plants produced per explant was related to the percent area of transformed tissue in transient expression. These results indicate that there is an apparent correlation between the GUS gene transient expression and the frequency of stable transformation.

With respect to bacterial strains, we again found that the nopaline strain A208SE was superior to EHA101 in giving stable transformants. The transformation frequency for A208SE after 4 and 6 days of cocultivation was 35.7% and 75% respectively, while the respective data for EHA101 was only 9.5 and 33.3%. The number of transgenic plants produced per explant for A208SE was also double that obtained for EHA101 (Table 3).

In conclusion, we have developed a reproducible, efficient transformation system for *K. laciniata*. The high frequency of transformation is probably due to the following three factors: firstly, the *Kalanchoe* sp. is highly susceptible to *A. tumefaciens*. Our observation on GUS gene transient expression demonstrated that *K. laciniata* was at least as susceptible to *A. tumefaciens* as *N. tabacum*; secondly, somatic cells of *K. laciniata* are highly totipotent and hundreds of shoots can be produced from one leaf explant; thirdly, the use of prolonged coculture and the appropriate agrobacterial strain, e.g., A208SE, remarkably increased the transformation frequency.

Table 2. Transformation frequency of *K. laciniata* obtained by different days of cocultivation of leaf explants with *A. tumefaciens* strain A208SE haboring the binary vector pROA93

Days of cocultivation	Total no. of explants inoculated	No. of explants producing TP[a]	Transformation frequency (%)	Total no. of TP[a] produced	No. of TP[a] produced per explant
Control	10	0	0	0	0
4	28	10	35.7	33	1.2
6	16	12	75.0	54	3.4
7	11	9	81.8	108	9.8

[a] TP = transgenic plants.

Table 3. Comparison of transformation efficiency in leaf explants of *K. laciniata* by cocultivation with *A. tumefaciens* A208SE and EHA101 both harboring the binary vector pROA93

Days of cocultivation	Bacteria strain	Total no. of explants inoculated	No. of explants producing TP[a]	Transformation frequency (%)	Total no. of TP[a] produced	No. of TP[a] produced per explant
4	A208SE	28	10	35.7	33	1.2
	EHA101	63	6	9.5	38	0.6
6	A208SE	16	12	75.0	54	3.4 .
	EHA101	12	4	33.3	14	1.2

[a] TP = transgenic plants.

4 Transformation Protocol

1. Grow plants of *K. laciniata* in a growth chamber under a photoperiod of 14-h light/10-h dark cycle and day/night temperature regime of 25/20 °C, respectively. The light intensity ranges from 200–300 μE/m^2/s.
2. Culture *Agrobacterium tumefaciens* strain A208SE carrying a disarmed Ti-plasmid pTiT37-SE and a plasmid pROA93 (for details see Jia et al. 1989) on opine medium containing 100 mg/l nopaline plus antibiotics spectinomycin, kanamycin, and chloramphenical at a concentration of 100, 25, and 25 mg/l, respectively for 3 days at 27 °C.
3. Suspend the agrobacteria in a liquid C1 medium to a density of $A_{600} = 0.2$.
4. Excise the second or third pair of leaves from the top of the plant grown in growth chamber and surface-sterilize by dipping them into 70% ethanol for 30 s, followed by continuous agitation for 10 min in a 10% (v/v) commercial bleach (Clorox) solution (final concentration 0.5% sodium hypochlorite). Rinse the leaves four times with sterile distilled water, then cut into 1×1 cm^2 pieces prior to inoculation with agrobacteria.
5. Inoculate leaf explants with bacterial suspension for 10 min in a 50-ml Falcon screw-capped tube. Invert the tube gently for agitation. Blot dry the explants between two layers of sterile filter paper and then transfer them onto a filter paper placed on solidified C1 medium for 6 days. To avoid bacterial overgrowth, the explants are rinsed with sterile distilled water at day 3 and 6 respectively.
6. Histochemical assay of GUS activity in cells of inoculated explants as described by Jefferson (1987). The wound sites of ten explants are sectioned individually by hand (two to three sections per explant) after cocultivation. Incubate sections overnight at 37 °C in 5-bromo-4-chloro-3-indolyl-β-glucuronide (X-Gluc, Research Organics, Inc., Ohio) solution consisting of 10 mM ethylene diamine tetra-acetic acid (EDTA), 100 mM sodium phosphate (pH 7.0), 0.5 mM potassium ferrocyanide, 0.5 mM potassium ferricyanide, and 0.1% (W/V) X-Gluc. Chlorophylls in the sections are removed by adding FAA (10 ml formaldehyde: 5 ml glacial acitic acid: 50 ml ethyl alcohol: 35 ml water) for 15 min, followed by sequential 2 min each in 70, 90, and 100% ethanol and water before sections are examined microscopically. Transformed cells expressing the chimeric 35S/GUS gene are blue in color. Calculate the TTF and PATT.
7. After 6 days coculture, transfer explants onto fresh C1 agar medium containing 40 mg/l of kanamycin and 500 mg/l of carbenecillin in a 9-cm Petri dish for shoot induction and selection of kanamycin-resistant shoots. Place cultures in a culture room under a 16-h light/8-h dark cycle with a photon fluence rate of 60 μE/m^2/s from cool white fluorescent lamps. The day and night temperatures are 27 and 23 °C, respectively.
8. After 2 weeks' culture on the above selection medium shoot bud differentiation is visible. Subculture explants once for another 2 weeks on the same medium. Score the number of explants producing shoots.
9. Cut the clumps of multiple shoots into several pieces and transfer them onto C2 medium containing 100 mg/l kanamycin and 500 mg/l carbenecillin for further selection and shoot elongation.

10. For rooting, transfer individual elongated, dark-green shoots at the two to three leaf pair stage onto C3 medium containing the same concentrations of antibiotics as in C2 medium in a Plantcon container.
11. Assay plantlets, which are rooted within 2 weeks on C3 medium, for GUS activity. Score the GUS positive ones as stable transformants. Calculate transformation frequency by the number of explants producing transgenic plants as a percentage of the total number of inoculated explants.
12. Transplant rooted transformants into pots and grow in a growth chamber.

5 Perspectives

We believe that after minor modification, the transformation system established for *K. laciniata* may be applicable to other kalanchoe species, e.g., *K. blossfeldiana.* In a preliminary experiment using leaf explants of different cultivars of *K. blossfeldiana* for in vitro culture, we found that some hybrids are highly regenerative.

For utilization of this system, a critical issue should be considered, which is that many kalanchoe cultivars marketed at present are interspecific hybrids in nature. Since kalanchoe species differ greatly in chromosome number (2n = 34, 36, 40, 68, 72, or 102, etc., Baldwin 1938), the hybrids usually exhibit male sterility (the normal pollen ranged from 0 to over 90% in five different cultivars of *K. blossfeldiana* tested in our experiment). The meiosis may also be abnormal due to the difficulty in chromosomal pairing which may result in failure of setting seeds, although *K. blossfeldiana* is self-pollinated. Therefore, for investigation of transgene expression in seed progeny, it is important to select the genotype with an ability of normal seed setting as a starting material. Furthermore, the seed germination of *K. blossfeldiana* is phytochrome-controlled. It has been found that seed germination of this species is enhanced by keeping the seeds in the dark and by addition of gibberellic acid (Dedonder et al. 1983; Fredericq et al. 1983; Rethy et al. 1987, 1990).

Anthocyanin synthesis and flower induction in *Kalanchoe* spp. are sensitive to photoperiod and known to be phytochrome-controled (Neyland et al. 1963; Lange et al. 1971; Schwabe 1985). Phytochrome gene from oat (Hershey et al. 1985), zucchini (Sharrock et al. 1986; Lissemore et al. 1987), pea (Tomizawa et al. 1986; Sato 1988), rice (Kay et al. 1989a, b), and *Arabidopsis thaliana* (Sharrock and Quail 1989) have been cloned and characterized. The transfer of phytochrome gene between dicotyledonous and monocotyledonous plants has been reported (Kay et al. 1989c; Boylan and Quail 1989; Keller et al. 1989). Recently we have also transferred a chimeric 35S/tobacco phytochrome gene into *K. laciniata.* Our preliminary results indicated that in transformants the percentage of plantlets with purple lower leaf surface was seven to eight fold higher than that of non-transformed control plants. The plant height and length of internodes were reduced by 20% in transgenic plants (Jia et al. unpubl.). This is in agreement with the results of Boylan and Quail (1989) in which they found that transgenic tomato plants, which highly expressed oat phytochrome, tended to be dwarfed and with a higher level of anthocyanin content. To date, the effect of foreign phytochrome gene on flowering response of the transgenic plants has not yet been reported. Interestingly, our preliminary data showed that after 4 weeks of short-day treatment, the flower induction in transgenic kalanchoe plants was significantly inhibited; flower number on each plant was reduced by 50%. The reduction was due to the number of inflorescences decreased which was only 45% of the control (Jia et al. unpubl.).

Many *Kalanchoe* species are characterized by Crassulacean acid metabolism (CAM), and have been extensively used in the study of CAM photosynthetic pathway (Bidwell 1983; Foyer 1984; Ting 1987). Therefore, it may be useful to incorporate the genetic transformation system into the study of expression and regulation of genes involved in the CAM pathway. In addition, it is also of great commercial value to incorporate this system in creating new kalanchoe varieties with novel flower color or pattern through genetic engineering.

Acknowledgements. The author is grateful to Dr. Eng-Chong Pua and Dr. Anil Kush for critically reading the manuscript.

References

Baldwin JT (1938) Kalanchoe: the genus and its chromosomes. Am J Bot 25:572–579

Bidwell RGS (1983) Crassulacean acid metabolism. In: Steward FC, Bidwell RGS (eds) Plant physiology – a treatise vol VII. Energy and carbon metabolism. Academic Press, New York pp 389–405

Boylan MT, Quail PH (1989) Oat phytochrome is biologically active in transgenic tomatoes. Plant Cell 1:765–773

Dedonder A, Rethy R, Fredericq H, de Greef JA (1983) Interaction between P_{fr} and growth substances in the germination of light-requiring *Kalanchoe* seeds. Physiol Plant 59:488–492

Everett TH (1981) Kalanchoe. The New York Botanical Garden illustrated encyclopedia of horticulture, vol. 6. Garland, New York, pp 1884–1889

Foyer CH (1984) Photosynthesis: crassulacean acid metabolism (CAM). Wiley, New York, pp 175–196

Fredericq H, Rethy R, van Onckelen H, de Greef JA (1983) Synergism between gibberellic acid and low P_{fr} levels inducing germination of *Kalanchoe* seeds. Physiol Plant 57:402–406

Garfinkel DJ, Nester EW (1980) *Agrobacterium tumefaciens* mutants affected in crown gall tumorigenesis and octopine catabolism. J Bacteriol 144:732–743

Hershey HP, Barker RF, Idler KB, Lissemore JL, Quail PH (1985) Analysis of cloned cDNA and genomic sequences for phytochrome: complete amino acid sequences for two gene products expressed in etiolated *Avena*. Nucleic Acids Res 13:8543–8559

Herwig R (1984) The Hamlyn encyclopedia of house plants: kalanchoe. Hamlyn, London, pp 188–189

Jefferson RA (1987) Assaying chimeric genes in plants: the β-glucuronidase as a versatile gene fusion marker in higher plants. EMBO J 6:3901–3907

Jia S-R, Yang M-Z, Ott R, Chua N-H (1989) High frequency transformation of *Kalanchoe laciniata*. Plant Cell Rep 8:336–340

Kay SA, Keith B, Shinozaki K, Chua N-H (1989a) The sequence of the rice phytochrome gene. Nucleic Acids Res 17:2865–2866

Kay SA, Keith B, Shinozaki K, Chye M-L, Chua N-M (1989b) The rice phytochrome gene: structure, autoregulated expression, and binding of GT-1 to a conserved site in the 5′ upstream region. Plant Cell 1:351–360

Kay SA, Nagatani A, Keith B, Deak M, Furuya M, Chua N-H (1989c) Rice phytochrome is biologically active in transgenic tobacco. Plant Cell 1:775–782

Keller JM, Shanklin J, Vierstra RD, Hershey HP (1989) Expression of a functional monocotyledonous phytochrome in transgenic tobacco. EMBO J 8:1005–1012

Klee HJ, Gordon MP, Nester EW (1982) Complementation analysis of *Agrobacterium tumefaciens* Ti plasmid mutations affecting oncogenicity. J Bacteriol 150:327–331

Lange H, Shropshire W, Mohr H (1971) An analysis of phytochrome-mediated anthocyanin synthesis. Plant Physiol 47:649–655

Lissemore JL, Colbert JT, Quail PH (1987) Cloning of cDNA for phytochrome from etiolated *Cucurbita* and coordinate photoregulation of the abundance of two distinct phytochrome transcripts. Plant Mol Biol 8:485–496

Love JW (1980) Kalanchoe. In: Larson RA (ed) Introduction to floriculture. Academic Press, New York, pp 409–434
Morton JF (1981) Atlas of medicinal plants of middle America. Charles C Thomas, Springfield Illinois, USA pp 258–260
Murashige T, Skoog F (1962) A revised medium for rapid growth and bioassays with tobacco cultures. Physiol Plant 15:473–497
Neyland M, Ng YL, Thimann KV (1963) Formation of anthocyanin in leaves of *Kalanchoe blossfeldiana* – a photoperiod response. Plant Physiol 38:447–451
Rethy R, Dedonder A, de Petter E, van Wiemeersch L, Fredericq H, de Greef J, Steyaert H, Stevens H (1987) Biphasic fluence-response curves for phytochrome-mediated *Kalanchoe* seed germination. Plant Physiol 83:126–130
Rethy R, Dedonder A, Cappelle M, Fredericq H, de Greef J (1990) Long-lasting light effects in imbibed *Kalanchoe blossfeldiana* seeds in the presence of gibberellic acid. Plant Physiol 93:333–338
Sato N (1988) Nucleotide sequence and expression of the phytochrome gene in *Pisum sativum*: differential regulation by light of multiple transcripts. Plant Mol Biol 11:697–710
Schwabe WW (1969) *Kalanchoe blossfeldiana* Poellniz. In: Evans LT (ed) The induction of flowering, Macmillan, Melbourne, Australia, pp 227–246
Schwabe WW (1985) *Kalanchoe blossfeldiana*. In: Halevy AH (ed) CRC Handbook of flowering, vol III. CRC Press, Boca Raton, Florida, pp 217–235
Sharrock RA, Quail PH (1989) Novel phytochrome sequence in *Arabidopsis thaliana*: structure, evolution, and differential expression of a plant regulatory photoreceptor family. Genes Dev 3:1745–1757
Sharrock RA, Lissemore JL, Quail PH (1986) Nucleotide and amino acid sequence of a *Cucurbita* phytochrome cDNA clone: identification of conserved features by comparison with *Avena* phytochrome. Gene 47:287–295
Spena A, Schmulling T, Koncz C, Schell JS (1987) Independent and synergistic activity of rol A, B and C loci in stimulating abnormal growth in plants. EMBO J 6(13):3891–3899
Ting IP (1987) Crassulacean acid metabolism. In: Newman DW, Wilson KG (eds) Models in plant physiology and biochemistry vol I. CRC Press, Boca Raton, Florida, pp 45–47
Tomizawa K, Komeda Y, Sato N, Nagatani T, Furuya M (1986) Isolation of cDNA for pea phytochrome using an expression vector. Plant Cell Physiol 27:1101–1108
White FF, Taylor BH, Huffman GA, Gordon MP, Nester EW (1985) Molecular and genetic analysis of the transferred DNA regions of the root-inducing plasmid of *Agrobacterium rhizogenes*. J Bacteriol 164:33–44

II.12 Transformation in *Linum usitatissimum* L. (Flax)

M.C. JORDAN and A. McHUGHEN[1]

1 Introduction

Flax (*Linum usitatissimum* L.) has been an important world crop for thousands of years. Its importance stems from its wide adaptation and distribution and its versatility; in fact, the name "usitatissimum" means "most useful". The stems can be used for fiber, the seeds for oil, and the meal for feed (Janick et al. 1974).

Due to the production of synthetic fibers the demand for fiber flax has been decreasing in recent years but the development of high yielding, shorter strawed oilseed flax has maintained the importance of flax as a crop. Flax (linseed) oil is used as an industrial oil in the production of paint, varnish, linoleum and as a concrete sealant (Kenaschuk 1975). The suitability of flax oil for these purposes is due to the presence of a high proportion of linolenic acid which results in a good drying oil. Unfortunately, linolenic acid causes the oil to turn rancid quite rapidly and thus it cannot be used for commercial edible oil purposes (Green 1986).

However, flax as an edible product is receiving a good deal of attention recently due to the health aspects of linolenic acid and other compounds. Properties such as cancer prevention and lowering of blood cholesterol have been ascribed to components in flax (Stitt 1986).

Flax is amenable to many tissue culture techniques including regeneration from protoplasts as well as from hypocotyl and leaf-derived callus (for a review see McHughen 1990). The ease with which plants can be regenerated from various flax seedling tissues, especially the hypocotyl and cotyledon, greatly facilitates its use in genetic transformation studies requiring the production of transgenic plants.

2 *Agrobacterium*-Mediated Flax Transformation

Agrobacterium tumefaciens is the causative agent in the formation of crown gall disease to which many dicotyledons and a few monocotyledons are susceptible (De Cleene and De Ley 1976). This disease involves the formation of galls which are capable of producing auxins and cytokinins as well as novel amino acids or sugar derivatives known as opines.

[1] National Research Council of Canada, Plant Biotechnology Institute, Saskatoon, Sask. S7N 0W9, and Crop Development Centre, University of Saskatchewan, Saskatoon, Sask. S7N 0W0, Canada

Biotechnology in Agriculture and Forestry, Vol. 22
Plant Protoplasts and Genetic Engineering III (ed. by Y.P.S. Bajaj)

The bacteria cause the plant cells in the gall to produce these opines by transferring a segment of DNA known as the T-DNA (tranfer DNA) from a large (~ 200 kb) tumor-inducing (Ti) plasmid carried by the bacteria into the plant cell. The T-DNA integrates into the plant genome and the transcriptional and translational machinery of the plant cell are used to produce the proteins which are responsible for opine production and gall formation (see Kado 1991 for a review of the molecular mechanisms of crown gall disease).

In 1983 Hepburn et al. demonstrated that wild-type nopaline strains of *Agrobacterium* are capable of inserting their T-DNA into the flax genome. The knowledge that flax is susceptible to *Agrobacterium* coupled with the hypocotyl regeneration system (see McHughen 1990), which fulfills the requirements for the rapid regeneration of multiple shoots, set the stage for the first efforts at obtaining transgenic flax plants.

Basiran et al. (1987) presented evidence for the production of transformed cells in callus derived from hypocotyl explants cocultivated with *Agrobacterium* strains carrying "disarmed" Ti plasmids (Ti plasmids with the tumor-causing genes deleted) and was able to produce shoots from this callus which tested positively for opines. However, in a similar experiment, Jordan and McHughen (1988a) demonstrated that such shoots are not necessarily transformed, as opines can be transported from transformed cells in the callus to shoots derived from nontransformed cells. The nontransformed cells are protected from the selective agent by the surrounding transformed cells. It is therefore possible to produce escape shoots from "transformed" callus. Later work showed that, contrary to common opinion at the time, the inoculated hypocotyl/callus tissue is composed of transformed cells and nontransformed cells. The regenerating cells in this system are exclusively or almost exclusively the nontransformed cells, which yield only nontransformed, "escape" regenerants (Dong and McHughen 1991).

Regeneration from hypocotyl explants occurs primarily from epidermal cells (Murray et al. 1977) while the transformed cells tend to occur and proliferate around the cut ends of the explant. Thus the vast majority of the regenerated shoots arise from nontransformed epidermis tissue. In order to avoid this problem the epidermis can be partially peeled off the explant with tweezers thus removing a major source of escapes as well as providing wounded tissue along the entire length of the explant as opposed to just the cut ends. Such a procedure led to the production of fully fertile transgenic flax plants which passed the foreign genes to their progeny in a normal Mendelian manner (Jordan and McHughen 1988b).

Even with epidermal peeling, a number of shoots are produced whose progeny do not contain the foreign genes or else a much smaller number of progeny contain the genes than would be expected on the basis of Mendelian inheritance. Although most of these "escapes" are most likely due to cross-protection in the original explant, at least some of them are chimeric. Fully transformed plants can be produced from these chimeric shoots either by testing more progeny than would (theoretically) be necessary (if the germ line cells are derived from at least some transformed cell lines) or by culturing leaf pieces from the chimeric plant on selective medium and regenerating shoots from any resulting callus (McHughen and Jordan 1989).

In order to overcome the problems of escapes associated with *Agrobacterium tumefaciens* inoculated hypocotyl explants Zhan et al. (1988) used *A. rhizogenes* to inoculate flax cotyledon explants and recovered transformed shoots from the resulting hairy roots. However, the shoots possessed the abnormal phenotype characteristic of shoots derived from hairy root cells – that is curled leaves, short internodes and abnormal root systems. Thus hypocotyl peeling combined with *A. tumefaciens* remains the method of choice for the production of phenotypically normal transgenic flax.

In an effort to increase the number of transgenic plants obtainable with this procedure the use of a preculture period was investigated (McHughen et al. 1989). It was found that when the explants were cultured for 9–12 days on regeneration medium prior to removal of the epidermis, significantly more transformed shoots were produced as compared to either no preculture or a 3–7-day preculture period. During the preculture stage the cells become competent for regeneration. Transformation after this period results in transformed cells which are already regeneration competent. It is possible that cells transformed without a preculture period are prevented from becoming regeneration competent – perhaps because of stress caused by the presence of the bacteria or the selective agent (in this case, kanamycin).

Another approach was recently investigated in an effort to improve the recovery of transgenic flax plants. Here, co-culture durations were investigated. Until this point, the inoculation period was generally 2–3 days, followed by transfer of the plant tissues to a selection medium containing anti-*Agrobacterium* agents to eliminate the vector organism. If the co-culture period was increased, the *Agrobacteria* tend to overgrow and smother the plant tissue (resulting in zero transgenic plants recovered). However, Dong developed a system where the hypocotyls were placed on a layer of sterile filter papers on top of the co-culture medium. Under this condition, the *Agrobacteria* did not overgrow the plant tissue and so much longer co-cultivation durations could be studied (Dong and McHughen 1991). Using an intron containing GUS as a reporter gene (Vancanneyt et al. 1990), it appears that the greatest amount of cellular transformation occurs with a 5–7 day co-culture period. This represents a 2–15-fold improvement over a 2–3-day period, depending on other conditions of the culture (Dong and McHughen 1991). In the same study, it was found that the preculture period detracted from cellular transformation (Dong and McHughen 1991). This indicates that the preculture period serves to convert at least some cells from being transformation competent to being regeneration- competent. What we are currently working on is inoculating transformation competent cells, then converting them to regeneration competent and subsequently recovering transgenic plants.

3 Genetic Engineering for Flax Crop Improvement

3.1 Herbicide Resistance

Genes conferring resistance to herbicides will be among the first agronomically useful genes inserted into crop species. Several classes of such genes have been isolated, cloned and modified for insertion into plant cells. We have been working on the transfer of genes conferring resistance to three types of herbicides. These are glyphosate (Roundup; Monsanto), glufosinate (Basta, Ignite; Hoechst), and sulfonylurea (e.g., Glean; DuPont).

Flax is known to be a poor competitor with many common broadleaf and grassy weeds, and therefore good weed control is essential for maximum crop production. Due to this poor competitive ability, even fairly low weed infestations can reduce flax yields considerably. The presence of weeds can reduce the quantity and quality of the oil, add to shipping costs, interfere with harvesting and render the straw unusable for paper making (Friesen 1988). Herbicides currently in use in western Canada provide effective weed control in flax; however, they are expensive and can also cause some crop damage especially under certain environmental conditions such as those causing plant stress (drought, heat) or those causing delayed germination and emergence (for soil incorporated herbicides) (Kneeshaw 1981).

3.2 Glyphosate Resistance

Glyphosate (n-[phosphonomethyl]glycine) is a broad spectrum herbicide which interferes with the enzyme 5-enolpyruvylshikimate-3-phosphate (EPSP) synthase by inhibiting the binding of the substrate phospoenolpyruvate to the active site of the enzyme (Steinrucken and Amrhein 1980). This enzyme is a member of the shikimate pathway which is involved in the formation of the aromatic amino acids and secondary products derived from chorismate. The gene is nuclear but the protein is functional in the chloroplasts. A transit peptide is responsible for transport of the enzyme to the chloroplast (Della-Cioppa et al. 1986).

Shah et al. (1986) isolated an EPSP synthase gene from a *Petunia* overproducer cell line and this gene was subsequently modified to confer enhanced resistance to glyphosate. The resistance gene driven by the 35S promoter was inserted into an *Agrobacterium* transformation vector and used to transform flax.

The resulting transgenic flax was shown to contain a functional glyphosate resistance gene by leaf callus assays where the leaves from the transformed shoots remained green and callused in the presence of 5 mM glyphosate whereas leaves from nontransformed control plants bleached white and died (Jordan and McHughen 1988b). When progeny from the transformed plants were grown in the greenhouse and sprayed with 25 g/ha glyphosate, the control plants died whereas the transformed plants survived. However, some damage to the meristematic tissue was observed leading to the death of the apical meristem and subsequent regrowth from axillary meristems. These results were confirmed by field tests where the plants survived but suffered delayed maturity when sprayed with levels of glyphosate needed to provide effective weed control (McHughen and Mitchell 1990). Since then, Monsanto Co. has isolated and modified EPSP synthase genes from other sources; these have been shown to confer field-level resistance to Roundup in tests of transgenic canola (*Brassica napus*) (Mitchell 1991, pers. comm.).

3.3 Glufosinate Resistance

Glufosinate is also a nonselective herbicide. The active ingredient, phosphinothricin, is an antagonistic analog of glutamine synthetase. Genes conferring resistance to phosphinothricin have been isolated and transferred to plant species (Deblock et al. 1987; Thompson et al. 1987). We are currently involved in a project (in collaboration with Hoechst) to transfer a gene coding for Phosphinothricin Acetyl Transferase (PAT) to flax. The PAT gene detoxifies PPT and thus represents a different method of herbicide resistance. The glyphosate resistance system of Monsanto is based on modifying the target enzyme for the herbicide, so that it is no longer recognized as a target. In the glufosinate resistance system of Hoechst, the gene codes for an enzyme which metabolizes the herbicide. Several transgenic lines of commercial cultivars of flax containing this PAT gene have been generated and are now in early testing stages.

3.4 Sulfonylurea Resistance

Haughn et al. (1988) isolated and cloned a mutant acetolactate synthase (ALS) gene from *Arabidopsis* which confers resistance to the sulfonylurea herbicide chlorsulfuron (Glean; DuPont). ALS is a critical enzyme in the branched-chain amino acid biosynthetic pathway, and it is inhibited by sulfonylurea herbicides. Application of the herbicide to a susceptible plant results in death by starvation. This mutant gene confers resistance by generating an enzyme with a reduced affinity for the herbicide while maintaining the normal enzymic function.

Using disarmed *Agrobacterium* as the vector system, this gene was transferred into several different commerical cultivars of flax (McHughen 1989). Subsequent analysis of the transgenic lines showed several of them to be fully resistant at the field level doses of two common sulfonylurea herbicides, chlorsulfuron (Glean; DoPont) and metsulfuron methyl (Ally; DuPont) (Fig. 1, Table 1) (McHughen and

Fig. 1. Wild-type cultivar NorLin and transgenic line 12115 seeded into soil previously treated with the sulfonylurea herbicide metsulfuron methyl (Ally; DuPont) at the full field rate of 4.5 g/ha. The wild-type row *in the foreground* exhibited poor germination and stunted chlorotic seedlings while the row of transgenic plants carrying the mutant ALS gene (*background*) had normal growth and development

Table 1. Percentage comparisons between normal soil and herbicide treatments for productivity of transgenic and wild-type flax. Degree of significance of differences is noted: ns = not significant; ** = $p < 0.01$; *** = $p < 0.001$ by t-test. (McHughen and Holm 1991, reprinted by permission of Kluwer Academic Publishers)

	Chlorsulfuron	sig.?	Metsulfuron	sig.?
Height				
12115	97.4	ns	93.8	ns
NorLin	30.7	***	31.2	**
Biomass				
12115	98.2	ns	91.6	ns
NorLin	20.0	***	29.8	***
Seed Yield				
12115	91.3	ns	83.3	ns
NorLin	11.5	***	20.8	***
1000 Seed wt.				
12115	98.8	ns	101	ns
NorLin	97.8	ns	100	ns

Holm 1991). These transgenic lines are also being tested for resistance to other sulfonylurea herbicides, such as triasulfuron (Ciba-Geigy). Assays of enzyme activities showed the transgenic lines to be 25- to 260-fold more resistant to chlorsulfuron than the parental, nontransgenic lines (McSheffrey et al. 1991).

Interestingly, flax is already resistant to post emergent application of sulfonylureas, which is a registered product for weed control in flax crops in some countries. Flax shoots and leaves can metabolize chlorsulfuron to nontoxic byproducts, but the roots cannot (Hutchison et al. 1984). The problem arises in that chlorsulfuron has a long soil residue activity, lasting up to several years in certain types of soil. Farmers who wish to grow flax (or many other types of broadleaf crops) have to ensure there is no sulfonylurea residue active in the soil. This can cause rotational problems for farmers who like to use sulfonylureas in their cereal crops. Therefore, the development of a sulfonylurea -resistant transgenic flax will be of interest to those farmers who might have sulfonylurea residue problems on the farm: not only will it permit them to grow a crop on land that otherwise might have to be left fallow, they might also reduce the net amount of chemical herbicide applied by taking advantage of the residual activity of the sulfonylurea on weeds that might otherwise have to be controlled in the flax crop (McHughen and Holm 1991).

3.5 Other Traits

Genetic engineering for crop improvement cannot be sustained on the basis of herbicide resistance. While herbicide resistant cultivars will be among the first transgenic products to hit the seed market, they will also be among the first to be phased out. Genes for other agronomically useful traits are being developed in many labs around the world. Insect resistance, using either *Bacillus thuringiensis* endotoxin gene (Delannay et al. 1989) or the cowpea trypsin inhibitor gene (Hilder et al. 1987) is being developed, along with virus resistance systems (Kaniewska et al. 1990; Slightom et al. 1990). While these traits will provide significant advances in some crops, they will probably not be inserted into flax, because flax does not suffer from viral disease or insect infestation to the same extent as, say, cotton for insects and potato for viral diseases. Instead, flax will have to wait to benefit from the isolation, modification, and improvement of other useful genes, such as those conferring salt tolerance (Boyd et al. 1991) or genes designed to modify fatty acid composition.

Most of the attention given to crop improvement using genetic engineering technology is based on the transfer of a useful trait to the host crop species. More recently, attention has been focused on the deletion of undesirable traits. This can be achieved by inserting a gene coding for the production of antisense mRNA or by the use of a "suicide time-bomb" to destroy tissues or organs responsible for the undesired trait. Mariani et al. (1990) used a ribonuclease structural gene driven by a tapetum-specific promotor to kill the tapetal cells as they were differentiating. Because the promotor is active only in tapetal cells, no other tissues of the plant were affected, and the transgenic plants, lacking a functional tapetum, were male sterile. Flax might be a good species to attempt to induce male sterility because there are no hybrid systems available at present. No doubt, this technique will provide many uses in addition to male sterility, limited only by the imagination. Similarly, the anti-sense technology can be applied to inactivating any dominant gene, so long as the gene is known and characterized.

A major question is the effect of transformation on normal agronomic performance. While farmers might like to have, and perhaps even pay a premium for a transgenic cultivar with improved insect or herbicide resistance, they will not be pleased if the line does not yield as well as conventional cultivars. It is possible that the insertion of T-DNA causes the inactivation of a major gene in the host plant genome, or that the gene product of the novel gene somehow interferes with the normal agronomic functioning of the plant enough to cause a reduction in vigor and therefore result in a reduction in yield. Because of this, all transgenic lines must be screened as carefully for normal agronomic function as for expression of the novel trait. Large scale field tests of transgenic material are only just beginning to be reported. Fortunately, it appears that the transformation process itself does not affect normal agronomic function in the majority of transgenic lines (McHughen and Rowland 1991).

4 Summary and Conclusions

The flax transformation system we have developed has produced transgenic flax lines carrying agriculturally valuable genes (herbicide resistance) which are now being field- tested. Evaluation of normal agronomic traits as well as for herbicide resistance established that transformation per se had no effect on normal agronomic performance. With this being the case, we can look forward to genetic engineering playing a major role in crop improvement. However, further advances will depend on the isolation of more genes which play a major role in economically important traits.

References

Basiran N, Armitage P, Scott RJ, Draper J (1987) Genetic transformation of flax (*Linum usitatissimum*) by *Agrobacterium tumefaciens*: regeneration of transformed shoots via a callus phase. Plant Cell Rep 6:396–399

Boyd LA, Adam L, Pelcher LE, McHughen A, Hirji R, Selvaraj G (1991) Characteristics of an *Escherichia coli* gene encoding betaine aldehyde dehydrogenase (BADH): structural similarity to mammalian ALDHs and a plant BADH. Gene 103:45–52

DeBlock M, Botterman J, Vandewiele M, Dockx J, Thoen C, Gossele V, Rao Movva N, Thompson C, Van Montague M, Leemans J (1987) Engineering herbicide resistance in plants by expression of a detoxifying enzyme. EMBO J 6:2513–2518

De Cleene M, De Ley J (1981) The host range of crown gall. Bot Rev 42:389–466

Delannay X, Lavallee B, Proksch R, Fuchs R, Sims S, Greenplate J, Marrone P, Dodson R, Augustine J, Layton J, Fischhoff D (1989) Field performance of transgenic tomato plants expressing the *Bacillus thuringiensis* var. *kurstaki* insect control protein. Bio/Technol 7:1265–1269

Della-Cioppa G, Bauer SC, Klein BK, Shah DM, Fraley RT, Kishore GM (1986) Translocation of the precursor of 5-enolpyruvylshikimate-3-phosphate synthase into chloroplasts of higher plants in vitro. Proc Natl Acad Sci USA 83:6873–6877

Dong J-Z, McHughen A (1991) Patterns of transformation intensity on flax hypocotyls inoculated with *Agrobacterium tumefaciens*. Plant Cell Rep 10:555–560

Friesen GH (1988) Weed competition in flax. Flax Newsletter 70. Flax Growers Western Canada, Regina, Saskatchewan

Green AG (1986) Development of an edible-oil flax genotype. In: Carter JF (ed) Proc 51st Annual Flax Institute of the United States. Fargo, ND, pp 43–47

Haughn G, Smith J, Mazur B, Somerville C (1988) Transformation with a mutant *Arabiodopsis* acetolactate synthase gene renders tobacco resistant to sulfonylurea herbicides. Mol Gen Genet 211:266–271

Hepburn AG, Clarke LE, Blundy KS, White J (1983) Nopaline Ti-plasmid, pTiT37, T-DNA insertions into a flax genome. J Mol Appl Genet 2:211–224

Hilder VA, Gatehouse AMR, Sheerman SE, Barker RF, Boulter D (1987) A novel mechanism of insect resistance engineered into tobacco. Nature 330:160–163

Hutchison J, Shapiro R, Sweetser PB (1984) Metabolism of chlorsulfuron by tolerant broadleaves. Pestic Biochem Physiol 22:243–247

Janick J, Schery RW, Woods FW, Ruttan VW (1974) Plant science. An introduction to world crops. Freeman, San Francisco, pp 518, 572–573

Jordan MC, McHughen A (1988a) Transformed callus does not necessarily regenerate transformed shoots. Plant Cell Rep 7:285–287

Jordan MC, McHughen A (1988b) Glyphosate tolerant flax plants from Agrobacterium mediated gene transfer. Plant Cell Rep 7:281–284

Kado CI (1991) Molecular mechanisms in crown gall tumorigenesis. Crit Rev Plant Sci 10:1–32

Kaniewska W, Lawson C, Sammons B, Haley L, Hart J, Delannay X, Tumer N (1990) Field resistance of transgenic Russet Burbank potato to effects of infection by Potato Virus X and Potato Virus Y. Bio/Technol 8:750–754

Kenaschuk E (1975) Flax breeding and genetics. In: Harapiak JT (ed) Oilseed and pulse crops in Western Canada. Western Canada Co-operative Fertilizers Ltd, Calgary, Alberta, pp 203–221

Kneeshaw PG (1981) Differential flax cultivar tolerance to registered herbicide treatments. MSc Thesis, Univ of Saskatchewan, Saskatoon

Mariani C, De Beuckeleer M, Truettner J, Leemans J, Goldberg RB (1990) Induction of male sterility in plants by a chimaeric ribonuclease gene. Nature 347:737–741

McHughen A (1989) *Agrobacterium*-mediated transfer of chlorsulfuron resistance to commerical flax cultivars. Plant Cell Rep 8:445–449

McHughen A (1990) Flax (*Linum usitatissimum* L.) in vitro studies. In: Bajaj YPS (ed) Biotechnology in agriculture and forestry vol 10. Legumes and oilseed crops 1. Springer, Berlin, Heidelberg, New York, pp 502–514

McHughen A, Holm FR (1991) Herbicide-resistant transgenic flax field test: Agronomic performance in normal and sulfonylurea-containing soils. Euphytica 55:49–56

McHughen A, Jordan MC (1989) Recovery of transgenic plants from "escape" shoots. Plant Cell Rep 7: 611–614

McHughen A, Mitchell A (1990) Preliminary field test results of Roundup resistant transgenic flax. In: J.F. Carter (ed) Proc 53rd Annual Flax Inst of the United States, Fargo, ND, USA, pp 105–108

McHughen A, Rowland GG (1991) The effect of T-DNA on the agronomic performance of transgenic flax plants. Euphytica 55:269–275

McHughen A, Jordan MC, Feist G (1989) A preculture period prior to *Agrobacterium* inoculation increases production of transgenic plants. J Plant Physiol 135:245–248

McSheffrey SA, McHughen A, Devine MD (1992) Characterization of transgenic sulfonylurea-resistant flax. Theor Appl Genet 84:480–486

Murray BE, Handyside RJ, Keller WA (1977) In vitro regeneration of shoots on stem explants of haploid and diploid flax (*Linum usitatissimum*). Can J Genet Cytol 19:177–186

Shah DM, Horsch RB, Klee HJ, Kishore GM, Winter JA, Tumer NE, Hironaka CM, Sanders PR, Gasser CS, Aykent S, Siegel NR, Rogers SR, Fraley RT (1986) Engineering herbicide tolerance in transgenic plants. Science 233:478–481

Slightom JL, Chee PP, Gonsalves D (1990) Field testing of cucumber plants which express the CMV coat protein gene: field plot design to test natural infection pressures. In: Nijkamp HJJ, Van Der Plas LHW, Van Aartrijk J (eds) Progress in plant cellular and molecular biology, Proc VIIth Int Congr Plant Tissue and Cell Culture. Kluwer, Dordrecht, pp 201–206

Steinrucken H, Amrhein N (1980) The herbicide glyphosate is a potent inhibitor of 5-enolpyruvylshikimic acid-3-phosphate synthase. Biochem Biophys Res Commun 94:1207–1212

Stitt P (1986) Nutritional importance of flax. In: Carter JF (ed) Proc 51st Annual Flax Institute of the United States, Fargo, ND, USA, pp 23–24

Thompson CJ, Rao Movva N, Tizard R, Crameri R, Davies JE, Lauwereys M, Botterman J (1987) Characterization of the herbicide-resistant gene *bar* from *Streptomyces hygroscopicus*. EMBO J 6: 2519–2523

Vancanneyt G, Schmidt R, O'Connor-Sanchez A, Willmitzer L, Rocha-Sosa M (1990) Construction of an intron-containing marker gene: splicing of the intron in transgenic plants and its use in monitoring early events in *Agrobacterium*-mediated plant transformation. Mol Gen Genet 220:245–250

Zhan X-C, Jones DA, Kerr A (1988) Regeneration of flax plants transformed by *Agrobacterium rhizogenes*. Plant Mol Biol 11:551–559

II.13 Transformation in *Lobelia inflata*

Masahiko Tanaka[1], Hiroshi Yonemitsu[1], Koichiro Shimomura[2], Kanji Ishimaru[3], Shunji Mochida[1], Tohru Edno[1], and Akira Kaji[4]

1 Introduction

Lobelia inflata L. (Campanulaceae) is an annual plant with an erect, angular, slightly pubescent stem of up to 50 cm. It is commonly known as Indian tobacco, since it is used as a substitute for tobacco by American Indians. Several dialkylpiperidine alkaloids have been isolated from this plant (Schöpf and Kauffmann 1957). The total alkaloid content of *L. inflata* was found to be 0.134–0.635%, and the quantity of alkaloid varies with the culture conditions (Kalashnikov 1939).

Among the alkaloids of *L. inflata* the major and most important one is lobeline (Fig. 1), which has some interesting pharmacological and therapeutic properties. Lobeline has a marked stimulatory effect on the respiratory center of the brain (Marshall 1928), and its hydrochloride salt is commercially used as a respiratory stimulant. Furthermore, it strongly inhibits the binding of nicotine to the brain, without showing nicotine-like behavioral effects (Reavill et al. 1990). Thus, lobeline has been studied as a deterrent to tobacco smoking (Davison and Rosen 1972).

The production of lobeline alkaloid has been reported in cultivated *L. inflata* (Krochmal et al. 1972), and in tissue cultures of this plant (Wysokinska 1977; Krajewska et al. 1988). Generally, hairy roots grow faster than the untransformed roots or cultured cells. Furthermore, hairy root cultures are genetically stable with respect to the production of secondary metabolites (Aird et al. 1988). Hence, it appeared desirable to produce lobeline alkaloid in the hairy root cultures of *L. inflata* by transformation with *Agrobacterium rhizogenes*.

2 Establishment of Hairy Root Culture and Production of Secondary Metabolites

Transformed hairy roots were obtained by the inoculation of the stems of *Lobelia inflata* L. with *Agrobacterium rhizogenes* 15834. The hairy roots were cultured in

[1] Research Institute for Molecular Genetics, Tsumura & Co., 3586 Yoshiwara, Ami-machi, Inashiki, Ibaraki 300-11, Japan
[2] Tsukuba Medicinal Plant Research Station, National Institute of Hygienic Sciences, 1 Hachimandai, Tsukuba, Ibaraki 305, Japan
[3] Genetic Engineering Laboratory, Faculty of Agriculture, Saga University, Honjo, Saga, 840 Japan
[4] Department of Microbiology, School of Medicine, University of Pennsylvania, Philadelphia, PA 19104-6076, USA

Biotechnology in Agriculture and Forestry, Vol. 22
Plant Protoplasts and Genetic Engineering III (ed. by Y.P.S. Bajaj)

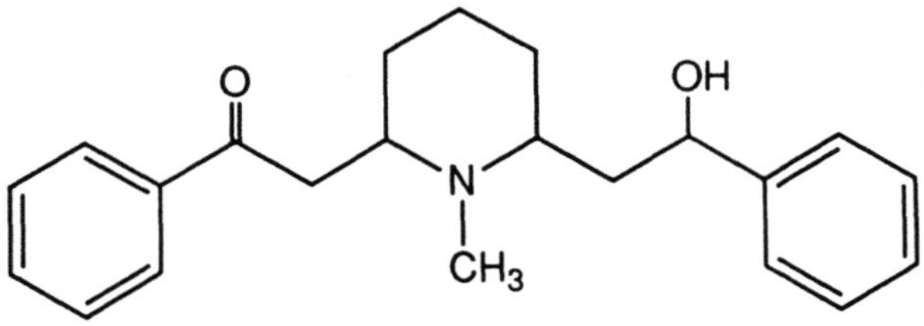

Fig. 1. Chemical structure of lobeline

hormone-free basal medium. Transformation was confirmed by the detection of opines and T-DNA, and lobeline alkaloid and polyacetylenes were present in these hairy roots.

2.1 Bacterial Strain

Agrobacterium rhizogenes strain ATCC 15834 was grown on YEB agar (Vervliet et al. 1975) and maintained by subculturing on the same medium every month.

2.2 Transformation of *L. inflata* and Establishment of Hairy Root Cultures

The stems of sterile 20-day-old plantlets were cut with a sterile scalpel and *A. rhizogenes* strain ATCC 15834 was inoculated on the cut ends. The infected plantlets were incubated on hormone-free MS solid medium (Murashige and Skoog 1962) at 25 °C under a 16-h photoperiod (6 $\mu E/m^2/s$). After 3–4 weeks, hairy roots appeared at the inoculated sites. Hairy root cultures were then established by transferring the root tips (ca. 1–2 cm in length) of the hairy roots into hormone-free MS liquid medium containing cefotaxime (500 μg/ml). This culture was maintained at 25 °C in the dark on a gyratory shaker (80 rpm). Root tips of the hairy roots were further subcultured in the same fresh medium every month.

Effects of media on growth rate and lobeline production of hairy roots were studied by inoculating segments of the hairy roots from hormone-free MS liquid medium into hormone-free MS, 1/2 MS (one-half strength standard Murashige and Skoog's medium), B5 (Gamborg et al. 1968), and NN (Nitsch and Nitsch, 1967) media, respectively. They were cultured at 25 °C in the dark on a gyratory shaker (80 rpm) for 4 weeks with two changes of medium at 2 weeks' interval. Subsequent to this adaptation period, root tips from conditioned hairy roots (0.2 g fresh weight; about 1–2 cm, ten tips) were inoculated into 50 ml of the same fresh medium and cultured at 25 °C in the dark on a gyratory shaker (80 rpm) for 4 weeks.

2.3 Detection of Opines

Extraction and detection of opines (agropine and mannopine) were performed by following the method described by Petit et al. (1983). Paper electrophoresis was according to the method of Otten and Schilperoort (1978). The electrophoretogram was visualized with alkaline silver nitrate reagent (Trevelyan et al. 1950).

2.4 Southern Hybridization Analysis

DNA hybridization was performed by the procedure of Southern (1975). Total DNAs were isolated from the hairy roots and nontransformed plants of *L. inflata* according to the method of Saito et al. (1990). The probe DNA (*Eco*RV digested TL-DNA region of pRiHR1; 2.7×10^9 cpm/μg) was labeled with [α–^{32}P]- dCTP using the multiprime DNA labeling kit (Amersham). Hybridized blots were autoradiographed with Kodak X-OMAT AR film for 30 hrs at -70 °C.

2.5 Extraction and Analysis of Lobeline

Lyophilized hairy roots were ground with pestle and mortar. The powdered sample (1 g) was sonicated in 10 ml of 0.01 N HCl for 5 min at room temperature. After filtration, the residue was resonicated in equal volume of 0.01 N HCl. The combined filtrates were dried under vacuum at 40 °C and dissolved in 1 ml of 0.01 N HCl. The identical procedure was applied to extraction of lobeline from the roots of mother plants.

The extracts were analyzed by high performance liquid chromatography (HPLC, Shimadzu LC6A) on a YMC-pack AM312 ODS column (6 mm i.d. × 15 cm, Yamamura Chemical Lab. Co.) at room temperature. Samples (20 μl) were eluted with the eluant [acetonitrile–0.1% trifluoroacetic acid (3:7, v/v)] at a flow rate of 1 ml/min. The eluate was monitored by a UV detector (Shimadzu SPD6A) at 254 nm. This extraction procedure gave 80–90% yield. A lobeline fraction from this procedure was further purified by HPLC and was analyzed by a mass spectrometer (JEOL JMS DX-300).

2.6 Analysis of Polyacetylenes

Extraction and isolation: lyophilized hairy roots (18 g, dry wt), cultured in MS liquid medium for 6 weeks in the dark, were mashed. The powdered sample was macerated with methanol (700 ml) at room temperature. The extract was concentrated under reduced pressure to ca. 50 ml. This sample was poured into 50 ml of H_2O and reconcentrated to ca. 35 ml. The extract was placed on Sephadex LH-20 column (3 × 35 cm) and eluted by 60% methanol to obtain Fr. 1 and 2. Fr. 2 was dried under reduced pressure and applied to silica gel 60 column chromatography (CC) (2 × 220 cm, $CHCl_3$–MeOH, 19:1–4:1) to give lobetyolin and Fr. 21. Fr. 21 was further purified by silica gel 60 CC (2 × 11 cm, C_6H_6–Me_2CO, 5:1–3:1) to give lobetyol.

HPLC analysis: column; Wakosil-II $5C_{18}$ (4.6 mm × 250 mm), sol.; 50 mM NaH_2PO_4–CH_3CN (80:20–20:80), detector; 270 nm (UV), column temp; 40 °C, Retention time; lobetyolin (20.04 min), lobetyol (24.39 min).

3 Results and Discussion

3.1 Establishment of Hairy Root Culture

Stems of *L. inflata* were inoculated with *A. rhizogenes* strain 15834 and incubated on hormone-free MS solid medium. After 3–4 weeks, numerous hairy roots appeared at the inoculated site (Fig. 2A). These hairy roots were separated from the stems and placed on hormone-free MS solid medium containing cefotaxime (500 μg/ml). Among 50 inoculants, 3 clones grew vigorously on hormone-free MS medium (Fig. 2B). These clones were designated Li-1, 2, and 3, respectively. They were transferred to hormone-free MS liquid medium and cultured. After 4–5 weeks, rapidly growing axenic hairy roots were obtained in hormone-free MS liquid medium.

In general, the transformation with *Agrobacterium* has been typically represented by the production of opines in these transformed plant tissues. The presence of agropine and mannopine in these hairy roots was examined (Fig. 3). The extracts of two hairy root clones (Li-1 and 2) gave two spots corresponding to the positions of mannopine and agropine (Fig. 3, lanes 2 and 3). On the other hand, neither agropine nor mannopine could be detected in the nontransformed root extract (Fig. 3, lane 5). It is interesting to note that the extract of Li-3 gave only one faint spot at mannopine position (Fig. 3, lane 4). Mano et al. (1986) have reported that the same opine producing phenotype was observed on the hairy root of *Scopolia japonica* which was induced by *A. rhizogenes* 15834.

Southern hybridization analysis was performed to confirm that the hairy root Li-2 contained the TL-DNA region of pRi15834 (Fig. 4). The distinct bands, at the position of about 0.87, 1.03, 1.4, 1.58, 2.0, 2.32, 2.55 kb, indicated by the arrows, revealed the presence of TL-DNA in this hairy root (lane 1) (Slightom et al. 1986). Other bands which showed slower mobility than 2.55 kb have been considered as composite bands containing plant DNA and TL-DNA (junction). No such bands

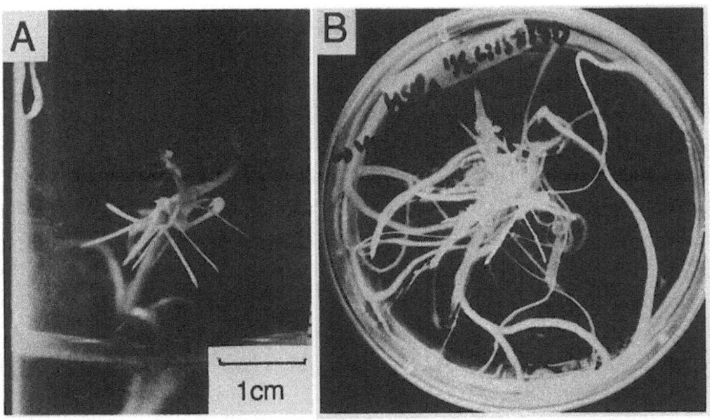

Fig. 2. A Hairy roots induced on a stem of *L. inflata* inoculated with *A. rhizogenes* strain 15834. **B** Hairy roots cultured on hormone-free MS solid medium

Fig. 3. Detection of agropine and mannopine in the extracts of *L. inflata* hairy roots. Extract from equivalent of 10 mg fresh weight of hairy root was applied (10 µl) on a paper and analyzed by paper electrophoresis. *Lane 1* authentic mannopine and agropine. Various extracts were analyzed as follows; *lane 2* transformed hairy roots (Li-1); *lane 3* transformed hairy roots (Li-2); *lane 4* transformed hairy roots (Li-3); *lane 5* roots of the plants grown in pots; *lane 6* authentic mannopine. **a** and **m** indicate positions where agropine and mannopine migrate, respectively

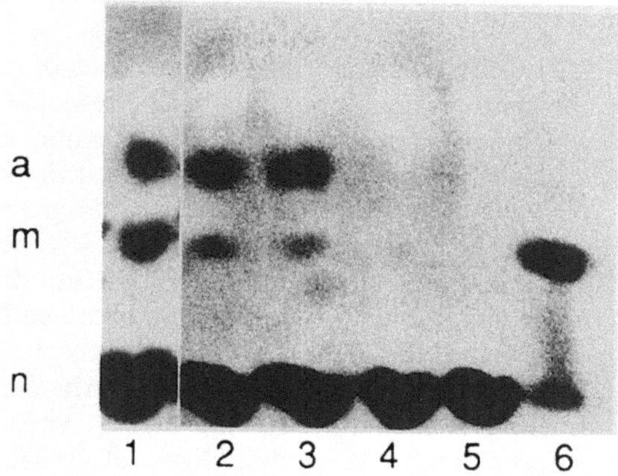

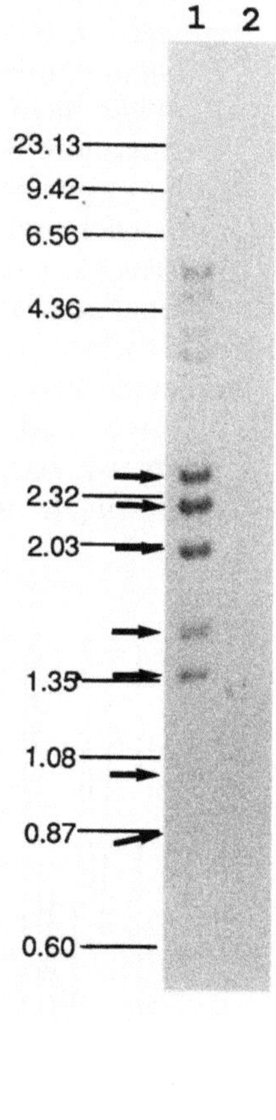

Fig. 4. Southern blot analysis of TL-DNA sequence in *L. inflata* hairy root. Total DNAs (10 µg) isolated from hairy root (Li-2) and nontransformed plant of *L. inflata* were digested with *Eco*RV and were fractionated in an 1% agarose gel, transferred to a nitrocellulose paper and hybridized with ^{32}P-labeled TL-DNA as described. *Lane 1* transformed hairy roots (Li-2); *lane 2* nontransformed plants of *L. inflata*. The *number at the left* show the molecular weight (in kb) of standard DNAs which migrated to the indicated positions

were detected in lane 2 where the DNA of the nontransformed plant was electrophoresed.

Since the presence of opines (agropine and mannopine) and TL-DNA in the hairy root Li-2 clone was demonstrated, this clone was used in further experiments.

3.2 Production of Lobeline in Hairy Root Cultures

HPLC analysis confirmed the presence of lobeline in the hairy root. The HPLC chromatogram of the extract of Li-2 revealed a peak at the retention time, 19.9 min which was identical to that of authentic lobeline (Fig. 5). Mass spectrometric analysis of this peak further supported the identification of lobeline.

Various media were tested for the growth of hairy roots as well as for the content of lobeline (Table 1). After 4 weeks of culture, more hairy roots were obtained from cultures in MS, 1/2 MS, and B5 medium (10–12 g, wet weight) while much less (4.2 g, wet weight) were obtained from cultures in NN medium. On the other hand, the highest lobeline content was observed in hairy roots cultured in NN medium. This point is more clearly shown from the last column of Table 1, where NN medium gave the highest yield of lobeline per unit wet weight. In agreement with this finding, Ko et al. (1988, 1989) reported that NN medium was suitable for the production of secondary metabolites in hairy root cultures of *Cassia occidentalis* and *Glycyrrhiza uralensis.* The lobeline content of hairy roots cultured in NN medium was the same order of magnitude compared with that of *L. inflata* cultivated in pots. From these results, we concluded that hairy root of *L. inflata* have the same capacity for lobeline synthesis as the mother plant roots. Furthermore, the lobeline content of hairy roots was similar to that of *L. inflata* callus (reported by Krajewiska et al. 1988). It should be noted that the growth rate of *L. inflata* hairy roots was much higher than that of *L. inflata* callus. Therefore the present hairy root culture of *L. inflata* appears to be a better system for production of lobeline than callus.

During serial cultivations of hairy roots in various media, morphological differences were observed between hairy roots cultured in NN medium and those cultured in other basal media. Hairy roots cultured in 1/2 MS, MS, and B5 medium produced extensive lateral branching and profuse root hairs (Fig. 6A). It is interesting to note that hairy roots in NN medium exhibited poor branching, short

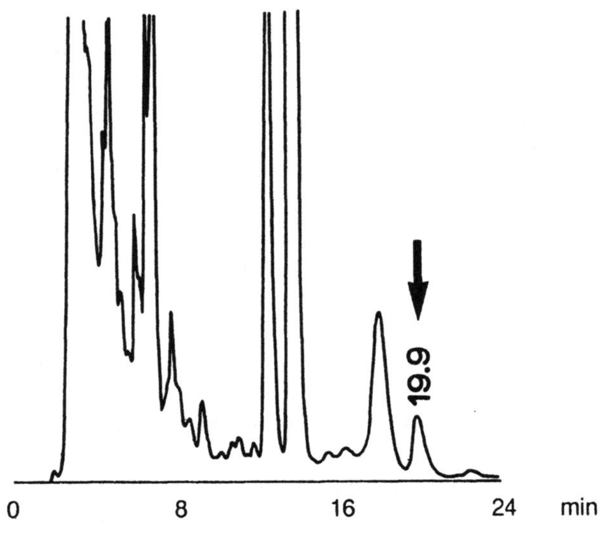

Fig. 5. HPLC profile of extract of *L. inflata* hairy roots. Extract from 20 mg dry weight of *L. inflata* hairy roots was applied on YMC-pack AM312 ODS column. Peak of lobeline is indicated by the *arrow*

Table 1. Growth and lobeline content of hairy roots of *Lobelia inflata*

Material	Medium	Fresh wt. (growth ratio) (g)	Dry wt. (growth ratio) (g)	Lobeline content per unit dry wt. (μg/g dry wt.)	Lobeline content per unit frest wt. (μg/g fr. wt.)
Hairy roots	MS	12.0 ± 0.7 (60.0)	0.62 ± 0.03 (60.2)	19 ± 2.0	1.0 ± 0.11
	1/2 MS	12.2 ± 0.3 (61.0)	0.74 ± 0.01 (74.7)	25 ± 0.5	1.5 ± 0.05
	B5	10.0 ± 0.1 (50.0)	0.65 ± 0.01 (55.1)	18 ± 0.4	1.2 ± 0.03
	NN	4.2 ± 0.1 (21.0)	0.66 ± 0.01 (21.1)	54 ± 1.6	8.6 ± 0.26
Roots of mother plants	–	–	–	24 ± 2.1	–

Hairy roots were precultured in each liquid medium at 25 °C in the dark on a gyratory shaker (80 rpm) for 4 weeks with medium change at 2 weeks of culture. After this adaptation period, approximately 0.2 g fresh wt. of each hairy root was further cultured in the same liquid medium under the same condition for 4 weeks and weighed. The value of the growth ratio are the quotient of the weight after 4-week culture and the weight of the inoculum. (On the basis of dry wt., approximately 30 mg of root tips were inoculated into NN medium and approximately 10 mg were inoculated into each of the other three media.) Values are means ± standard error of three replicates. *L. inflata* mother plants were cultured in pots at 25 °C under a 16-h photoperiod (98 μE/m²/s) for 2–3 months.

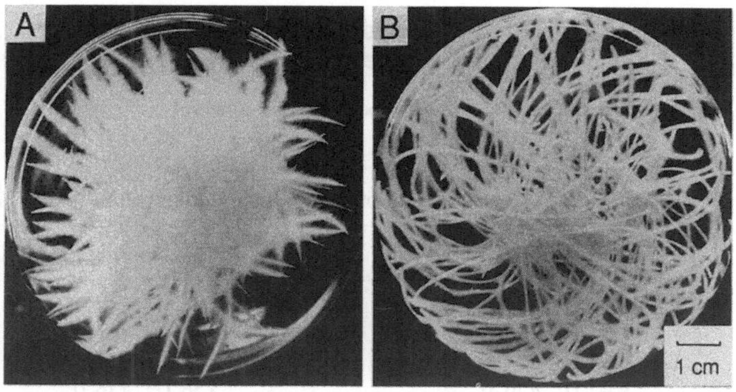

Fig. 6A, B. Morphological differences between hairy roots cultured in 1/2 MS and NN liquid media. **A** Hairy roots cultured in 1/2 MS medium for 2 weeks. **B** Hairy roots cultured in NN medium for 2 weeks

root hairs, and somewhat harder tissue compared with others (Fig. 6B). When a part of the *L. inflata* hairy root grown in NN medium was cultured in MS medium, the root morphology reverted to that grown in MS medium. From the appearances of hairy root in our experiment, NN medium seems to produce the hairy root with less water content and more tightly packed tissues, resembling those of the mother plant. This is probably related to higher production of lobeline by these tissues. Variations of morphology due to different culture media have been reported in other systems such as hairy roots of *Datura* (Hamill et al. 1987) and *Duboisia* (Mano et al. 1989). Together with our finding these reports suggest that the production of secondary product and morphology of hairy roots are strongly affected by the culture medium.

3.3 New Polyacetylenes in the Hairy Root Culture

During the course of chemical studies on the hairy root of *L. inflata*, two new polyacetylene compounds, lobetyolin (9-0-β-D-glucopyranosyl-2,10-tetradecadien-4,6-diyne-8,14-diol) and lobetyol (2,10-tetradecadien-4,6-diyne-8,9,14-triol) were also isolated (Fig. 7). Polyacetylenes are generally found in plants of the Compositae and Umbelliferae etc., and rarely appear in other families. These two compounds were isolated from methanol extract of the hairy root with Sephadex LH-20 and silica gel column chromatographies. The yield was 200 and 8 mg for lobetyolin and lobetyol, respectively. For determination of the structure of these polyacetylenes, mass spectrometry (MS), and nuclear magnetic resonance (NMR) (^{1}H and ^{13}C) spectra were used. The ^{1}H-NMR spectral data of lobetyolin, lobetyol,

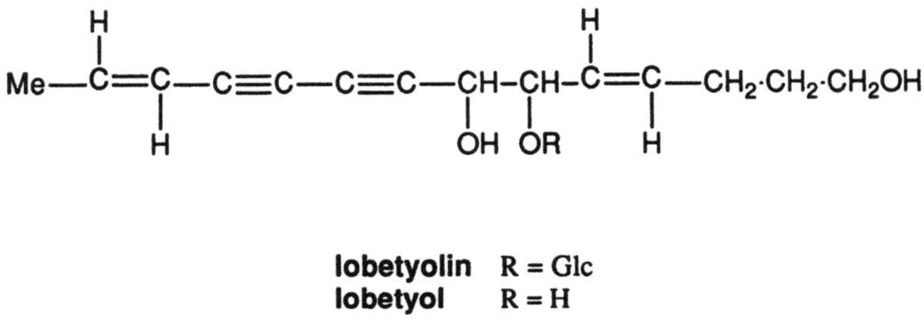

Fig. 7. Chemical structures of lobetyolin and lobetyol

Table 2. ^{1}H NMR spectral data of lobetyolin and lobetyolin acetate (270 MHz, δ values)

H	Lobetyoiln (ME_2CO-d_6)	Lobetyolin acetate ($CDCl_3$)
1	1.73 3H, *dd* (6.8, 2.3)	1.82 3H, *dd* (7.0, 2.3)
2	6.29 1H, *dq* (16.2, 6.8)	6.34 1H, *dq* (15.9, 7.0)
3	5.55 1H, *dd* (16.2, 2.3)	5.51 1H, *dd* (15.9, 2.3)
8	4.40 1H, *d* (6.8)	5.56 1H, *d* (6.7)
9	4.20 1H, *t* (6.8)	4.30 1H, *dd* (6.7, 7.6)
10	5.40 1H, *dd* (16.2, 6.8)	5.42 1H, *dd* (15.3, 7.6)
11	5.86 1H, *dt* (16.2, 6.8)	5.83 1H, *dt* (15.3, 7.6)
12	2.09 2H, *br dd* (13.5, 6.8)	2.22 2H, *br dd* (13.8, 7.6)
13	1.56 2H, *quin* (6.8)	1.76 2H, *quin* (7.6)
14	3.49 2H, *t* (6.8)	4.10 2H, *br t* (7.6)
Glc		
1′	4.32 1H, *d* (7.5)	4.59 1H, *d* (8.0)
2′	3.24 1H, *m*	4.98 1H, *dd* (9.2, 8.0)
3′	3.24 1H, *m*	5.19 1H, *t* (9.2)
4′	3.34 1H, *t* (8.3)	5.09 1H, *t* (9.2)
5′	3.24 1H, *m*	3.66 1H, *m*
6′	3.57 1H, *dd* (12.1, 6.8)	4.25 1H, *dd* (12.2, 4.6)
	3.77 1H, *dd* (12.1, 2.7)	4.14 1H, *m*
Ac		2.00, 2.02, 2.04, 2.07, 2.08 2.10 each 3H,*s*

Coupling constants (*J* in Hz) in parentheses.

Table 3. ^{1}H NMR spectral data of compounds lobetyol and lobetyol acetate (270 MHz, δ values in $CDCl_3$)

H	Lobetyol	Lobetyol acetate
1	1.82 3H, *dd* (6.8, 2.3 Hz)	1.83 3H, *dd* (6.8, 2.3 Hz)
2	6.34 1H, *dq* (16.2, 6.8 Hz)	6.34 1H, *dq* (16.2, 6.8 Hz)
3	5.52 1H, *dd* (16.2, 2.3 Hz)	5.50 1H, *m*
8	4.29 1H, *d* (6.4 Hz)	5.50 1H, *m*
9	4.13 1H, *t* (6.4 Hz)	5.40 1H, *m*
10	5.57 1H, *dd* (16.2, 6.4 Hz)	5.50 1H, *m*
11	5.88 1H, *dt* (16.2, 6.4 Hz)	5.87 1H, *dt* (16.2, 6.8 Hz)
12	2.18 2H, *br dd* (13.5, 6.4 Hz)	2.16 2H, *br dd* (13.8, 6.8 Hz)
13	1.68 2H, *quin* (6.4 Hz)	1.73 2H, *quin* (6.8 Hz)
14	3.67 2H, *t* (6.4 Hz)	4.06 2H, *t* (6.8 Hz)
Ac		2.05, 2.08, 2.09 each 3H.s

and their acetates are shown in Tables 2 and 3. The ^{13}C-NMR spectral data of these compounds are also given in Table 4. The conjugated diyne structure of these polyacetylenes was also confirmed by the specific absorptions in the infrared (νKBr: 2240 cm^{-1}) and ultraviolet (λH_2O: 207.2, 241.2, and 267.6 nm) spectra.

Lobetyolin: an off-white amorphous powder, $[\alpha]^{28}$ $-8.9°$ (Me_2CO; Conc. 0.12). ^{1}H-NMR see Table 2; ^{13}C-NMR: see Table 4; FAB-MS *m/z* (rel. int.): 397 $[M+H]^+$ (28), 419 $[M+Na]^+$ (15), 435 $[M+K]^+$ (8); IRν KBr cm^{-1}: 3375 (OH), 2940 (Me), 2870 (CH_2), 2240 ($C\equiv C$), 1675 ($C=C$), 1390 (CH_3), 1050 (CH_2OH), 980 ($C=C$); UV λ H_2O nm (log ε): 207.2 (4.51), 215.0 (4.59), 241.2 (3.63), 253.6 (3.92), 267.6 (4.30), 283.6 (3.94), (Found: C, 58.63; H, 7.63. $C_{20}H_{28}O_8$ $1/2H_2O$ requires: C, 59.24; H, 7.21).

Acetylation of lobetyolin: lobetyolin (27 mg) was acetylated with Ac_2O (1 ml) and pyridine (1 ml) overnight at room temperature. Usual work-up gave the product, which was chromatographed over silica gel 60 CC (2 × 16 cm). Elution with C_6H_6-Me_2CO (1:0–15:1) furnished the hexaacetyl lobetyolin (40.9 mg) as an off-white amorphous powder, $[\alpha]^{26}$ $-71.4°$ ($CHCl_3$; Conc. 0.55). ^{1}H NMR: see Table 2; ^{13}C NMR: see Table 4; FAB-MS *m/z* (relative intensity): 649 $[M+H]^+$ (32), 671 $[M+Na]^+$ (100), 687 $[M+K]^+$ (23); UV λ $CHCl_3$ nm (log ε): 256.0 (3.57), 270.0 (3.70), 286.2 (3.58).

Hydrolysis of lobetyolin with β-glucosidase: a solution of lobetyolin (70 mg) in 0.02 M NH_4OAc (6 ml) was incubated with β-glucosidase (Sigma, No. G-8625) overnight at 35 °C. The reaction mixture was poured into H_2O (20 ml) and extracted with EtOAc (3 × 5 ml). The EtOAc layer was dried under vacuum and subjected to silica gel 60 CC (2 × 14 cm, C_6H_6–Me_2CO, 9:1 – 3:1) to yield the aglycone (35.9 mg) which was identical to lobetyol.

Lobetyol: an off-white amorphous powder, $[\alpha]^{26}$ $+18.8°$ (Me_2CO; Conc. 0.13). ^{1}H NMR: see Table 3; ^{13}C NMR: see Table 4; FAB MS *m/z* (relative intensity): 257 $[M+Na]^+$ (47), 273 $[M+K]^+$ (32).

Acetylation of lobetyol: lobetyol (35 mg) was acetylated with Ac_2O (1 ml) and pyridine (0.5 ml) overnight at room temperature. This sample was purified by silica gel 60 CC (2 × 15 cm, C_6H_6–Me_2CO, 19:1–18:1) to give triacetyl lobetyol

Table 4. ^{13}C NMR spectral data of lobetyolin, lobetyol and their acetate form (67.5 MHz, values)

C	Lobetyolin[a]	Lobetyolin acetate[b]	Lobetyol[b]	Lobetyol acetate[b]
1	19.2	18.9	18.9	18.9
2	145.8	144.7	144.6	144.9
3	110.4	109.4	109.4	109.4
4	82.4	78.0	79.1	78.3
5	78.2	75.6	78.1	74.9
6	72.9	71.6	71.4	72.0
7	70.8	71.5	71.4	71.3
8	66.5	65.6	66.9	65.1
9	81.8	78.3	75.5	73.6
10	126.4	124.5	127.5	123.8
11	138.6	138.1	135.2	136.7
12	29.7	28.8	28.9	28.7
13	33.0	28.0	31.8	27.7
14	61.9	63.5	62.3	63.6
Glc				
1′	100.8	97.1		
2′	74.7	71.2		
3′	77.9	72.9		
4′	71.5	68.5		
5′	77.8	71.7		
6′	62.7	62.0		
Ac		20.6		20.7
		20.7		20.9
		21.0		
C=O		169.1		169.2
		169.3		169.5
		169.4		171.0
		170.3		
		170.7		
		171.1		

[a] In acetone-d_6.
[b] In chloroform-*d*.

(49.6 mg), an off-white amorphous power, $[\alpha]^{26}$ – 79.2° ($CHCl_3$; Conc. 0.28). ^1H NMR; see Table 3; ^{13}C NMR: see Table 4; FD MS *m/z* (relative intensity): 360 $[M]^+$ (100), UV λ $CHCl_3$ (log ε); 256.2 (3.52), 270.2 (3.65), 286.2 (3.52).

Polyacetylenes isolated from plants are generally not conjugated with highly hydrophilic molecules (such as sugars). It was therefore unexpected that the polyacetylene (lobetyolin) found in *L. inflata* hairy root existed as glycoside. This result suggests the presence of a high level of glycosylation enzyme for this compound in this plant.

The content of these polyacetylene (lobetyolin and lobetyol) in mother plant, plantlet *in vitro* and hairy root were also determined by the HPLC analysis (Table 5). The content of lobetyolin (19.51 mg/g dry wt.) in the hairy root cultured in hormone free MS liquid medium in the dark for 4 weeks was about 1.5-fold of that of the roots of the mother plant (12.38 mg/g dry wt.) which was grown in the

Table 5. Content of polyacetylenes (lobetyolin and lobetyol) in *Lobelia inflata*

Material	Content (mg/g dry wt.)	
	Lobetyolin	Lobetyol
Hairy root	19.51	1.44
Root of mother plant	12.38	n.d.
Leaf of mother plant	9.91	n.d.
Root of plantlet in vitro	7.11	2.42
Leaf of plantlet in vitro	0.32	n.d.
Stem of plantlet in vitro	1.25	+

(n.d.: not detected, + : < 0.2 mg/g dry wt.)

field and harvested in June. The content of lobetyolin in the plantlet cultured in MS solid medium for 2 months under a 16-h photoperiod condition was fairly low (especially in the leaf and stem) compared to those of the mother plant and hairy roots. The content of lobetyol (2.42 mg/g dry wt.) in the root portion of the plantlet was slightly higher than that of the hairy root. However, considering the rapid growth of the hairy root culture of this plant, the hairy root system seems to be the most effective method for production of these polyacetylenes.

4 Summary and Conclusions

Hairy roots of *Lobelia inflata* were obtained following infection with *Agrobacterium rhizogenes* 15834. The lobeline content of the hairy roots was of the same order of magnitude as that of the root of the mother plants.

Various media were tested for the growth of hairy roots as well as for lobeline content in hairy roots. These results suggest that the physiological conditions of transformed root cells of *L. inflata* can be changed under the influence of its culture medium. Although hairy roots cultured in NN medium grow relatively slowly, they produced the secondary metabolite lobeline in much higher yield.

Two new polyacetylene compounds (lobetyolin and lobetyol) were isolated from hairy roots. Lobetyol was detected in cultured roots (hairy roots and roots of plantlet), but not detected in mother plants grown in the field. The tissue culture under the controlled conditions might bring the capability of the production or accumulation of the new compound lobetyol in the roots of *L. inflata*. Hence, the hairy root system provides the possibility for production of a new compound which has hitherto escaped detection in the plants due to its low content.

Acknowledgment. We thank Dr. H. Kamada (University of Tsukuba) for providing the authentic agropine and mannopine, and T. Ogata (Research Institute for Molecular Genetics, Tsumura) for helping with HPLC analysis. This work was supported by a grant from the Japanease Health Science Foundation, Ministry of Health and Welfare.

References

Aird ELH, Hamill JD, Rhodes MJC (1988) Cytogenetic analysis of hairy root cultures from a number of plant species transformed by *Agrobacterium rhizogenes*. Plant Cell Tissue Organ Cult 15:47–57

Davison GC, Rosen RC (1972) Lobeline and reduction of cigarette smoking. Psychol Rep 31:443–456

Gamborg OL, Miller RA, Ojima K (1968) Nutrient requirements of suspension cultures of soybean root cells. Exp Cell Res 50:151–158

Hamill JD, Parr AJ, Rhodes MJC, Robins RJ, Walton NJ (1987) New routes to plant secondary products. Bio/Technol 5:800–804

Kalashnikov VP (1939) The contents of alkaloids in the cultivated and in the wild types of lobelia. Farmatsiya 1:20–28

Ko KS, Ebizuka Y, Noguchi H, Sankawa U (1988) Production of secondary metabolites by hairy roots and regenerated plants transformed with Ri plasmids. Chem Pharm Bull 36:4217–4220

Ko KS, Noguchi H, Ebizuka Y, Sankawa U (1989) Oligoside production by hairy root cultures transformed by Ri plasmids. Chem Pharm Bull 37:245–248

Krajewska A, Szoke E, Szarvas T (1988) The influence of the new synthetic cytokinins on *Lobelia inflata* L. tissue culture. Herba Pol 34:27–34

Krochmal A, Wilken L, Chien M (1972) Plant and lobeline harvest of *L. inflata* L. Econ Bot 26:216–220

Mano Y, Nabeshima S, Matsui C, Ohkawa H (1986) Production of tropane alkaloids by hairy root cultures of *Scopolia japonica*. Agric Biol Chem 50:2715–2722

Mano Y, Ohkawa H, Yamada Y (1989) Production of tropane alkaloids by hairy root cultures of *Duboisia leichhardtii* transformed by *Agrobacterium rhizogenes*. Plant Sci 59:191–201

Marshall WR (1928) Alpha lobelin as a respiratory stimulant. Arch Int Med 42:180–188

Murashige T, Skoog F (1962) A revised medium for rapid growth and bio assays with tobacco tissue cultures. Physiol Plant 15:473–497

Nitsch C, Nitsch JP (1967) The induction of flowering *in vitro* in stem segments of *Plumbago indica* L. Planta (Berl.) 72:355–370

Otten LABM, Schilperoort RA (1978) A rapid micro scale method for the detection of lysopine and nopaline dehydrogenase activities. Biochem Biophys Acta 527:497–500

Petit A, David C, Dahl GA, Ellis JG, Guyon P, Casse-Delbart F, Tempé J (1983) Further extention of the opine concept: Plasmids in *Agrobacterium rhizogenes* cooperate for opine degradation. Mol Gen Genet 190:204–214

Reavill C, Walther B, Stolerman IP, Testa B (1990) Behavioural and pharmacokinetic studies on nicotine, cytisine, and lobeline. Neuropharmacology 29:619–624

Saito K, Kaneko H, Yamazaki M, Yoshida M, Murakoshi I (1990) Stable transfer and expression of chimeric genes in licorice *Glycyrrhiza uralensis*, using an Ri plasmid binary vector. Plant Cell Rep 8:718–721

Schöpf C, Kauffmann T, (1957) Über die stärker hydrophilen Nebenalkaloide aus *Lobelia inflata* L.; ein Beitrag zur Biogenese der Lobelia-alkaloide. Annalen der Chemie 608:88–127

Slightom JL, Durand-Tardif M, Jouanin L, Tepfer D (1986) Nucleotide sequence analysis of TL-DNA of *Agrobacterium rhizogenes* agropine type plasmid. J Biol Chem 261:108–121

Southern EM (1975) Detection of specific sequence among DNA fragments separated by gel electrophoresis. J Mol Biol 98:503–517

Trevelyan WE, Procter DP, Harrison JS (1950) Detection of sugars on paper chromatograms. Nature 166:444–445

Vervliet G, Holsters M, Teuchy H, Montagu M van, Schell J (1975) Characterization of different plaque-forming and defective temperate phages in *Agrobacterium* strains. J Gen Virol 26:33–48

Wysokinska H (1977) Wystepowanie alkaloidow w hodowli tkankowej *Lobelia inflata* L. Farm Pol 33:725–727

II.14 Transformation in *Lycopersicon esculentum* L. (Tomato)

C. BELLINI[1]

1 Introduction

Cellular genetic manipulations offer new possibilities for breeding tomato. Until the last few years, the introduction of favorable characteristics such as resistance to pathogens, modification of plant habit, and improvement of fruit quality have been carried out by crossing with wild species (Rick 1978). Nevertheless, because of the unilateral or interspecific incompatibility between tomato and its wild relatives, a vast reserve of genetic resources remains unexploited (Rick 1982). Protoplast fusion can be an alternative method for mixing genomes and the development of plant genetic transformation offers new possibilities. Transformation of tomato cultivars was achieved using Ti or Ri plasmid vectors and stable transformants were selected on medium containing kanamycin (Koorneef et al. 1986; McCormick et al. 1986; Shahin et al. 1986). This could be achieved because of the relative facility to regenerate tomato plants from primary explants. Direct gene transfer procedures are useful for inserting new genes into the genome, as it is not necessary to clone the gene to be transferred into vectors derived from *Agrobacterium*. In addition, they offer the possibility of performing transient gene expression assays that facilitate comparisons between the strength of promotors that regulate the expression of structural genes (Ou-Lee et al. 1986; Boston et al. 1987), and permit investigation into the interactions between several genes in the same cell (Ecker and Davis 1986). This approach requires an efficient procedure of protoplast culture and plant regeneration. Regeneration of plants from tomato protoplasts still presents several difficulties and makes the use of direct gene transfer techniques less attractive. Direct gene transfer in tomato was first achieved using a calcium-phosphate DNA transformation procedure or a PEG treatment (Koorneef et al. 1986; Jongsma et al. 1987). Toyoda et al. (1989) adapted the microinjection technique to tomato callus cells, but did not succeed in regenerating transformed plants. More recently, Tsukuda et al. (1989) compared the efficiency of promoters in the transient expression of foreign genes introduced in tomato protoplasts by electroporation.

Electroporation of plant protoplasts has already led to the transformation of several crop plants such as oilseed rape (Guerche et al. 1987b), lettuce (Chupeau et al. 1989) and potato (Masson et al. 1989) in our laboratory, and now has been extended to the transformation of tomato protoplasts. This, combined with the optimization of tomato protoplast culture (Bellini et al. 1990), enabled us to obtain

[1] Laboratoire de Biologie Cellulaire, I.N.R.A., Route de Saint-Cyr, 78026 Versailles Cédex, France

Biotechnology in Agriculture and Forestry, Vol. 22
Plant Protoplasts and Genetic Engineering III (ed. by Y.P.S. Bajaj)

stable transformants from both *L. esculentum* and *L. peruvianum* (Bellini et al. 1989a). In this chapter the optimized methodologies and the factors influencing the success of electroporation of tomato protoplasts are discussed.

2 Parameters Important for the Preparation and Culture of Tomato Protoplasts

In the transformation experiments that will be described later, we chose to use as a model system *Lycopersicon peruvianum* (LA2172), because of its good regeneration ability, and two cultivars of *Lycopersicon esculentum*, Manapal and Golden Sunrise. The most relevant parameters for the preparation and culture of tomato protoplasts are summarized. Tables 1 and 2 describe the protocol and the different optimized media, respectively.

2.1 Light Intensity and Photoperiod

Several reports dealing with the preparation and culture of tomato protoplasts have already been published (Shahin 1985; Niedz et al. 1985; Tan et al. 1987). The authors dwelt on the preconditioning of the donor plant. Under our conditions, it was important to grow the donor plants under low light intensity (63 $\mu E/m^2/s$). These results are in agreement with previous studies, which showed that growing plants under low light intensity was a prerequisite to obtain a high yield of viable protoplasts from tobacco (Chupeau et al. 1974), lettuce (Chupeau et al. 1989), and various species of *Solanum* (Hassanpour-Estahbani and Demarly 1984; Tan et al. 1987) and *Lycopersicon* (Cassells and Barlass 1976, Tabaeizedeh et al. 1984). In these conditions, plants synthesize cell wall with fewer pectate components (Cassells and Barlass 1976), and therefore lower concentrations of enzymes can be used for protoplast isolation. The toxic effect of enzyme preparations is then décreased and the production of viable protoplasts is facilitated. The best results were obtained when plants were grown at 63 µE/m/s, with 16 h of light in the case of *L. esculentum* and 8 h light in the case of *L. peruvianum*.

2.2 Influence of the pH of the Digestion Medium for Obtaining *L. esculentum Protoplasts*

In the first experiments, very few viable protoplasts could be obtained when a digestion medium with the standard pH 5.8 was used, even though the enzyme concentration in medium GSG was lower than that used by other groups (Zapata et al. 1981; Shahin 1985; Tan et al. 1987), the degradation of the cell wall seemed to be too rapid. This probably damages the plasma membrane and then decreases the yield of viable protoplasts. Preplasmolysis of leaf tissue, before protoplast isolation, was not found to be efficient for obtaining viable protoplasts. This can be explained by the fact that donor plants were grown in a humid environment (90% humidity)

Table 1. Protocol for isolation and culture of *L. esculentum* and *L. peruvianum* protoplasts

Operation	Description
Growth of the donor plants	Germinate seeds on a wet filter paper, in a Petri dish. Transfer seedlings in soil and grow in the greenhouse under 200 μE/m²/s, until they have four expanded leaves. Place the plants in a controlled environment chamber, under low light intensity (63 μE/m²/s) and a photoperiod of 8 h for *L. peruvianum* and 16 h for *L. esculentum*, with a temperature of 25 °C during the day and 17 °C during the night. Hygrometry 90%. After 1 week the first subapical expanded leaves developed in these conditions can be used for protoplast isolation
Preparation of protoplasts	
[a] *L. esculentum*	Sterilize leaves for 30 min in a chloride solution: 1 tablet of Bayrochlor (sodium dichloroisocyanurate releasing 1.5 g of active chloride) + 800 ml distilled water + 5×10^{-3} (w/v) Tween 80. Then wash three times with 800 ml of sterile distilled water. Slice leaves into 1 mm strips and put them in Petri dishes containing 9/10 of medium I_{10}^{a}, pH 6.5, and 1/10 of the enzymatic mixture[b]. Place in the dark, at 22 °C for 16 h without shaking. Separate protoplasts from the leaf fragments by slightly shaking the dishes. Filter through a stainless sieve (mesh size = 80 μm) and pellet by centrifugation at 70 g for 6 mn. Resuspend the pellet in the washing medium (WM). Wash twice with this medium. Count the protoplasts and adjust density to 2×10^6 protoplast/ml.
[b] *L. peruvianum*	Use the same procedure but replace medium I_{10} by medium GSG[c]
Protoplast culture and plant regeneration	
[a] *L. esculentum*	Plate protoplasts at a density of 10^5 p/ml in medium C_1 and place in the dark at 28 °C. One week after the first division dilute the protoplast-derived colonies four times with medium C_1D. Leave them in the dark for a further 2 weeks and then progressively increase light intensity to 63 μE/m²/s
[b] *L. peruvianum*	Plate protoplasts at a density of 6.5×10^4 p/ml in medium MC and incubate in the dark at 28 °C. Three to 4 days later place the protoplasts derived colonies in the light (63 μE/m²/s) at 28 °C. After 10 days dilute the colonies four times in the same medium containing only 0.2 mg/l BAP and 0.1 mg/l NAA. After 1 month, transfer microcalli on RCP[d] medium for regeneration. Buds appear after 1 to 2 months. Cut them off from the calli and transfer them to medium B[e] for root induction

[a] Bellini et al. (1989).
[b] Enzyme mixture (10 X): 0.2% (w/v) Macerozyme R_{10}, 1% (w/v) Onozuka R_{10} (Yakult Biochemicals, Japan), 0.5% Driselase (FLUKA).
[c] Chupeau et al. (1990).
[d] Masson et al. (1987).
[e] Bourgin et al. (1979).

Table 2. Protoplast preparation and culture media. (Bellini et al. 1990)

	I_{10}	GSG	WM	EB	MC	C_1	C_1D
Macronutrients mM							
NH_4NO_3					5	5	5
KNO_3	2.5	2.5			15	40	15
$CaCl_2$, $2H_2O$	0.5	0.5	14		3	3	3
$MgSO_4$, $7H_2O$	0.1	0.1			3	3	3
KH_2PO_4					0.5	0.5	0.5
$(NH_4)2SO_4$	0.1	0.1					
NaH_2PO_4, H_2O	0.11	0.11					
KCl			260	5			
Micronutrients mM							
KI	4.5	4.5					
H_3BO_3	50	50			100	100	100
$MnSO_4$, H_2O	60	60					
$MnCl_2$					10	10	10
$ZnSO_4$, $7H_2O$	7	7			10	10	10
Na_2MoO_4, $2H_2O$	1	1			0.1	0.1	0.1
$CuSo_4$, $5H_2O$	0.1	0.1			0.1	0.1	0.1
$CoCl_2$, $6H_2O$	0.1	0.1			0.1	0.1	0.1
Fe Citrate	150	150			150	150	150
Vitamins (mg/l)							
Ca Pantothenate	1	1			1	1	1
Biotine	0.01	0.01			0.01	0.01	0.01
Niacine	1	1			1	1	1
Pyridoxine	1	1			1	1	1
Thiamine	1	1			1	1	1
Adenine sulfate						10	10
Osmoticum (g/l)							
Inositol	100				5	5	5
Sucrose					20	60	60
Mannitol		45		90	65	25	25
Sorbitol		25					
Glucose		15					
Growth substances (mg/l)							
BAP	0.5	0.5			0.5	0.5	0.5
2,4-D					0.5	0.5	
NAA	1	1			1	1	1
Other							
BCP mg/l	8	8	8	8	8	8	8
MES mM	35	35	3.5		3.5	3.5	3.5
MOPS μM				200			
Tween 80 μl/l	10	10	10	10	10	10	10

Media were sterilized by autoclaving for 20 min at 120 °C. Growth factors and iron citrate were added after autoclaving.

and regularly watered. In these conditions the intra-cellular osmotic pressure could be maintained at a low level. This renders the plasmolysis in media with a high sugar concentration unnecessary. The increase of the pH of media GSG and I_{10} was the most efficient way to reduce the enzyme activity.

2.3 Use of myo-Inositol as an Osmoticum for the Preparation and Culture of Protoplasts

The use of myo-inositol as osmoticum in the digestion medium enhanced the stability of the tomato protoplasts. It acted differently on protoplasts relative to the other osmotica. Ettlinger and Lehk (1987) have shown that, after labeling *Catharanthus roseus* cells with ^{3}H-myo-inositol with a 3-h incubation, 5% of the radioactivity was found in the lipid fraction, 2% in the cell wall polymers, and less than 1% was composed of inositol phosphates, the rest being unmetabolized myo-insitol. In the experiment described here, the role of myo-inositol can be explained by the fact that after 16-h incubation a certain amount of myo-inositol has accumulated in the tomato cells. Part of the myo-inositol led to the formation of phosphatidylinositol, which is a component of the plasma membrane. This could explain the improvement of the stability of protoplasts. Myo-inositol seems also to be an important element for tomato protoplast culture, since it is often found at concentrations higher than 1 g/l in the different culture media described in the literature (Shahin 1985; O'Connel and Hanson 1985; Tan et al. 1987). Masson et al. (1987) have also shown that myo-inositol enhances the division of potato protoplasts. In our conditions the best results were obtained with 5 g/l of myo-inositol. Higher concentrations rapidly became toxic, probably because of the accumulation of myo-inositol and of the products of its metabolization.

2.4 Role of Calcium in Protoplast Culture

An appreciable improvement of the plating efficiency was obtained when 40 mM $CaCl_2$ was added to the culture medium (Fig. 1). This improvement can be

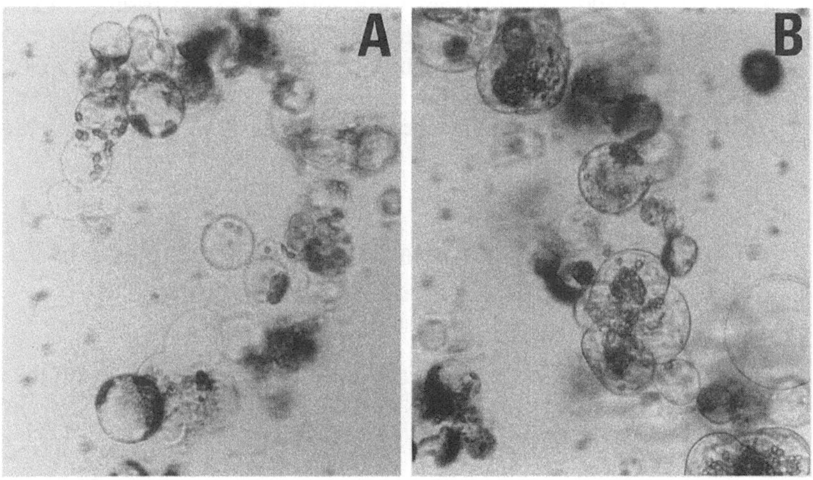

Fig. 1A, B. Influence of calcium concentration on plating efficiency. **A** protoplasts of genotype Golden Sunrise 10 days after plating in a medium without calcium. **B** protoplasts of genotype Golden Sunrise 10 days after plating in a medium containing 40 mM $CaCl_2$. (Bellini et al. 1990)

explained by the stabilization of protoplasts due to interactions between the plasma membrane and the calcium, but also by an enhancement of cell division induced by the increase of the intracellular level of calcium. The role of calcium has already been implicated in cell division and proliferation in higher plants (Trewavas 1985; Tanimoto and Harada 1986; Das et al. 1987). This may be correlated with the importance of myo-inositol described above. Das et al. (1987) have shown the possible involvement of the phosphoinositide cycle and calcium ions in cell proliferation. Concentration of calcium higher than 40 mM inhibited the growth of colonies. This can be explained by the fact that a too high concentration of calcium inhibits the polymerization of microtubules and then stops the cell division.

2.5 Influence of Ammonium Nitrate on Protoplast Culture

Ammonium nitrate was earlier shown to be toxic to tomato mesophyll protoplasts (Zapata et al. 1981). Under our conditions, 5 mM ammonium nitrate was found to be necessary for tomato protoplast culture. Similar results were obtained by Masson et al. (1987), who found that 10 mM NH_4NO_3 enhanced the plating efficiency of potato protoplasts over that obtained on media containing only nitrate. Ammonium ions are generally an important nitrogen source for plant tissue culture. Meyer and Abel (1975) have shown that a stimulating effect on cell wall formation of tobacco protoplasts may be due to the accumulation of NH_4^+ in the medium, as glutamine alone did not stimulate cell wall formation.

2.6 Influence of Sugar Concentration

The use of 60 g/l of sucrose led to a significant increase of the plating efficiency (Bellini et al. 1990). No divisions were observed in the absence of sucrose. Sucrose and glucose were found to be necessary for tomato protoplast culture (Zapata et al. 1981). Medium TM2 described by Shahin (1985) contains 68 g/l of sucrose, and Mühlbach (1980) has shown that using glucose as osmoticum, instead of mannitol, greatly enhanced plating efficiency of protoplasts of tomato cv. Hilda 72.

3 Direct Gene Transfer into Mesophyll Protoplasts of *L. esculentum* and *L. pervianum* by Electroporation

The experimental protocol is described in Table 3. Comments on the relevant steps are reported here.

3.1 Chloramphenicol Acetyl Transferase Transient Activity Assay

The CAT transient activity assay has been proved to be an efficient technique for optimizing electroporation conditions on plant protoplasts (Fromm et al. 1986;

Table 3. Protocol for transformation of *L. esculentum* and *L. peruvianum* protoplasts by electroporation

Operation	Description
Preparation of protoplasts	Prepare protoplasts of *L. esculentum* and *L. peruvianum* as described in Table 1 Wash once with the electroporation buffer (EB, pH 7.2) Resuspend the protoplasts in EB, at a density of 2×10^6 p/ml Maintain the protoplast suspension in ice
Preparation of plasmids	Just before the electroporation resuspend the plasmid[a] to be used, in the electroporation buffer, at a concentration of 1 μg/ml
Electroporation	Mix 50 μg of sonicated calf thymus DNA + 10 μg of plasmid + 2×10^6 protoplasts. Put this mixture in the electroporation chamber and deliver three pulses at 250 V in the case of *L. peruvianum* protoplasts or at 300 V in the case of *L. esculentum*[b]. Then culture the protoplasts as described in Table 1
CAT transient activity assay	Three days after the elctroporation harvest the colonies derived from 2×10^6 protoplasts and pellet them by centrifugation at 70 g for 6 min Rinse with washing medium and resuspend in 500 μl of CAT buffer [c] containing 2.5% Polyclar powder. Then proceed as described by Guerche et al. (1987a)
Selection of stable transformants	For the culture of electroporated protoplasts proceed as described in Table. 1. When diluting the colonies add the selective agent[d] in the dilution medium. Maintain the selective pressure until buds appear

[a] Plasmids were extracted according to the procedure of Holmes (1982) and purified by a centrifugation in a CsCl gradient and stored at − 20 °C, precipitated in ethanol.
[b] Electric pulses were delivered at 5 s intervals, by a 16 μF capacitor charged with an ISCO 490 generator. The pulse length was 100 ms.
[c] CAT buffer: 1 μM leupeptine, 30 mM ascorbic acid, 20 mM EDTA, 250 mM Tris-HCl, pH 8.
[d] Paromomycin was used at a concentration of 20 mg/l to select kanamycin-resistant colonies as described by Bellini et al. (1990).
Chlorsulfuron was used at a concentration of 50 nM to select chlorsulfuron-resistant colonies.

Guerche et al. 1987a). Protoplasts of *L. esculentum* were electroporated in presence of the plasmid pCaMVCAT (Fromm et al. 1985). Different voltages and number of pulses were tested. Voltage was critical to obtain a maximum CAT activity. An optimum of transient activity was obtained with 250 or 300 V (Fig. 2). When higher voltages were used, protoplasts did not divide or died very soon after elctroporation (results not shown). Generally, no significant difference was observed when 1, 2, or 3 pulses were delivered. In the nonelectroporated controls (Fig. 2), a signal corresponding to acetylated derivatives of chloramphenicol was detected. This signal is due to the presence of an endogenous CAT activity, since it is detected with the same intensity in all conditions tested. A similar situation has been described previously for tobacco (Guerche et al. 1987a). These conditions are very close to those developed, in our laboratory, for the electroporation of tobacco protoplasts.

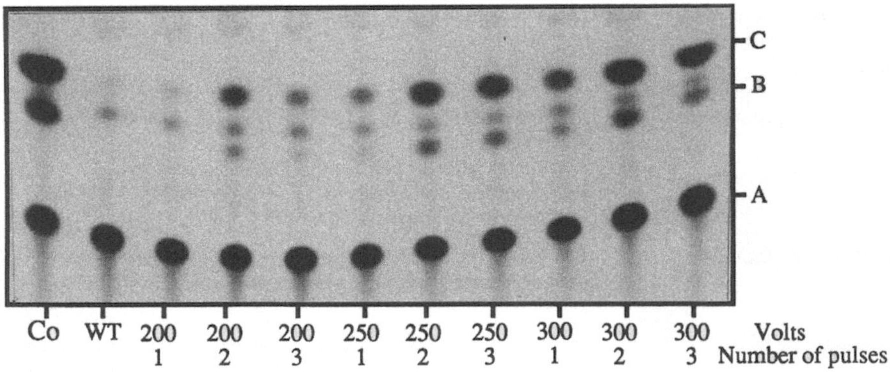

Fig. 2. CAT transient activity after electroporation of protoplasts of *L. esculentum* cv. Manapal. *Co* control (cellular extract of tobacco protoplasts electroporated with plasmid pCaMVCAT and expressing a high CAT activity was used as control); *WT* wild type protoplasts (i. e., nonelectroporated); *A* chloramphenicol; *B* 1-acetylchloramphenicol; *C* 3-acetylchloramphenicol. (Bellini et al. 1989a)

3.2 Selection of Stable Transformants

Protoplasts of tomato were electroporated in the presence of plasmid pGH1 (Haughn and Summerville 1986) which contains the entire gene coding for a mutant form of the acetolactate synthase of *Arabidopsis thaliana.* This mutation confers resistance to chlorsulfuron, a herbicide of the sulfonylureas family. Transformed colonies were selected as described in Table 3. Under the optimized conditions described above the plating efficiency after electroporation was decreased by only 50%. Chlorsulfuron was found to be an efficient selective agent. The growth of wild type colonies was stopped a few days after adding chlorsulfuron to the culture medium (Fig. 3). The relative transformation efficiency (number of

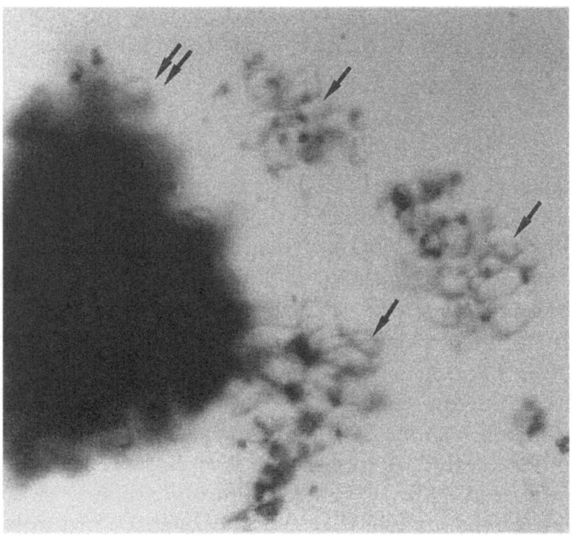

Fig. 3. Selection of chlorsulfuron resistant colonies after electroporation of protoplasts of *L. esculentum* cv. Golden Sunrise. → sensitive colony ten days after adding 50 mM chlorsulfuron in the culture medium; ⇉ resistant colony. (Bellini et al. 1989a)

Table 4. Transformation efficiency after electroporation of protoplasts of *L. esculentum*. (Bellini et al. 1989a)

Treatment	No. of protoplasts treated	Plating efficiency (%)	No. of colonies submitted to selection	No. of colonies selected as resistant	Relative transformation efficiency (%)
Manapal					
Non elect.	2×10^6	0.2	4.0×10^3	0	0
Elect. + pGH1	6×10^6	0.1	6.0×10^3	150	2.5
Golden Sunrise					
Non elect.	2×10^6	15	1.5×10^4	0	0
Elect. + pGH1	8×10^6	6	5.0×10^4	1549	0.3

After electroporation protoplasts were cultured as described in Table 1. Ten days later part of the protoplast-derived colonies were submitted to selection, the other part cultured without the selective agent in order to determine the plating efficiency and the relative transformation efficiency. The plating efficiency was the number of protoplast derived colonies scoring 15 after plating out of the number of protoplasts plated. The relative transformation efficiency was the number of resistant colonies out of the colonies submitted to selection.

resistant colonies/number of colonies) could reach 2.5% in the case of cultivar Manapal (Table 4).

L. peruvianum protoplasts were electroporated in the presence of plasmid pABD1 (Paszkowski et al. 1984), which carries a chimeric gene that confers resistance to kanamycin.

Paromomycin, which is an antibiotic analogous to kanamycin was used as selective agent. Previous experiments have shown that it has a more drastic effect on colonies derived from susceptible tobacco protoplasts (Bellini et al. 1989b). The toxicity of paromomycin was also observed with *L. peruvianum* protoplasts. In these conditions it was possible to select paromomycin resistant colonies. The relative frequency of transformation varied from 0.3 to 0.9% (Bellini et al. 1989a).

When resistant calli were transferred on regeneration medium containing paromomycin they continued to grow and regenerated buds. Plantlets were transferred on medium B (Bourgin et al. 1979), with no selective pressure, for root induction. Resistance to kanamycin was confirmed by transferring cuttings on medium B supplemented with 100 mg/l of kanamycin. In this test, susceptible plantlets were unable to root and bleached whereas the resistant ones grew normally (Fig. 4). Those which have developed roots were transferred to the greenhouse.

3.3 Evidence for the Presence of the Foreign Gene

The presence of the foreign gene in the genome of two clones, Lp1Kr and Lp6Kr, was confirmed by Southern blot analysis. When genomic DNA from a normal plant and from the two resistant clones was digested by Hind3 and EcoR5 restriction enzymes, bands of respectively 0.8 kb and 1.2 kb, corresponding to the coding sequence of the NPT II gene and to the entire chimaeric gene, were

Fig. 4A, B. Confirmation of the resistance to kanamycin of clone Lp1Kr. Cuttings of wild type (**A**) plants and of Lp1Kr (**B**) were transferred onto medium **B** supplemented with 100 mg/l kanamycin. (Bellini et al. 1989a)

observed in plants Lp1Kr and Lp6Kr but not in the control (Fig. 5). Reconstruction experiments indicated that the number of integrated copies ranged from one to three copies per genome. This experiment demonstrated that sequences originating from plasmid pABD1 were present in the genome of both clones.

The kanamycin resistance trait was transmitted to the progeny in a Mendelian manner, as expected for a monogenic dominant nuclear marker (Bellini et al. 1990).

4 Conclusion

The technique for the production and culture of tomato protoplasts described here led to reproducible results. It is stressed that the growth conditions of the donor plants are very important for obtaining viable protoplasts. We have shown the importance of calcium and myo-inositol for protoplast preparation and culture, and also the importance of ammonium, that was previously found to be toxic for tomato protoplasts. The protocol described here allowed direct gene transfer by electroporation and selection of stable transformants calli from both *L. esculentum* and *L. peruvianum* protoplasts. The relative transformation efficiency ranged from 0.2 to 2.5%. This corresponds to an increase of 10 to 100 fold compared to the

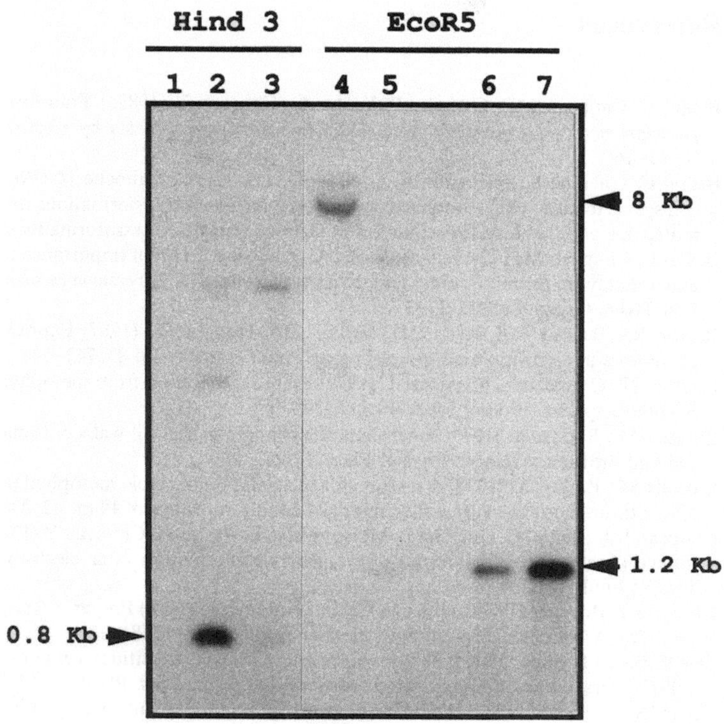

Fig. 5. Southern blot analysis of plant DNA. Total DNA of wild type (WT) plant (10 µg) and of two transgenic plants resistant to kanamycin, Lp1Kr (10 µg) and Lp6Kr (5 µg), were digested with the restriction enzymes HindIII or EcoRV, and hybridized with the 0.8 Kb HindIII fragment of pABD1 (containing the coding sequence of the NPTII gene) used as a probe. *Lane 1* DNA from wild type plant, *lanes 2* and *4* DNA from Lp1Kr; *lanes 3* and *5*: DNA from Lp6Kr; *lanes 6* and *7*: the plasmid pABD1 was used for a reconstruction experiment at a concentration of 1 and 3 copies respectively. (Bellini et al. 1989a)

results presented by Jongsma et al. (1987), who used polyethyleneglycol or the calcium phosphate DNA transformation procedures. However the transformation efficiency is not as high as that obtained after microinjection of tomato cells derived from calli, as described by Toyoda et al. (1989). In their case nearly 20% of the cells, in which a plasmid containing a kanamycin resistant gene was microinjected, could be selected on a kanamycin containing medium. In *L. esculentum* transformation, the relative efficiency of transformation was as high as that obtained with *L. peruvianum* protoplasts, but the use of this technique is still restricted in this approach by the low rate of regeneration.

References

Bellini C, Chupeau MC, Guerche P, Vastra G, Chupeau Y (1989a) Transformation of *Lycopersicon peruvianum* and *Lycopersicon esculentum* mesophyll protoplasts by electroporation. Plant Sci 65 (1):63–76

Bellini C, Guerche P, Spielmann A, Goujaud J, Lessaint C, Caboche M (1989b) Genetic analysis of transgenic tobacco plants obtained by liposome-mediated transformation: absence of evidence for the mutagenic effect of inserted sequences in sixty characterised transformants. J Hered 80:361–367

Bellini C, Chupeau MC, Gervais M, Vastra G, Chupeau Y (1990) Importance of myo-inositol, calcium and ammonium for the viability and division of tomato (*Lycopersicon esculentum*) protoplasts. Plant Cell Tissue Organ Cult 23:27–37

Boston RS, Becwar MR, Ryan RD, Larkins BA, Hodges TK (1987) Expression from heterologous promotors in electroporated carrot protoplasts. Plant Physiol 83:742–746

Bourgin JP, Chupeau Y, Missonier C (1979) Plant regeneration from mesophyll protoplasts of several *Nicotiana* species. Physiol Plant 45:288–292

Cassells AC, Barlass M (1976) Environmental changes in the cell walls of tomato leaves in relation to cell and protoplast release. Physiol Plant 37:239–246

Cassells AC, Barlass M (1978) A method for the isolation of stable mesophyll protoplasts from tomato leaves throughout the year under standard conditions. Physiol Plant 42:236–242

Chupeau MC, Bellini C, Guerche P, Maisonneuve B, Vastra G, Chupeau Y (1989) Transgenic plants of lettuce (*Lactuca sativa*) through direct gene transformation after electroporation of protoplasts. Bio/Technol 7:503–507

Chupeau Y, Bourgin JP, Missonier C, Dorion N, Morel G (1974) Préparation et culture de protoplastes de divers *Nicotiana*. CR Acad Sci Paris Ser D 278:1565–1568

Das R, Bagga S, Sopory SK (1987) Involvement of phosphoinositides, calmoduline and glyoxalase-I in cell proliferation in callus cultures of *Amaranthus paniculatus*. Plant Sci 53:45–51

Ecker JR, Davis RW (1986) Inhibition of gene expression in plant cells by expression of antisens RNA. Proc Natl Acad Sci USA 83:5372–5376

Ettlinger C, Lehle L (1987) Auxin induces rapid changes in phosphatidylinositol metabolites. Nature 331:176–178

Fromm ME, Taylor LP, Walbot V (1985) Expression of genes transferred into monocot and dicot plant cells by electroporation. Proc Natl Acad Sci USA 82:5824–2828

Fromm ME, Taylor LP, Walbot W (1986) Stable transformation of maize after gene transfer by electroporation. Nature 319:791–793

Guerche P, Bellini C, Le Moullec JM, Caboche M (1987a) Use of a transient expression assay for the optimization of direct gene transfer into tobacco mesophyll protoplasts by electroporation. Biochimie 69:621–628

Guerche P, Charbonnier M, Jouanin L, Tourneur C, Paszkowski J, Pelletier G (1987b) Direct gene transfer by electroporation in *Brassica napus*. Plant Sci 52:111–116

Haberlach GT, Cohen BA, Reichert NA, Baer MA, Towill LE, Helgeson JP (1985) Isolation, culture and regeneration of protoplasts from potato and several related *Solanum* species. Plant Sci Lett 39:67–74

Haughn GW, Somerville C (1986) Sulfonylurea-resistant mutants of *Arabidopsis thaliana*. Mol Gen Genet 211:430–435

Hassanpour-Estahbani A, Demarly Y (1984) Plant regeneration from protoplasts of *Solanum pennellii*: effect of photoperiod applied to donor plants. J Plant Physio. 121:171–174

Holmes DS (1982) Rapid purification of bacterial plasmids and coliphage M13 RF without CsCl centrifugation. Anal Biochem 127:428–433

Jongsma M, Koornneef M, Zabel P, Hille J (1987) Tomato protoplasts DNA transformation: physical linkage and recombination of exogenous DNA sequences. Plant Mol Biol 8:383–394

Koornneef M, Hanart C, Jongsma M, Toma I, Weide R, Zabel P, Hille J (1986) Breeding of a tomato genotype readily accesible to genetic manipulation. Plant Sci 45:201–208

Masson J, Lancelin D, Bellini C, Lecerf M, Guerche P, Pelletier G (1989) Selection of somatic hybrids between diploid clones of potato (*Solanum tuberosum* L.) transformed by direct gene transfer. Theor Appl Genet 78:153–159

Masson J, Lecerf M, Rousselle P, Perennec P, Pelletier G (1987) Plant regeneration from protoplasts of diploid potato derived form crosses of *Solanum tuberosum* with wild *Solanum* species. Plant Sciences 53:167–176

Maxon Smith JW, Ritchie DB (1983) A collection of near-isogenic lines of tomato: research tool of the future? Plant Mol Biol Rep 1:41–46

McCormick S, Niedermeyer J, Fry J, Barnason A, Horsch R, Fraley R (1986) Leaf disc transformation of cultivated tomato (*Lycopersicon esculentum*) using *Agrobacterium tumefaciens*. Plant Cell Rep 5:81–84

Meyer Y, Abel WO (1975) Budding and division of tobacco protoplasts in relation to pseudo-wall and wall formation. Planta 125:1–13

Muhlbach HP (1980) Different regeneration potentials of mesophyll protoplasts from cultivated and wild species of tomato. Planta 148:89–96

Negrutiu I, Shillito R, Potrykus I, Biasini G, Sala F (1987) Hybrid genes in the analysis of transformation conditions. Plant Mol Biol 8:366–373

Niedz RP, Rutter SM, Handley LW, Sink KC (1985) Plant regeneration from leaf protoplasts of six tomato cultivars. Plant Sci 39:199–204

O'Connel MA, Hanson MR (1985) Somatic hybridisation between *Lycopersicon esculentum* and *Lycopersicon penellii*. Theor Appl Genet 70:1–12

Ou-Lee TM, Turgeon R, Wu R (1986) Expression of a foreign gene linked to either a plant virus or a *Drosophila* promoter, after electroporation of protoplasts of rice, wheat and sorghum. Proc Natl Acad Sci USA 84:4870–4874

Paszkowski J, Shillito RD, Saul M, Mandak V, Hohn T, Hohn B, Potrykus I (1984) Direct gene transfer to plants. EMBO J 3:2717–2722

Rick CM (1978) The tomato. Sci Am 239:76–87

Rick CM (1982) The potential of exotic germplasm for the tomato improvement. In: Vasil I, Scowcroft W, Frey K (eds) Plant improvement and somatic cells genetics. Academic Press, London New York, pp 1–28

Seguin M, Lalonde M (1988) Gene transfer by electroporation in Betulaceae protoplasts: *Alnus incana*. Plant Cell Rep 7:367–370

Shahin EA (1985) Totipotency of tomato protoplasts. Theor Appl Genet 69:235–240

Shahin EA, Sukhapinda K, Simpson RB, Spivey R (1986) Transformation of cultivated tomato by a binary vector in *Agrobacterium rhizogenes*: transgenic plants with normal phenotype harbor binary vector T-DNA, but no Ri-plasmid T-DNA. Theor Appl Genet 72:770–777

Shillito RD, Saul MW, Paszkowski J, Muller M, Potrykus I (1985) High efficiency of direct gene transfer to plants. Bio/Technol 3 1099–1105

Tabaeizedeh Z, Bunisset-Bergounioux C, Perennes C (1984) Environmental growth conditions of protoplast source plants: effect on subsequent protoplast division in two tomato species. Physiol Vég 22 (2):223–229

Tan M, Rietveld E, van Marrewijk GA, Kool AJ (1987) Regeneration of leaf mesophyll protoplasts of tomato cultivars (*Lycopersicon esculentum*): factors important for efficient protoplast culture and plant regeneration. Plant Cell Rep 6:172–175

Tanimoto S, Harada H (1986) Involvement of calcium in adventitious bud initiation in *Torenia* stem segments. Plant Cell Physiol 27 (1):1–10

Toyoda H, Matsuda Y, Utsumi R, Ouchi S (1989) Intranuclear microinjection for transformation of tomato callus cells. Plant Cell Rep 7:293–296

Trewavas AJ (1985) Growth substances, calcium and the regulation of cell division. In: Bryant and Francis (eds) The cell division cycle in plants. Cambridge University Press, Cambridge, pp 133–156

Tsukuda M, Kusano T, Kitayawa Y (1989) Introduction of foreign genes ınto tomato protoplasts by electroporation. Plant Cell Physiol 30 (4):599–603

Zapata FJ, Sink KC, Cocking EC (1981) Callus formation from leaf mesophyll protoplasts of three *Lycopersicon* species: *L. esculentum*, cv. Walter, *L. pimpinellifolum* and *L. hirsutum*, F. *glabratum*. Plant Sci Lett 23:41–46

II.15 Transformation in *Mentha* Species (Mint)

A. Spencer[1], J.D. Hamill[2], and M.J.C. Rhodes[3]

1 Introduction

1.1 Distribution and Importance of *Mentha*

The genus *Mentha* (Labiateae) occurs in all five continents; however, there is only one species native to North America and its occurrence in the tropics is infrequent. *Mentha* occurs in Australia, but there are a number of species whose relationship with the rest of the genus is uncertain (Harley and Brighton 1977). *Mentha* species are important because of their production of essential oils which find wide use as food flavors and as fragrances and include spearmint, *M. spicata*, and peppermint *M. piperita*.

In the section *Mentha*, plants are self-compatible, set seed on self-pollination, yet are capable of a fair degree of outbreeding. While many of the hybrids are infertile, the capacity of *Mentha* species for vigorous vegetative propagation due to a highly invasive rhizome system leads to many subfertile hybrids being retained in the population. This is of interest as many of the potentially commercially important species are hybrids; for instance, peppermint is a hybrid between *M. aquatica* and *M. spicata*.

The major components of the essential oils of mints are monoterpenes, C-10 isoprenoid compounds which arise biosynthetically from acetate via mevalonate, isopentenyl pyrophosphate, and geranyl pyrophosphate. Extensive chemical analyses of the essential oils of mints (Alston and Turner 1963) has indicated the presence of "chemical races" producing distinctive types of essential oil, each largely consisting of a group of structurally related monoterpenes. The main types are those oils consisting solely of acyclic monoterpenes typified by linalool and linalyl acetate which are found in *M. citrata* (Bergamot mint), oils containing 2-oxygenated monoterpenes as in *M. spicata* (spearmint) typified by carvone, and oils containing 3-oxygenated monoterpenes such as menthol as found in peppermint, *M. piperita*. (Loomis 1967).

1.2 Objectives in the Genetic Engineering of *Mentha*

Since the commercial value of an oil is critically dependent on the composition of its monoterpene components, it is desirable to be able to modify oil composition to

[1] Department of Biological Sciences, University of East Anglia, Norwich, Norfolk, United Kingdom
[2] Department of Genetics and Developmental Biology, Monash University, Melbourne, Australia
[3] Genetics and Microbiology Department, AFRC Institute of Food Research, Norwich Laboratory, Norwich, United Kingdom

Biotechnology in Agriculture and Forestry, Vol. 22
Plant Protoplasts and Genetic Engineering III (ed. by Y.P.S. Bajaj)

improve its quality by changing the balance between these constituents. However, another desirable end may be to direct oil synthesis to the predominant formation of an individual component such as menthol which has commercial value in its own right. Evidence from many breeding experiments suggests that differences in oil components may result from single gene effects, presumably related to expression or repression of the structural genes coding for individual biosynthetic enzymes (see early evidence reviewed by Loomis 1967). The biochemistry of the production of monoterpenes from their immediate precursor, geranyl pyrophosphate, is known in outline (Gershenzon and Croteau 1990). There is increasing knowledge of the enzymology of at least some of the steps involved (Kjonaas et al. 1982; Croteau and Venkatachalam 1986; Suga et al. 1986; Gershenzon and Croteau 1990, Karp et al. 1990) and of at least some of the factors involved in the regulation of the pathway to form both monoterpenes (Gershonzon and Croteau 1990) and other related isoprenoid compounds (Gray 1987). Genetic modification would be aimed at increasing the flux through the pathway either by increasing the in vivo activity of key enzymes by over-expressing the relevant genes behind strong promoters or by decreasing the activity of enzymes involved in the diversion of intermediates to unwanted side-products using anti-sense RNA or ribozyme technology. This approach is at present limited by the unavailability of many of the relevant genes. cDNAs encoding an enzyme, hydroxymethylglutaryl CoA reductase (HMG-CoA reductase), playing an important role in regulating the flux through the pathway, have been isolated from two plant sources, *Arabidopsis* (Caelles et al. 1989) and *Hevea* (Chye et al. 1991); a genomic clone has also been isolated from *Hevea*. Progress in this area has been limited in the lack of suitable in vitro systems for study of the biosynthesis of monoterpenes. One of our initial objectives was to develop transformed cultures which produced oils at levels and in composition similar to the whole plant. Previous work with cell cultures of *Mentha* species led to cultures which produced only very low levels of the desired end products (Paupardin 1976) although many retain the capacity to catalyze biotransformation of intermediates in the pathway (Aviv et al. 1981). Much of this low productivity was attributed to the toxicity of these compounds to cells in culture which lack the capacity to store monoterpenes extracellularly (Charlwood et al. 1988). In a new approach, we investigated the use of transformed shoot cultures as biochemically and genetically stable systems in which to study monoterpene biosynthesis. To this end we have developed transformation procedures based on wild-type (Spencer et al. 1990) and genetically modified strains (Spencer 1991; Rhodes et al. 1992) of *Agrobacterium tumefaciens* for a range of *Mentha* species representing different chemical races. This work represents the first successful transformation of *Mentha*. More recently, there has been a preliminary report of the transformation of *Mentha citrata* by the virulent *Agrobacterium tumefaciens* stain A281 (Berry et al. 1990).

2 Procedures for the Genetic Manipulation of *Mentha* Species

An important contribution to our understanding of the interactions between the plant growth regulators and plant organ differentiation comes from the work of Skoog and Miller (1957) on *Nicotiana tabacum*, who showed that increasing the ratio of cytokinin to auxin in the culture medium induced callus tissue to differentiate to produce shoots. Subsequently, media containing high exogenous cytokinins concentrations have been used to maintain shoot tip and auxillary meristem cultures of many species including *M. piperita* and *M. spicata* (Bricout et al. 1978; Heble 1985; Hirata et al. 1990). Shoot differentiation from calli submerged in liquid medium, with exogenous supplies of cytokinins at levels greater than 1 mg/l has been demonstrated in a number of species, including *Chrysanthemum cinerariaefolium* (Zieg et al. 1983). However, attempts to develop shoot cultures of *Mentha* from callus using a range of concentrations of cytokinins in the culture medium have proved less successful (Bricout and Paupardin 1975; Paupardin 1976).

To overcome these problems, we have developed hormone-independent shoot cultures (shooty teratomas) of *Mentha*. The approach was intended to parallel the use of transformed root culture based on transformation with the T-DNA of *A. rhizogenes*, which synthesize and accumulate high levels of root-produced secondary metabolites (Hamill et al. 1986, 1987).

However, there is no comparable strain of *Agrobacterium* that has evolved the ability, following infection, to induce shoot formation on a wide range of species. Genes in the transferred-DNA (T-DNA) of *A. tumefaciens* code for the synthesis in the plant cell of specific bacterial metabolites, opines, and for factors inducing the tissue to undergo disorganised growth to form a mass of callus-like tissue, the gall. Strains of *A. tumefaciens* are classified on the basis of the type of opines they produce, the common ones being strains producing octopine or nopaline. The processes that lead to oncogenesis and gall development are not fully understood, but the two genes, *tms1* and *tms2*, coding for two enzymes of auxin biosynthesis, tryptophan mono-oxygenase and indole acetamide hydrolase, respectively, and one (*ipt*) for cytokinin biosynthesis, coding for dimethylallyl transferase, play an important role. The integration of the *tms* and *ipt* genes leads to increased levels in the auxin and cytokinin in the transformed tissue (Akiyoshi et al. 1983). However, other genes present (such as gene 5 and 6b) in the T-DNA probably also play an important role in modifying the action of the *tms* and *ipt* genes (Tinland et al. 1989; Korber et al. 1991).

2.1 Approaches to the Transformation of *Mentha* to Produce Shoot Cultures

N. tabacum cv. Xanthi has been used in this work as the model species to confirm the shoot-inducing capacity of the different strains of *A. tumefaciens*. From information in the literature there are four types of strains of *A. tumefaciens* which can induce shoot formation in *N. tabacum* (see Rhodes et al. 1992). These are: octopine strains of *A. tumefaciens* bearing mutations in the *tms* genes (Ooms et al. 1980, 1981; Garfinkel et al. 1981), disarmed strains bearing the wild-type cytokinin gene (*ipt*) isolated from the T-DNA of *A. tumefaciens* (Smigoki and Owens 1988; Schmulling et al. 1989), strains designed to overexpress the *ipt* coding sequence under the control of the powerful Cauliflower Mosaic Virus 35S (CaMV35S) promoter (Smigoki and Owens 1988), and surprisingly, some wild-type nopaline strains such as T37 and C58 (Braun and Wood 1976; Gresshoff et al. 1979) albeit at low frequency.

These four transformation strategies were tested on a range of *Mentha* species particularly *M.* × *piperita citrata* Ehrh. Brig. (Bergamot mint) referred to as *M. citrata* and *M.* × *piperita vulgare* (Scotch black mint), referred to subsequently as *M. piperita*. Those two species of *Mentha* produce a very different spectrum of monoterpene products in their essential oils: *M. citrata* accumulates a simple oil consisting of over 90% of two acyclic monoterpenes, linalool and linalyl acetate, while *M. piperita* makes the typical peppermint cyclic monoterpenes, menthone and menthol. Shooty teratomas developed from galls derived from transformations with the *A. tumefaciens* wild-type nopaline strains, pTiT37 and pTiC58, and from transformations with expression constructs containing the coding sequence of cytokinin gene (*ipt*) derived from pTiT37, and under the control of the powerful CaMV35S promoter, with or without duplicated 5′ enhancer sequences. A further construct containing the *onc*-genes (*tms1*, *tms2*, and *ipt*) from pTiC58 was also tested on the two *Mentha* species. In some experiments, a spearmint variety, *M. spicata*, was also tested.

2.2 Plant Materials and Bacterial Strains Used

Nicotiana tabacum cv. *Xanthi*: gift from J. Nichols, John Innes Institute. *Mentha spicata* and *Mentha citrata*: obtained from a commercial source, *Mentha piperita* (Scotch black mint) a gift from Dr. K. Svoboda, West of Scotland Agricultural College, Auchincruive, Ayr. The wild-type and mutant *Agrobacterium* strains used in this work are described in Table 1. Cloning of the *ipt* open reading frame into expression cassettes was described in detail by Spencer (1991). The *ipt* open reading frame was isolated from the wild-type gene of pTiT37 and Xba1 restriction sites were inserted at both ends of the ORF using the M13 bacteriophage M13K19 (Carter et al. 1985) together with two synthetic oligonucleotides which contained an Xba1 sequence at their 5′ ends in order to clone the *ipt* ORF into pBin19- based expression cassettes (Hamill et al. 1988).

Three expression cassettes were used to express the *ipt* ORF in transformed plant tissue; these are (1) a vector bearing the *ipt* gene under the control of the nopaline synthase promoter using the pea rubisco E9 small subunit poly A signal (pNos *ipt*), (2) the *ipt* gene with CaMV35S promoter and the CaMV35S poly A signal (p35S *ipt*), and (3) the *ipt* gene with CaMV35S promoter bearing duplicate upstream enhancer sequences (Kay et al. 1987) and the CaMV35S poly A signal (pE35S *ipt*). These vectors were chosen to provide a range of promoter strengths for driving expression of the *ipt* coding sequence following introduction into the plant genome. On the basis of other studies, the transcriptional strengths of these promoters were expected to be: nopaline synthase < CaMV35S < CaMV35S with duplicated 5′ enhancer sequences (Saunders et al. 1987; Kay et al. 1987; Hamill et al. 1988). These constructs were introduced into two disarmed strains of *A. tumefaciens* LBA 4404 (Hoekema et al. 1983) and C58/3 (Dale et al. 1989) by the methods of Hofgen and Willmitzer (1988).

2.3 Transformation and Production of Galls

Freshly harvested *Mentha* stems (5 cm) with all leaves except those of the apical bud removed, were surface sterilized with a 10% (v/v) solution of a commercial

Table 1. Summary of strains and constructs used in present work

Strain	Agrobacterium background	Plasmid	Insert	Inactivated gene(s)	Opine produced	Reference
A348 *tms* mutants	C58	pTiA6NC/*tms*	Tn5	*aux*-1	Octopine	a
T 37	T37	pTiT37	–	–	Nopaline	b
C 58	C58	pTiC58	–	–	Nopaline	b
LBA 4404	Ach5	pTiAch5/LBA4404	–	T-DNA	–	c
C 58/3	C58	pTiC58	–	T-DNA	–	d
SSI	Ach5	pTiAch5/pSSI	–	T-DNA	–	e
Constructs						
pNos-*ipt*	4404	pFIE10[a]	*ipt*	–	–	f
	C58/3	pFIE10	*ipt*	–	–	
p35S-*ipt*	4404	pFIE20[a]	*ipt*	–	–	f
	C58/3	pFIE20	*ipt*	–	–	
pE35S-*ipt*	4404	pFIH10[a]	*ipt*	–	–	f
	C58/3	pFIH10	*ipt*	–	–	
p-*onc*	4404	pIFR*onc*	*aux*1, *aux*2, *ipt*			f
	C58/3	pIFR*onc*	*aux*1, *aux*2, *ipt*			

[a] See Hamill et al. (1988).
a: Garfinkel et al. (1981). b: Depicker et al. (1980). c: Hoekema et al. (1983). d: Dale et al. (1989). e: S. Scholfeld (pers. comm.). f: Spencer (1991).

bleach (Domestos, Unilever) for 10 min followed by thorough rinsing in sterile demineralized water. Growth of *Agrobacterium* and the infection of plant stems was essentially as described by Ooms et al. (1985). *Agrobacterium* strains were maintained on TY agar medium (Ooms et al. 1980). Ten-ml TY broth was inoculated with the *Agrobacterium* strain and grown on a shaker overnight to an A_{600} of about 0.6. *Mentha* stems were inoculated by allowing a small volume of the *Agrobacterium* culture to flow into the wounded area. The basal end of the inoculated stem was inserted into 50 ml of MSO agar medium (MS salts; Murashige and Skoog 1962), containing 30 g/l sucrose and 10 g/l agar, in 200-ml jars, and were maintained at 25 °C under a 16-h day, illuminated by daylight fluorescent tubes at 840 lx.

2.4 Establishment of Transformed Shoots in in Vitro Culture

After a period of about 4 weeks, galls which had developed on the inoculated stems were excised and cultured in Petri dishes on MSO medium containing 0.8% agar and 3% sucrose, with 10% conditioned medium, and a suitable antibiotic to eliminate any free-living *Agrobacterium*. Galls were transferred to fresh medium after 4 days and maintained in a 16-h day with a light intensity of 840 lx, at 25 °C. The galls grew rapidly and were maintained subsequently by transferring to fresh medium in Petri dishes at 21-day intervals. Shoot initiation on *Mentha* galls became apparent after about 5 weeks in culture. Single stem tips were removed and grown as individual cultures on MSO medium containing 10% conditioning factor.

2.4.1 Conditioned Medium

The presence of conditioned medium was an important factor in the survival of the excised galls during the early stages of tissue culture. Conditioned medium was obtained by culturing a fine cell suspension of *Nicotiana tabacum* (grown in liquid MS salts containing 30 g/l sucrose and 0.5 mg/l 2,4-D) for 7 days on a shaker at 90 rpm. The cells were allowed to settle, and the spent medium was transferred to a sterile container and stored at − 20 °C until required for use as a conditioning factor. The conditioned medium was added at 10% (v/v) to MSO agar medium, under sterile conditions immediately prior to pouring the medium into Petri dishes. Conditioning factor was added to MSO medium preparations to culture freshly excised galls, but was not included in medium preparations for established galls and teratoma tissue.

2.4.2 Verification of Asepsis

The possible presence of contaminating *Agrobacterium* cells in the shoot cultures was investigated by crushing samples of approximately 500 mg of teratoma under sterile conditions, and plating the crushed tissue onto TY medium. In the case of some T37-derived shooty teratomes, initially the presence of low levels of bacteria was detected by this method, but after a treatment high concentrations of ampicillin (3–5 mg/ml), bacteria were shown to be absent from the shooty teratoma tissue.

3 Phenotypes of *Mentha* Galls Transformed by Different Strains of *A. tumefaciens*

3.1 Auxin Mutants

Transformation of the two *Mentha* species with the auxin mutant strains of *A. tumefaciens*, (pTiA6NC::Tn5/tms-337, LBA 4060 and LBA 1501) (Garfinkel et al. 1981; Ooms et al. 1981) resulted in gall formation, but shoot differentiation was not observed. Galls which developed on *M. citrata* following transformations with a *tms1* mutant of pTiA6NC, (tms-337::Tn5) were excised and grown on MSO medium without hormones with monthly subculturings for over 2 years. These auxenic galls were green and friable and grew readily in culture, but did not undergo shoot differentiation.

3.2 Expression of Endogenous *ipt* Gene

Transformations of three *Mentha* species (*M. spicata*, *M. citrata*, and *M. piperita*) with the *ipt* gene isolated from pTiT37 HindIII fragment 22 (Goldberg et al. 1984), as a 2kbp HindIII-BamHI fragment and cloned into pBin19 housed in the disarmed strain LBA4404 (Schofield, pers. comm.) initiated gall formation in *M. piperita*. With other mints, LBA4404 showed very low transformation efficiency. The galls which developed on *M. piperita* stems from the endogenous *ipt* gene were cultured axenically on MSO agar medium without hormones. These galls had a similar friable texture to galls induced by the *tms1* mutants on the *Mentha citrata* but although galls derived from transformation of *M. piperita* with the *ipt* gene were darker green, they did not undergo differentiation into shoots.

3.3 Use of Wild-Type Nopaline Strains

Transformation of three *Mentha* species, *M. spicata*, *M. citrata*, and *M. piperita* with the wild-type nopaline strains (T37, C58, and N273) of *A. tumefaciens* initiated galls which developed at the inoculation sites of the stems of each *Mentha* species. Galls excised from the stems of *M. spicata* were cultured on MSO medium without hormones and developed as undifferentiating galls.

However, the galls which developed on the stems of *M. citrata* from inoculation with the nopaline strains of *Agrobacterium*, when excised and cultured on MSO agar medium with high levels of ampicillin to eradicate free-living *Agrobacterium* differentiated into shoots during subsequent tissue culture. These shooty teratomas which developed from inoculation with either T37 and C58, were initially hard galls, but after differentiation they became more friable and consisted of shooty clusters completely lacking root development (Fig. 1). The shooty teratomas which developed from inoculation with either T37 or C58 were similar in appearance but because the T37 strain of *A. tumefaciens* was more easily controlled with the antibiotics available, the T37-induced shooty teratomas of *M. citrata* were used for further work to investigate their growth and secondary metabolite production in culture. Nopaline assays confirmed that the shoot cultures which developed from

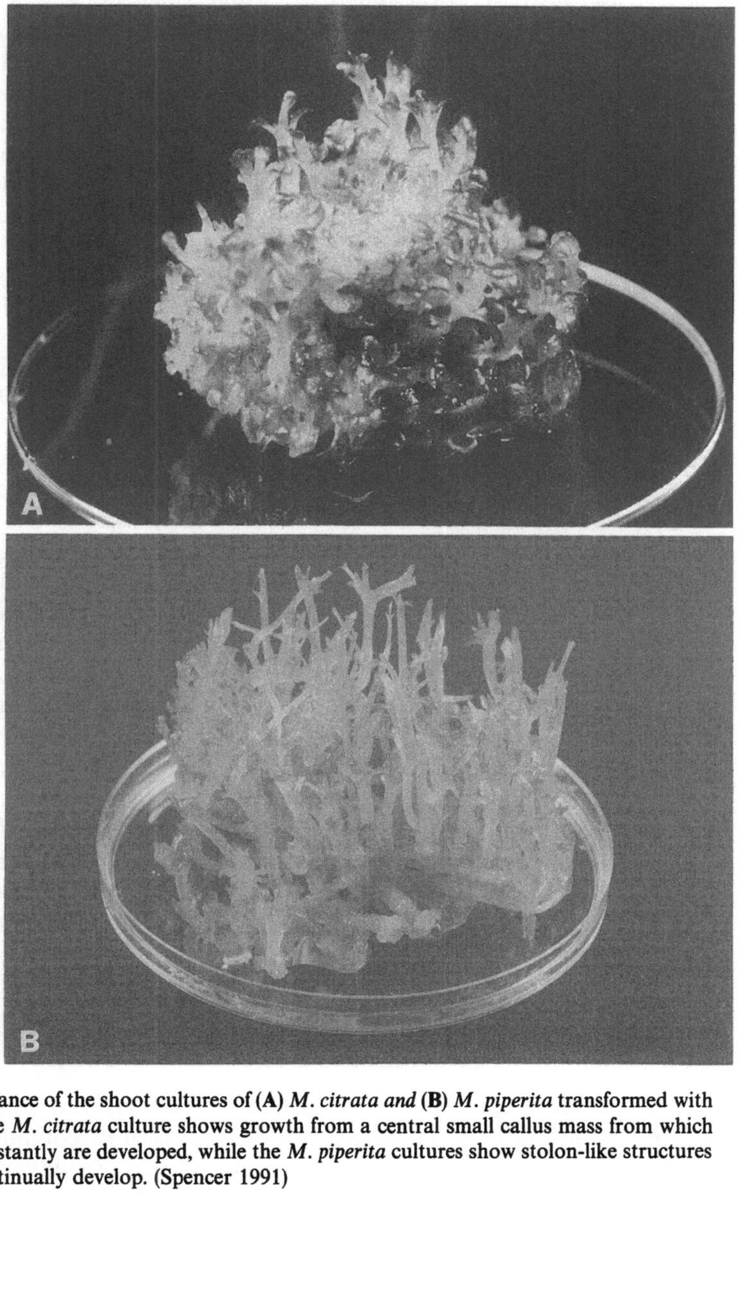

Fig. 1A, B. The appearance of the shoot cultures of **(A)** *M. citrata and* **(B)** *M. piperita* transformed with *A. tumefaciens* T37. The *M. citrata* culture shows growth from a central small callus mass from which shooty outgrowths constantly are developed, while the *M. piperita* cultures show stolon-like structures from which shoots continually develop. (Spencer 1991)

transformation of *M. citrata* with strain T37 contained nopaline, and were thus transformed and expressing the nopaline synthase gene of the T-DNA. Untransformed tissue from the parent plant was used as the control and was shown to lack nopaline. The elimination of bacteria is an important step in this procedure and the failure to induce shoot formation in *M. spicata* could, in part, be due to this factor. The galls which developed from inoculations of *M. citrata* with the nopaline strain of *A. tumefaciens* N2/73 (Anderson and Moore 1979) also produced shooty teratomas during subsequent tissue culture. Two distinctive shooty teratoma phenotypes were developed with this strain; a friable differentiating teratoma with leaves and no stem development, and one which grew as horizontal stolon-like structures, with vestigial root production from the leaf axial nodes.

The galls excised from *M. piperita* stems also developed into shooty teratomas, under culture conditions similar to those described for *M. citrata.* Shooty teratomas which developed on the *M. piperita* galls from transformations with either T37 or C58, were similar in appearance to the teratomas derived from *M. citrata* when grown on MSO agar medium. However, when the shooty teratoma cultures of *M. citrata* and *M. piperita* were grown on liquid B50 medium (B5 medium, 3% sucrose) without hormones, their phenotypic development was quite different. Shooty teratomas of *M. citrata* grown on B50 liquid medium developed as a large friable mass of shooty tissue, whereas the shoot cultures of *M. piperita* were more highly differentiated into stems without root development. Nopaline assays confirmed that the T37-derived shoot cultures of both species which developed from the galls were transformed with the *nos* gene of the T-DNA, while the untransformed parent plant was shown not to produce nopaline. Galls on *M. piperita* induced by N2/73 when cultured axenically were leafy but without stems. These teratomas did not develop the stolon-like shoots that developed from the N273 cultures of *M. citrata.*

A major factor that influenced the development of shooty teratomas in both *N. tabacum* and two *Mentha* species from transformation with the nopaline strains of *A. tumefaciens*, was the eradication of free-living *Agrobacterium* from the galls. The presence of *Agrobacterium* on the gall tissue appeared to result in death of the galls following subculturing, and in many instances a layer of *Agrobacterium* could be seen growing over the galls, but not on the surface of the agar, which contained antibiotics. Another factor which necessitated the complete elimination of free-living *Agrobacterium* in the galls was the endogenous expression of the *tzs* gene, responsible for bacterial cytokinin production in the nopaline strains of *Agrobacterium* (Fillatti et al. 1987). The presence of free-living *Agrobacterium* could influence the phenotype of the *Mentha* galls in a way that could not be attributed to transformation with the nopaline T-DNA. For these reasons it was necessary to use high levels of ampicillin, up to 5 mg/ml, to eradicate the contaminating *Agrobacterium.*

3.4 Use of Strains Which Over-Express the *ipt* Gene

Inoculation of the stems of *M. citrata* and *M. piperita* was carried out by the standard procedures. The three expression cassettes containing the *ipt* ORF in both *Agrobacterium* vectors LBA 4404 and C58/3, induced gall formation at inoculation sites on the stems of both *Mentha* species, but none of these galls differentiated into shoots on the plant stems. Control inoculations with LBA 4404 did not induce callus formation. However, prolific callusing did develop at several wound sites from inoculation with C58/3. Experiments showed that *M. piperita* is more responsive to the LBA 4404 binary vectors, while *M. citrata* is more responsive to the C58/3 binary vectors. The *ipt* gene under the control of the more powerful promoters, induced gall formation on the *Mentha* species more successfully than pSS1 and pNos *ipt*, C58/3 enhances the response of the *Mentha* species to the weaker promoters, which may again be due to the activity of *tzs* gene present in the virulence region of the nopaline Ti plasmid.

Galls which developed from inoculations with each of the expression vectors were excised from the *Mentha* stems, and transferred to MSO agar medium containing 10% conditioning factor, and 250 μg/ml ampicillin to eradicate free-living *Agrobacterium.* A stronger mixture of antibiotics containing ampicillin, cefotaxime, and cephalosporin, with each at 250 μg/ml, was required to eradicate free-living *Agrobacterium* from galls induced by the C58/3 binary vector strains. After axenic galls were established on MSO agar medium, antibiotics and conditioning factor were omitted from subculture medium. Galls isolated from inoculation of both *Mentha* species with pNos *ipt* were not successfully established on agar medium. These failures may in some cases be due to the phytotoxicity of the antibiotic cocktail used to eradicated C58/3.

Galls excised from *M. piperita* stems after inoculation with the wild-type *ipt* gene (pSSI), developed into green friable undifferentiating tissue on MSO medium, which were readily propagated and were frequently resistant to kanamycin at 50 mg/l. Although inoculations of *M. citrata* with the wild-type *ipt* gene induced gall formation, these galls were never successfully established on MSO agar.

Galls which developed on *M. piperita* stems from inoculation with the p35S *ipt* and the pE35S *ipt* strains were easily established in axenic culture on MSO agar medium, and after establishment became prolifically shooty. These shoots were apically dominant and produced roots only to a very limited extent.

Galls derived from inoculation of *M. citrata* stems with the p35S *ipt* and the pE35S *ipt* vectors were more difficult to establish on MSO agar medium than those of *M. piperita.* This again may have been due in part to the toxicity of the antibiotics used to control C58/3. However, galls which were successfully established on MSO medium differentiated into chlorotic leafy friable tissue which was very similar in appearance to the shooty teratomas that developed from transformation of *M. citrata* with the nopaline strain N2/73. Initially, areas of chlorophyll formation occurred on the galls, and then large green horizontally growing shoots developed, which again were very similar in appearance to the second phenotype reported from transformations of *M. citrata* with N2/73, although in this case they were more rooty. Galls of *M. citrata* derived from transformation with pE35S*ipt* appeared to generate more apically dominant shoots, when established on agar, than those derived from transformations with p35S*ipt.*

Transformations of *M. citrata* and *M. piperita* with the *ipt* coding sequence under the control of the four different promoters, induced gall formation on the stems of both *Mentha* species. However, shoot initiation did not occur directly from galls on plant stems, but developed only after the galls derived from transformation with the more powerful promoter had been excised from the stems and successfully established in axenic culture. It is clear that shooty teratoma development is influenced by both excision from the plant, resulting in a changed hormonal and nutritional environment and by elimination of free-living bacteria.

The above result suggests that these two species of *Mentha* behave in a manner similar to *N. tabacum*, in undergoing shoot differentiation in response to elevated levels of endogenous cytokinins. However, shoot initiation in the two *Mentha* species appear to require higher levels of expression of the *ipt* gene than in *N. tabacum*, although under these conditions the apical dominance in the *Mentha* shoots is not repressed to the same extent as in *N. tabacum*. The analysis of the endogenous cytokinin content of the various *Mentha* transformations is necessary to confirm these suggestions.

3.5 Transformations with the *onc*-Gene Construct

Transformations of stems from both *Mentha* species with an *onc*-gene construct which contained *tms1*, *tms2*, and *ipt* (Xba1 fragment 7 of pTiC58, Depicker et al. 1980) housed in LBA 4404, induced prolific gall development. Axenic galls which were established on agar medium were very firm and grew rapidly yet did not differentiate into shoots.

4 Analysis of Transformed Shoot Cultures

4.1 Genetic Analysis of Shoot Cultures

DNA was extracted from plant and shooty teratoma tissue by the method of Martin et al. (1985) and the presence of the *ipt* gene in the teratoma tissue was confirmed by PCR amplification and ethidium bromide agarose gel electrophoresis. Using appropriate primers for the *ipt* gene a 720 bp fragment was amplified (Hamill et al. 1991). Southern transfer of DNA and hybridization with a ^{32}P-labeled *ipt* probe (Feinberg and Vogelstein 1984) confirmed the identity of the PCR amplification product, to be the *ipt* gene using the methods and conditions described by Hamill et al. (1989, 1991). DNA was also extracted from shoot cultures derived from transformations of *M. piperita* with the *ipt* expression cassettes, by the method of Lassner et al. (1989), and the presence of the *ipt* gene was once again confirmed by PCR amplification. In addition, the DNA extracts of the cultures were probed with a ^{32}P-labeled fragment of pTiC58, (Hind III fragment 7), which had been isolated from pGV 0361 (Depicker et al. 1980), and is homologous to the virulence region of *A. tumefaciens* pTiT37 and pTiC58. This region of the Ti plasmid would not be present in axenic transformed tissue, and the absence of this region was used to demonstrate that the signals obtained by Southern blotting of genomic DNA were due to integration of T-DNA into the plant genome, rather than to bacterial contamination. In the PCR experiments, control amplifications were set up using primers to the *vir* region of the Ti plasmid again to confirm that positive signals obtained using *ipt* primers were due to integration of the *ipt* gene into the plant genome rather than due to bacterial contamination (Hamill et al. 1991).

4.2 Growth of *Mentha* Shoot Cultures in vitro

Shooty teratoma lines propagated from single stems of both *M. citrata* and *M. piperita* transformed with strain T37 were used for detailed studies of growth. A shooty teratoma line of *M. citrata* (MCI) was grown in a high light intensity of 840 lx at 25 °C with a 16-h day on 50 ml of either solid or liquid B50 medium. The results showed that this shooty teratoma grew more rapidly in the liquid medium, on a'rotary shaker at 90 rpm, achieving a dry matter of approximately 0.8 g in 18 days from an inoculum of equivalent to 0.05 g dry weight.

A shoot teratoma line of *M. piperita* (BM1) was grown in two liquid media (MSO and B50) under two conditions, one on a shaker at 90 rpm in a 16-h day at 25 °C, in a poorly illuminated room with a light intensity of 200 lx: and the second in a Gallenkamp orbital shaker at 90 rpm in a 16-h day at 25 °C with a higher light intensity of 840 lx. The results showed that this shoot culture grew more rapidly, and achieved a greater ultimate dry weight under the higher light intensities in both media. In MSO media the growth rate of the BM1 teratoma was doubled when grown in the higher illumination, achieving about 0.8 g dry weight in 10 days compared to about 0.75 g in 20 days at 200 lx. The *M. piperita* (BM1) in B50 liquid medium under the higher light intensity achieved 0.75 g in 28 days compared to about 0.65 g in 20 days in the lower light intensity. At high light intensity, the *M. citrata* teratoma (MC1) culture grew equally well in either media, achieving approximately 0.8 g dry weight in 18 days.

Shoots lines that were propagated from initial axenic single shoots, derived from transformations of *M. piperita* with p35S*ipt* (BM35S) or pE35S*ipt* (BME35S), were inoculated (0.1–0.3 g wet weight of shoot tissue) in 50 ml MSO medium to study their growth kinetics. The shoot culture of *M. piperita* derived from transformation with p35S*ipt* (BM35S), achieved a 2.5 times greater growth rate than phenotypically similar shoots derived from transformation with pE35S*ipt* (BME35S). The BM35S shoots achieved a maximum fresh weight of 7.8 g by day 44 compared to the BME35S shoots, which had achieved a fresh weight of only 3.2 g by day 44.

4.3 Analysis of Monoterpenes in Transgenic Shoots of *M. citrata* and *M. piperita*

The major components of the essential oils in each of the shoot cultures were extracted by the procedures of Spencer (1991) and the major monoterpene components were identified using GC/MS under electron ionization (EI) conditions (MS 80 RFA mass spectrometer Kratos UK), and compared with published MS data (McLafferty and Stauffer 1989).

The two major components in the essential oil extracted from the parent plant of *M. citrata* are linalool and linalyl acetate, which accounted for over 90% of the oil. Linalool and linalyl acetate also made up over 90% of the essential oil extracted from the T-37-derived culture (MC1) *M. citrata*. The maximum yield of

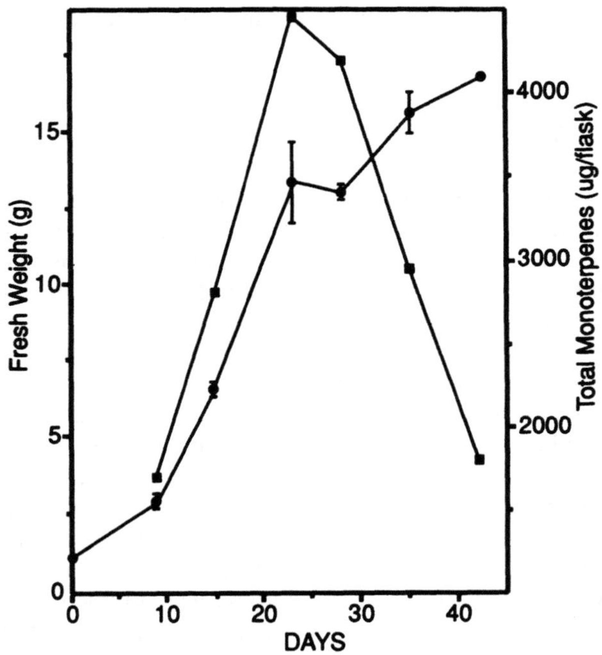

Fig. 2. Time course of growth and total monoterpene production of the T 37-derived culture of *M. piperita* (BM1) growing on MSO medium. This shows that the peak is monoterpene accumulation at the mid-point of the growth phase. (Spencer 1991)

the two monoterpenes linalool and linalyl acetate extracted from MC1 when grown on B50 medium was 195 μg/g fresh wt. and when grown on MSO medium was 102 μg/g fresh wt.

The major monoterpenes in the essential oil extracted from the parent plant of *M. piperita* (Scotch black mint) were menthone and menthol which accounted for up to 80% of the essential oil. The major component in the essential oil extracted from T37-derived shooty cultures of the *M. piperita* variety was menthol. However, these cultures have a much lower level of menthone and a higher level of menthofuran than the parent plant (Rhodes et al. 1992). The maximum total yield of essential oil extracted from the two shoot teratoma cultures (BM1 and BM2) (including menthone, menthofuran, menthol, and menthyl acetate) were similar, BM1 accumulated 635 μg/g fresh wt. and BM2 accumulated 577 μg/g fresh wt. BM2 shows a maximum yield of monoterpene at about the mid-point of the growth phase and yield was substantially reduced by the time the culture entered stationary phase (see Fig. 2).

The reduction in the level of monoterpenes in the *M. piperita* cultures observed late in the culture cycle may be due either to re-metabolism or to losses from the teratomas into the culture medium. Preliminary investigations were carried out to determine whether the decrease in monoterpene levels in the tissues could be accounted for by their export to the medium. Hexane extracts of the medium analyzed by G.C. showed 1-menthol to be present in the medium but only at low levels (about 1 μg/ml). This concentration of 1-menthol in the medium was insufficient to account for the decrease in total monoterpene in the culture, and suggests that a large amount of the monoterpenes must either be re-metabolized by the shooty teratomas, or that monoterpenes are being lost by metabolism in the medium.

Disorganized cultures resulting from transformation of *M. piperita* with the wild-type *ipt* gene (pSSI) were extracted by the standard procedure and analyzed by G.C. The major components identified in extracts of the essential oil from the parent plant were not present at detectable levels.

5 Summary and Conclusion

The induction of shoot formation in *Mentha*, which was the objective of the present study, can be achieved by two routes, the use of wild-type nopaline strains of *Agrobacterium* or expression of the *ipt* gene under the control of strong promoters. Stronger promoters are required to achieve shoot formation in *Mentha* than in tobacco. The *ipt* gene under the control of its own promoter, whether in isolation or within auxin mutants, is not in itself able to induce shoot formation in mint. *Mentha* appears to require high levels of endogenous cytokinin expression for shoot formation to occur. The efficacy of nopaline strains on inducing shoot formation in either mint or tobacco is difficult to explain. The fact that the T37 T-DNA in its entirety but not the isolated *ipt* gene or the *onc* gene fragment (Xba1 fragment 7) is able to induce shoot formation suggests that genes outside the *onc* region may play an important role in shoot formation effect. The reason why shoot differentiation in *Mentha* occurs only after excision of the initial gall from the explant is also unknown but is a feature of shooty teratoma function derived in either fashion. However, once the cultures are axenic, and the shooty phenotype is expressed, this is retained over extended periods. Each of the cultures discussed here has maintained its phenotype over repeated passages for more than 2 years. This work also demonstrates that transformation of some genotypes of *Mentha* can be achieved using conventional *Agrobacterium*-based vectors. Given the genetic heterogeneity within the genus, it is perhaps not surprising that some lines such as the *M. spicata* cultivar used are recalcitrant to transformation. Clearly where initial difficulties with a particular genotype are confronted, detailed investigation of the effects of different *Agrobacterium* host strains, and of the precise conditions for transformation are essential.

References

Akiyoshi DE, Morris RO, Hinz R, Mischke BS, Kosugo T, Garfinkel DJ, Gorden MP, Nester EW (1983) Cytokinin/auxin balance in crown gall tumors is regulated by specific loci in the T-DNA. Proc Natl Acad Sci USA 80:407–411

Alston RE, Turner BL (1963) In: Alston RE, Turner BL (eds) Biochemical systematics. p. 231 Prentice Hall, New Jersey, USA

Anderson AR, Moore LW (1979) Host specificity in the genus *Agrobacterium*. Phytopathology 60:320–323

Aviv A, Krochmal E, Dantees A, Galun E (1981) Biotransformation in monoterpenes by *Mentha* cell lines: conversion of menthothone to neomenthol. Planta Med 42:236–243

Berry C, Van Eck J, Kitto S, Smigocki A (1990) Transformation of orangemint (*Mentha citrata*) with *Agrobacterium tumefaciens*. Hort. Science 25: 1165 (Abstr)

Braun AC, Wood HN (1976) Suppression of the neoplastic state with the acquisition of specialised functions in cells, tissues, and organs of crown gall teratomas of tobacco. Proc Natl Acad Sci USA 73:496–500

Bricout J, Paupardin C (1975) Sur la composition de l'huile essentielle de *Mentha piperita* L. cultivé in vitro: influence de quelques facteurs sur sa synthase. CR Acad Sci Paris 281:383–386
Bricout J, Garcia-Rodriguez M, Paupardin C (1978) Action de la colchine sur la synthase d'huile essentielle par des tissues de *Mentha piperita* cultivés in vitro CR Acad Sci Paris 286:1585–1588
Caelles A, Ferrer A, Balcells L, Hegardt FG, Boronat A (1989) Isolation and characterisation of a cDNA encoding *Arabidopsis thaliana* 3-hydroxy-3-methylglutaryl CoA reductase. Plant Mol Biol 13:627–638
Carter P, Bedouelle H, Winter G (1985) Improved oligonucleotide site-directed mutagenesis using M13 vectors. Nucleic Acids Res 13:4431–4443
Charlwood BV, Brown JT, Mouston C, Morris GS, Charlwood KA (1988) The accumulation of isoprenoid flavour compounds in plant cell cultures. In: Schreier P (ed) Bioflavour '87. Gruyter, Berlin, pp 303–314
Chye M-L, Kush A, Tan C-T, Chau N-H (1991) Characterisation of cDNA and genomic clones encoding 3 hydroxy-3-methylglutaryl CoA reductase from *Hevea brasiliensis*. Plant Mol Biol 16:567–577
Croteau R, Venkatachalam KV (1986) Metabolism of monoterpenes: demonstration that (+)-cis-isopulegone, not piperitenone, is the key intermediate in the conversion of (−)-isopiperitenone to (+)-pulegone in peppermint (*Mentha piperita*). Arch Biochem Biophys 249:306–315
Dale PJ, Marks MS, Brown MM, Woolston CJ, Gunn HV, Mullineaux PM, Lewis DM, Kemp JM, Chen DF, Gilmour DM, Flavell RB (1989) Agroinfection of wheat: inoculation of in vitro grown seedlings and embryos. Plant Sci 63:237–245
Depicker A, De Wilde M, De Vos G, De Vos R, Van Montagu M, Schell J (1980) Molecular cloning of overlapping segments of the nopaline Ti-plasmid pTiC58 as a means to restriction endonuclease mapping. Plasmid 3:193–211
Feinberg AP, Vogelstein B (1984) A technique for radio labelling DNA restriction endonuclease fragments to high specific activity. Anal Biochem 137:266–267
Fillatti JJ, Sellmer J, McCown B, Haissid B, Comai L (1987) *Agrobacterium*-mediated transformation and regeneration of *Populus*. Mol Gen Genet 206:192–199
Garfinkel DJ, Simpson RB, Ream LW, White FF, Gordon MP, Nester EW (1981) Genetic analysis of crown gall: fine structure map of the T-DNA by site-directed mutagenesis. Cell 27:143–153
Gershenzon J, Croteau R (1990) Regulation of monoterpene biosynthesis in higher plants. Recent Adv Phytochem 24:99–160
Goldberg SB, Flick JS, Rogers SG (1984) Nucleotide sequence of the *tmr* locus of *Agrobacterium tumefaciens* pTi T37 T-DNA. Nucleic Acids Res 12:4665–4678
Gray JC (1987) Control of isoprenoid biosynthesis in higher plants. Adv Bot Res 14:25–91
Gresshoff PM, Skitnicki ML, Rolfe BG (1979) Crown gall teratoma formation is plasmid- and plant-controlled. J Bacteriol 137:1020–1021
Hamill JD, Parr AJ, Robins RJ, Rhodes MJC (1986) Secondary product formation by cultures of *Beta vulgaris* and *Nicotiana rustica* transformed with *Agrobacterium rhizogenes*. Plant Cell Rep 5:111–114
Hamill JD, Parr AJ, Rhodes MJC, Robins RJ, Walton NJ (1987) New routes to plant secondary products. Bio/Technol 5:800–804
Hamill JD, Evans DM, Robins RJ, Rhodes MJC (1988) Foreign gene insertion into transformed roots with binary vectors and *Agrobacterium rhizogenes* – potential for genetic manipulation of plant secondary metabolism. In: Robins RJ, Rhodes MJC (eds) Manipulating secondary metabolism in culture. Cambridge University Press, Cambridge, pp 145–153
Hamill JD, Rounsley S, Spencer A, Todd G, Rhodes MJC (1991) The use of the polymerase chain reaction in plant transformation studies. Plant Cell Rep 10:221–224
Harley RM, Brighton CA (1977) Chromosome numbers in the genus *Mentha* L. Bot J Linn Soc 74:91–96
Heble MR (1985) Multiple shoot cultures: a viable alternative in vitro system for the production of known and new biologically active plant constituents. In: Neumann KH, Barz W, Reinhard E (eds) Primary and secondary metabolism of plant cell cultures. Springer, Berlin Heidelberg New York, pp 281–289
Hirata T, Murakami S, Ogihara K, Suga T (1990) Volatile monoterpenoid constituents of the plantlets of *Mentha spicata* produced by shoot tip culture. Phytochemistry 29:493–495
Hoekema A, Hirsch PR, Hooykaas PJJ, Schilperoort RA (1983) A binary plant vector strategy based on separation of vir- and T-region of the *Agrobacterium tumefaciens* Ti-plasmid. Nature 303:179–180
Hofgen R, Willmitzer L (1988) Storage of competent cells for *Agrobacterium* transformation. Nucleic Acids Res 16:9877

Karp F, Mihaliak CA, Harris J-L, Croteau R (1990) Monoterpene biosynthesis. Specificity of the hydroxylation of levo-limonene by enzyme preparations from peppermint, *Mentha piperita*, spearmint *Mentha spicata* and perilla, *Perilla frutescens*. Arch Biochem Biophys 276:219–226
Kay R, Chan A, Daly M, McPherson J (1987) Duplication of CaMV35S Promoter sequences creates a stronger enhancer for plant genes. Science 236:1299–1302
Kjonaas R, Martinkus-Taylor C, Croteau R (1982) Metabolism of monoterpenes: conversion of 1-menthone to 1-menthol and d-neomenthol by stereospecific dehydrogenases from peppermint (*Mentha piperita*) leaves. Plant Physiol 69:1013–1017
Korber H, Strizhov N, Staiger D, Fledwisch J, Olsson O, Sandberg G, Palme K, Schell J, Koncz C (1991) T-DNA gene 5 of *Agrobacterium* modulates auxin response by autoregulated synthesis of a growth hormone antagonist in plants. EMBO J 10:3983–3991
Lassner MW, Peterson P, Yoder JI (1989) Simultaneous amplification of multiple DNA fragments by polymerase chain reaction in the analysis of transgenic plants and their progeny. Plant Mol Biol Rep 7:116–128
Loomis WD (1967) Biosynthesis and metabolism of monoterpenes. In: Pridham JD (ed) Terpenes in plants. Academic Press, London, pp 59–82
Maniatis T, Fritsch EF, Sambrook J (1982) Molecular cloning. A laboratory manual. Cold Spring Harbour Laboratory
Martin C, Carpenter R, Sommer H, Saedlet H, Coen ES (1985) Molecular analysis of instability in flower pigmentation of *Antirrhinum majus*, following isolation of the pallida locus by transposon tagging. EMBO J 4:1625–1630
McLafferty FW, Stauffer DB (1989) The Wiley/NBS registry of mass spectral data. Wiley, New York
Murashige T, Skoog F (1962) A revised medium for rapid growth and bioassays with tobacco tissue cultures. Physiol Plant 15:473–497
Ooms G, Klapwijk PM, Poulis JA, Schilperoort RA (1980) Characterization of Tn904 insertions in octopine Ti-plasmid mutants of *Agrobacterium tumefaciens*. J Bacteriol 144:82–91
Ooms G, Hooykaas PJJ, Moolenaar G, Schilperoort RA (1981) Crown gall plant tumours of abnormal morphology, induced by *Agrobacterium tumefaciens* carrying mutant octopine Ti plasmids; analysis of T-DNA functions. Gene 14:33–50
Ooms G, Basins A, Burrell M, Karp A, Twell D, Wilcox E (1985) Genetic manipulation in cultivars of oilseed rape (*Brassica napus*) using *Agrobacterium*. Theor Appl Genet 71:325–329
Paupardin C (1976) Sur la differentiation d'un tissu secreteur et la formation d'huile essentielle par des tissue végétaux cultivés in vitro. Comptes rendus du Congrès National Sociétés Savantes Section des Sciences vol 101(1), pp 619–628
Rhodes MJC, Spencer A, Hamill JD, Robins RJ (1992) Flavour improvement through plant cell culture. In: Charlwood BV, Patterson RLS, MacLeod G, Williams AA (eds) Bioformation of flavour. Royal Society of Chemistry, London pp 42–64. Cambridge University Press, Cambridge
Saunders PR, Winter JA, Barnason AR, Rogers SG, Fraley RT (1987) Comparison of cauliflower mosaic virus 35S and nopaline synthase promoters in transgenic plants. Nucleic Acids Res 15:1543–1558
Schmulling T, Beinsberger S, De Greef J, Schell J, Van Onckelen H, Spena A (1989) Construction of a heat-inducible chimeric gene to increase the cytokinin content in transgenic plant tissue. FEBS Lett 249:401–406
Skoog F, Miller CO (1957) Chemical regulation of growth and organ formation in plant tissues cultured in vitro. Symp Soc Exp Biol 11:118–131
Smigoki AC, Owens LD (1988) Cytokinin gene fused with a strong promoter enhances shoot organogenesis and zeatin levels in transformed plant cells. Proc Natl Acad Sci USA 85:5131–5135
Spencer A (1991) The development of shooty teratomas in *Mentha* species by genetic manipulation and studies on their growth and terpene production in vitro. PhD Thesis, Univ of East Anglia
Spencer A, Hamill JD, Rhodes MJC (1990) Production of terpenes by differentiated shoot cultures of *Mentha citrata* transformed with *Agrobacterium tumefaciens* T37. Plant Cell Rep 8:601–604
Suga T, Hirata T, Aoki T and Shishib T (1986) Inter-conversion and cyclisation of acylic allylpyrophosphates in the biosynthesis of cyclic monoterpenes in hydro plants Phytochem. 25:2769–2776
Tinland B, Huss B, Paulus F, Bonnard G, Otten L (1989) *Agrobacterium tumefaciens* 6b genes are strain-specific and affect the activity of auxin as well as cytokinin genes. Mol Gen Genet 219:217–224
Zieg RG, Zito SW, Staba EJ (1983) Selection of high pyrethrin-producing tissue cultures. Planta Med 48:88–81

II.16 Transformation in *Nicotiana edwardsonii*

J.M. Kiernan, Fang C. Wu, K.-B. Goldberg, S. Gowda and R.J. Shepherd[1]

1 Introduction

1.1 *Nicotiana edwardsonii*

Nicotiana edwardsonii Christie and D.W. Hall (Christie and Hall 1979) obtained by crossing *N. clevelandii* A. Gray with *N. glutinosa* L. is vigorous, fertile, and hypersensitive to tobacco mosaic virus (TMV). It possesses wide susceptibility to plant viruses like its *N. clevelandii* parent but has the vigor and size of its *N. glutinosa* parent.

Christie and Hall, who developed this hybrid, noted susceptibility to at least 11 plant viruses (Christie 1969). This wide range of susceptibility makes *N. edwardsonii* a useful plant for many aspects of plant virus research. In addition, the systemic susceptibility of *N. edwardsonii* to figwort mosaic virus (FMV), (Shepherd et al. 1987) and strain D4 of cauliflower mosaic virus (CaMV) (Schoelz et al. 1986) have made this an attractive plant for research efforts which focus on the molecular biology of caulimoviruses.

1.2 The Caulimoviruses

The caulimoviruses are a 12-member group of icosahedral viruses, averaging about 50 nm in diameter (Harrison et al. 1971) and containing circular double-stranded DNA. CaMV and FMV each have six major and two minor open reading frames and are approximately 8000 base pairs in size but appear to be only distantly related. Strong sequence homology is observed only in open reading frames I and V. The protein of gene I may participate in cell to cell movement (Linstead et al. 1988), gene V produces a reverse transcriptase involved in DNA replication (Hull and Covey 1983; Volovitch et al. 1984; Takatsuji et al. 1986). Gene VI, which we use in transformations reported in this chapter, encodes the inclusion body protein (Covey and Hull 1981; Xiong et al. 1982), and is believed to be largely responsible for host range determination and induction of virus symptoms in plants (Daubert et al. 1984; Schoelz and Shepherd 1988). Functions such as aphid transmission, DNA binding and coat protein production have been assigned to genes II, III, and

[1] Department of Plant Pathology, University of Kentucky, Lexington, KY 40546, USA

Biotechnology in Agriculture and Forestry, Vol. 22
Plant Protoplasts and Genetic Engineering III (ed. by Y P S. Bajaj)

IV, respectively. No products have been identified for the small open reading frames VII and VIII of CaMV and VII and IX of FMV.

1.3 Use of *Nicotiana edwardsonii* in Genetic Transformations and Transient Gene Expression Studies

We have developed two systems utilizing *N. edwardsonii*. One is the production of transgenic plants containing various genes of caulimoviruses by a leaf disc transformation method using *Agrobacterium tumefaciens* as a vector (Horsch et al. 1985). The other is the use of plant protoplasts made from suspension cultures of *N. edwardsonii* to observe transient expression of genes using the excised promoters of caulimoviruses coupled to reporter genes such as chloramphenicol acetyltransferase (CAT) and β-glucuronidase (GUS). In this chapter, are described experiments where gene I of FMV and gene VI of CaMV are introduced into plants via genetic transformation. The protocols for obtaining suspension culture protoplasts and their electroporation with caulimoviral DNA carrying the CAT and GUS markers are also outlined.

2 Leaf Disc Transformation and Regeneration

Attempts by us to introduce genetic material of CaMV or FMV into *N. edwardsonii* using the standard protocol for *Nicotiana tabacum*, which uses MS medium (Murashige and Skoog 1962) with 1 mg/l 6-benzylaminopurine (BAP) and 0.1 mg/l a-naphthaleneacetic acid (NAA) (Horsch et al. 1985), were not successful. Upon examining the shoot regeneration capabilities of several published media for *Nicotiana* species, we found a medium which gave good results. This medium, MS4 (Malmburg et al. 1985) contains MS salts (Gibco Laboratories, Grand Island, NY Cat No 500–1117), 10 mg/l (2-isopentenyl) adenine (2-iP), 0.3 mg/l indole-3-acetic acid (IAA), B5 vitamins (Gamborg et al. 1968), 30 g/l sucrose and 10 g/l agar (Difco Bacto–agar, Difco Labs., Detroit, MI) with an adjusted pH of 6.0 before autoclaving. This medium was also superior to MS medium with 1.13 mg/l 6-benzylaminopurine (BAP), a commonly used medium for shoot regeneration of *Nicotiana* species (Flick et al. 1983). MS4 medium (MS, 10 mg/l 2-iP and 0.3 mg/l IAA) resulted in 23% of the explants yielding shoots 28 days from initiation of culture and 45% at 42 days.

It was also found in initial experiments that shoot production was strongly inhibited when MS4 medium was used in conjunction with *A. tumefaciens* and the generally recommended levels of the appropriate antibiotics, kanamycin, or Geneticin (G418, Gibco Labs., Gaithersburg, MD) for the selection of the neomycin phosphotransferase gene and Carbenicillin for inhibiting *Agrobacterium*. Experiments with lowering the level of Carbenicillin (Geopen, manufacturer Roerig/Pfizer, New York, NY) indicated that a level of 500 mg/l was needed to control the growth of *Agrobacterium* sufficiently, so the experiments reported here deal only with differing levels of kanamycin or Geneticin (G418).

2.1 Transformation with Gene I of Figwort Mosaic Virus (FMV)

Region I of FMV exhibits a high degree of homology to region I of CaMV with 54% of the amino acid sequence being identically conserved (Richins et al. 1987). Hence, these genes probably have a similar function. Introduction of gene I of FMV into *N. edwardsonii* by genetic transformation was undertaken in order to study the possible role of this gene in cell-to-cell movement (discussed later).

Gene I of FMV was excised and ligated into the plasmid vector pKYLX7 (Schardl et al. 1987). Single and tandem constructs were made (Fig. 1). The resultant constructs were transferred into a wild-type *Agrobacterium tumefaciens* strain, A281 (Hood et al. 1984) using triparental mating (Ditta et al. 1980). The *Agrobacterium* culture was grown overnight in 35 ml YT medium (8 *g* Bacto–tryptone (Difco), 5 *g* Bacto yeast extract (Difco) and 2.5 *g* NaCl in 1 l of water) containing 10 μg/ml kanamycin and tetracycline and 100 μg/ml Rifampicin.

Plant material was from 6-week-old *Nicotiana edwardsonii* grown in a greenhouse under natural lighting conditions during the early part of summer with an ambient temperature range of 21–28 °C. The leaf material was sterilized by immersing in 70% ethanol for 20 s, then transferring to 10% bleach (diluted strength was 0.525% sodium hypochlorite) for 16 min. The leaf tissue was then given two consecutive washes by immersing in sterile deionized water for about 2 min each time. Small squares were cut with a sterile scalpel and immersed for about 30 s to 1 min in a solution of the *Agrobacterium* containing the relevant gene, blotted on sterile filter paper and placed on MS4 medium in sterile 100 × 20 mm Petri dishes. The squares were incubated for 48 h at room temperature in dim light

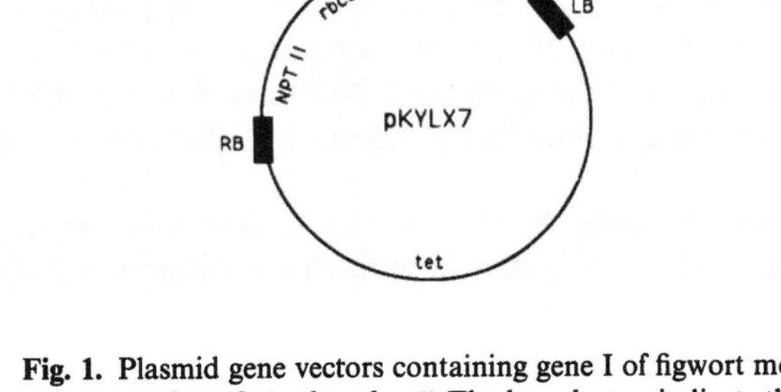

Fig. 1. Plasmid gene vectors containing gene I of figwort mosaic virus (FMV) (*12/1, 12/2*) used for the transformation of *N. edwardsonii*. The *boxed areas* indicate the right (*RB*) and left (*LB*) borders of the T-DNA. The neomycin phosphotransferase II (*NPTII*) and *tet* genes which provide kanamycin and tetracycline resistance, respectively, are indicated. The pKYLX7 vector provides a 35S CaMV promoter and a rbcS 3′ terminator for the FMV gene I open reading frame. The complete open reading frame of gene I is indicated by the *shaded areas* and is a single insert in construct 12/1 and a tandem insert in 12/2. (Goldberg et al. 1988)

and then transferred to sterile 100 × 20 mm Petri dishes containing the same medium and the following combinations of antibiotics: (1) 500 mg/l Geopen and no kanamycin, (2) 500 mg/l Geopen plus 10 mg/l G418, (3) 500 mg/l Geopen and 50 mg/l kanamycin, (4) 500 mg/l Geopen and 300 mg/l kanamycin. Dishes were kept at room temperature (24–28 °C) with a 16-h photoperiod of 2800 lx.

Two experiments were conducted, one using the single insert construct (12/1) and the other the tandem insert (12/2) of gene I of FMV. In each experiment, 15 dishes containing 7 leaf squares per dish were made for each antibiotic level. As controls, noninoculated squares (no *Agrobacterium*) were placed on medium without antibiotics and also with the same level of antibiotics as used in each experiment. The leaf squares were transferred to fresh medium every 3 weeks to further suppress growth of *Agrobacterium.* Shoots formed were placed in Magenta GA7 containers (Magenta Corp., Chicago, IL) on a rooting medium of one-half strength MS salts without plant hormones and containing 30 g/l sucrose, B5 vitamins, 10 mg/l agar (Difco Bacto-agar), pH 5.8 to which 500 mg/l Geopen had been added. Screening for insertion of the neomycin phosphotransferase gene was done by taking some leaf tissue and placing it on MS shoot medium containing 100 mg/l kanamycin and observing for the growth of shoots or callus.

In the experiment where the single insert of gene I (12/1) of FMV was used, no *N. edwardsonii* shoots were obtained on media containing Geopen and 50 mg or 300 mg/l kanamycin or 10 mg/l of G418. The medium containing Geopen alone (no kanamycin) resulted in 3 leaf squares out of 105 producing shoots. From these squares, 11 plants were eventually obtained. Three of these plants were found to be positive for the insertion of gene I of FMV by Southern blot analysis. None were kanamycin-resistant.

In the experiment where the plasmid with a tandem insert of gene I of FMV (12/2) was used, procedures were as when the single sequence (12/1) was used. No *Nicotiana edwardsonii* plants were obtained from either the medium containing Geopen and 10 mg/l G418 or that with Geopen alone (no kanamycin or G418). Leaf squares placed on medium with Geopen and 50 mg/l kanamycin yielded one square out of 105 with shoots. Squares on Geopen and 300 mg/l kanamycin resulted in four squares out of 105 with shoots. Thirteen plants were obtained from the one square that produced shoots on the 50 mg/l screening level of kanamycin and 7 from the squares on the 300 mg/l screening level. Only one plant from the 50 mg/l screening level was transgenic for gene I and it was not kanamycin-resistant. Three from the higher level of 300 mg/l were found to be transformed with gene I by Southern analysis (Fig. 2) and were also kanamycin-resistant. The time span from start of the experiment to transfer of rooted plants to the greenhouse was 26 weeks for construct 12/1 and 17 weeks for construct 12/2. In both experiments, the transformed plants were of normal size and appearance.

Transgenic plants are being used to investigate the function of gene I of the caulimoviruses. Gene I protein of caulimoviruses is found in the cell walls of infected plants, a feature of proteins involved in cell-to-cell movement of viruses. Our experiments will consist of using transgenic plants in complementation assays with gene I lethal mutants of the cloned genomes of figwort mosaic virus and peanut chlorotic streak virus. If gene I functions only in cell-to-cell movement of these viruses and has no role in replication or assembly, genomes of gene I lethal mutants should multiply and move systematically in gene I transgenic plants.

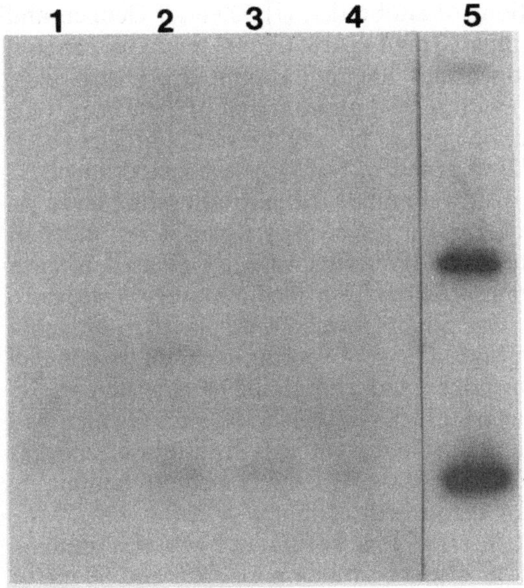

Fig. 2. Autoradiogram of a Southern hybridization analysis of *N. edwardsonii* transformed with the tandem insert (12/2) of gene I of figwort mosaic virus (FMV). Fifteen μg of total genomic DNA digested with *SacI* was subjected to agarose gel electrophoresis then transferred to a nylon membrane and hybridized to a ^{32}P-labeled probe containing gene I of FMV. *Lane 1* contains DNA from mature healthy *N. edwardsonii* plants. *Lanes 2, 3, and 4* contain DNA from plants 35, 27, and 25, respectively, which have the tandem insert (12/2) of gene I of FMV. *Lane5* contains plasmid DNA of the single insert construct of gene I (12/1) which has been digested with *SacI*. The *arrow* indicates the position of the single insert fragment of gene I. (Goldberg et al. 1988)

2.2 Transformation with Gene VI of Strain D4 of Cauliflower Mosaic Virus (CaMV)

Gene VI of CaMV is responsible for producing the inclusion body protein and is a determinant of virus host range and disease expression. We have found that when *Nicotiana tabacum* cv. Burley 21 tobacco was transformed with gene VI of strain D4 of CaMV, chlorosis, stunting and/or a mottling syndrome was observed in transformed plants similar to symptoms produced by virus infected plants (Goldberg et al. 1990, 1991). We were interested to see if this effect would be observed in *N. edwardsonii.*

Gene VI of CaMV strain D4 was excised from cloned viral DNA and ligated into a modified version (Kiernan et al. 1989) of the plasmid vector pGA472 (An et al. 1985) and the resultant construct (pJS65) (Fig. 3A) (Kiernan et al. 1989) was transferred into *Agrobacterium tumefaciens* strain A281 using triparental mating. In construct pJS65, gene VI retained its homologous 19S promoter. The *Agrobacterium* culture was grown overnight in YT medium containing 10 μg/ml kanamycin and 10 μg/ml tetracycline.

One hundred and five sterile leaf squares were inoculated with *A. tumefaciens* containing pJS65, incubated on MS4 medium for 48 h and then placed on MS4 medium with 500 mg/l Geopen and no kanamycin. Another 105 sterile squares were inoculated with pJS65, incubated on MS4 medium for 48 h, and then transferred to the same medium with 500 mg/l Geopen and 50 mg/l kanamycin. Light and temperature conditions were as in the previous experiment. Five leaf squares yielded shoots on the medium without kanamycin and three squares on the medium with 50 mg/l kanamycin. Eleven plants were regenerated from the squares placed on medium without kanamycin and ten plants from the 50 mg/l level of

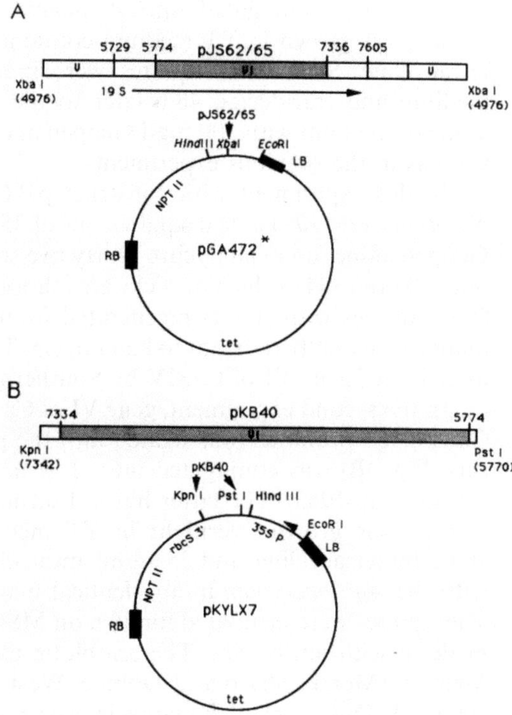

Fig. 3A, B. Plasmid gene vectors with gene VI of cauliflower mosaic virus (CaMV). **A** Constructs labeled pJS62 and pJS65 with CaMV gene VI in the vector pGA472 (An et al. 1985) used to transform *N. edwardsonii.* The *boxed areas* indicate the right (*RB*) and left (*LB*) borders of the T-DNA. The neomycin phosphotransferase II (NPTII) and *tet* genes which provide kanamycin and tetracycline resistance, respectively, are indicated. The 19S transcript of pJS62 and pJS65, using a modified version of pGA472, is indicated by the *arrow* and includes the homologous promoter and poly(A) signal. The *shaded areas* indicate the coding region of gene VI. **B** The plasmid pKB40 used for the transformation of *N. edwardsonii. Right* and *left borders* of the T-DNA and the NPTII and tet genes are indicated. The pKYLX7 vector provides a 35S promoter and a rbcS 3′ terminator for the CaMV gene VI open reading frame. The complete open reading frame of gene VI for the vector was obtained by olignucleotide mutagenesis arld is indicated by the *shaded areas.* (Goldberg et al. 1991)

kanamycin. The experiment resulted in six plants, three from medium without selection and three from 50 mg/l kanamycin selection expressing gene VI product. Expression of gene VI (Kiernan et al. 1989) was determined by Southern blot and western blot analysis. Only one plant was kanamycin-resistant. The time span of the experiment was about 15 weeks from initiation to transfer of rooted plants to the greenhouse. All plants were normal in size and appearance.

2.3 Transformation with Gene VI of Strain CM1841 of Cauliflower Mosaic Virus (CaMV)

Insertion of gene VI of strain CM1841 of CaMV into *N. tabacum* cv. Burley 21 resulted in expression of similar symptoms to those seen in Burley 21 transformed with gene VI of strain D4 of CaMV (Goldberg et al. 1991). We therefore wished to determine if similar effects would develop in *N. edwardsonii* plants transformed with these genes.

Two experiments were conducted. One using a construct (pJS62) (Kiernan et al. 1989) of gene VI of strain CM1841 with its homologous promoter inserted into a modified version of pGA472 and the other used a construct (pKB40) (Goldberg et al. 1990) with a CaMV 35S promoter inserted into a pKYLX7 gene vector (Schardl et al. 1987).

In the first experiment, gene VI of CaMV CM1841 was excised from viral DNA, ligated into the plasmid vector pGA472 and the resultant construct (pJS62)

(Fig. 3A) was conjugated into *A. tumefaciens* strain A281. The *Agrobacterium* culture was grown in YT medium containing 10 µg/ml kanamycin and 10 µg/ml tetracycline. Sterile leaf squares were inoculated with pJS62, placed on MS4 medium and transferred 48 h later to the same medium with 500 mg/l Geopen alone or medium with 500 mg/l Geopen and 10 mg/l G418. Light and temperature were as in the previous experiments.

In this experiment with construct pJS62 in strain A281 of *A. tumefaciens*, 4 *Nicotiana edwardsonii* leaf squares out of 35 formed shoots on medium containing Geopen alone (no kanamycin). Thirty five squares placed on medium with Geopen plus 10 mg/l G418 did not yield any shoots. Twenty three plants were obtained from squares with shoots regenerated from nonselective medium. Seven of these plants were mildly resistant to kanamycin. Two of the seven were found positive for insertion of gene VI of CaMV by Southern and western blot analysis.

In the second experiment, gene VI of CaMV strain CM1841 under control of a CaMV 35S promoter was ligated into the plasmid vector pKYLX7 (pKB40) and this (Fig. 3B) was conjugated into *A. tumefaciens* strain C58C1:pGV3850 (Zambryski et al. 1983). The latter has a disarmed plasmid vector. The *Agrobacterium* culture was grown overnight in YT medium containing 10 µg/ml kanamycin, 10 µg/ml tetracycline, and 25 µg/ml ampicillin. Sterile leaf squares were inoculated with the *Agrobacterium* in an identical manner as in the experiment with pJS62. The squares were incubated for 48 h on MS4 medium and then placed on the same medium with antibiotics. The antibiotic used to control the *Agrobacterium* was Mefoxin (Merck, Sharp and Dohme, West Point, PA) used at a concentration of 500 mg/l. One level of kanamycin was used for selection, 100 mg/l. Light and temperature were as in the other experiments. Ten out of 161 *Nicotiana edwardsonii* squares yielded shoots and 12 plants were eventually obtained from these. Nine of these plants were kanamycin-resistant. The presence of gene VI was detected in ten plants. Western analysis of four of these is shown in Fig. 4. The time span of the experiment from initiation to transfer of rooted plants to the greenhouse was 12 weeks for construct pJS62 and 18 weeks for pKB40. Very slight stunting and

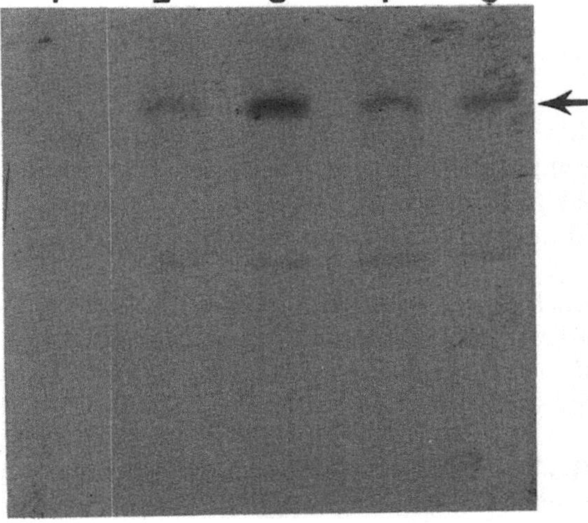

Fig. 4. Western blot analysis for gene VI protein from *N. edwardsonii* plants after transformation with gene VI of cauliflower mosaic virus (pKB40) (Goldberg et al. 1991). *Lanes 2 to 5* contain plants 11, 7, 3, and 1, respectively. *Lane 1* contains healthy *N. edwardsonii*. For each plant, 1 *g* of leaf tissue was suspended in 1.5 ml Laemmli buffer after pulverizing in liquid nitrogen. Fifty µl of this preparation was loaded per lane. The *arrow* indicates the position of the gene VI protein. The *lower band* indicates an immunologically related protein found in healthy and transformed *N. edwardsonii*

chlorosis were observed in transformed plants from experiments with both constructs, but the striking morphological effects of this gene observed in Burley 21 were not present in *N. edwardsonii.* However, levels of gene VI protein were much lower than those found in Burley 21. (Goldberg et al. 1991).

3 Transient Expression of Introduced Genes in *Nicotiana edwardsonii*

The following procedures lead to the establishment of suspension cell cultures, the successful isolation of protoplasts and electroporation of DNA into these protoplasts and detection of expression of reporter genes.

3.1 Production of Callus Cultures

Leaves from 6–8-weeks-old plants were sterilized, the midrib excised leaving about 0.25 cm green leaf tissue each side, cut into sections 0.5 cm long, and placed on LS medium (Linsmaier and Skoog 1965) containing 1 mg/l naphthaleneacetic acid (NAA), 0.2 mg/l 2,4-D and 0.2 mg/l kinetin. Dishes were stored at room temperature (24–28 °C) in the dark. The callus was white and friable, but rather slow-growing. Four to five subculturings at 4–5-week intervals were required before sufficient callus was available for suspension culture.

3.2 Initiation of Suspension Cultures

One to 2 g callus from LS medium was transferred to 75 ml modified liquid DM-1 medium containing 1 mg/l 2,4-D (Ranch and Giles 1980) in 250-ml Erlenmeyer flasks. Modified DM-1 contains 1360 mg/l $(NH_4)_2SO_4$, 3980 mg/l KNO_3 and remaining macronutrients as in standard MS medium except for NH_4NO_3, which is omitted. Micronutrients were one-tenth standard MS strength with the exception of $MnSO_4 \cdot H_2O$, which was standard strength. Thirty g/l sucrose and 1 g/l casein hydrolysate were included and in addition to 0.5 mg/l of the vitamins thiamine, nicotinic acid, and pyridoxine, 100 mg/l myo-inositol was added. Succinic or L-malic acid used in the original formulation were not included. The pH was adjusted to 5.5 before autoclaving.

Suspension cultures were maintained at 26 °C with 1000 lx of continuous light, on a G10 Gyrotory shaker (New Brunswick Scientific, Edison NJ) at 150 rpm. Suspensions were transferred every 4 days by pipeting 15 ml of suspension into 75 ml of fresh medium in 250-ml long-neck Erlenmeyer flasks.

3.3 Isolation of Protoplasts

Suspension cells of 2-day cultures were distributed in 50-ml tubes and centrifuged out of solution using a swing bucket rotor in a low speed clinical centrifuge at 100 *g* for 2 min. One volume of packed suspension cells was resuspended in 2 volumes of filter sterilized enzyme solution containing (w/v) 1.5% Cellulose Onozuka RS (Yakult Honsha Co., Ltd., Japan) and 0.15% Pectolyase Y23 (Seishin Pharmaceutical Co., Ltd., Japan) in MMC (0.5 M mannitol, 10 mM MES and 10 mM $CaCl_2$, pH 5.6) and the mixture was incubated in 250-ml Erlenmeyer flasks at room temperature under continuous gentle shaking at 30 rpm on a G2 gyrotory shaker (New Brunswick Scientific, N.J.) for 3–4 h. When virtually all cells were digested into protoplasts, the incubation mixture was filtered through eight layers of sterile cheesecloth and 25 ml aliquots were pelleted at 100 *g* for 90 s, then gently resuspended with 25% sucrose solution containing 10 mM MES and 10 mM $CaCl_2$, pH 5.6 to a total volume of 12 ml and transferred into a 15 ml tube. One ml of MMC was carefully layered on top of the solution and the tubes were centrifuged immediately in a swinging bucket rotor at 500 *g* for 2 min. The purified protoplasts were carefully removed with a sterile pipet from the interface and resuspended in 50 ml MMC. The concentration of protoplasts was determined using a hemocytometer. All procedures were performed under sterile conditions using sterile containers and solutions.

3.4 Transfection of Protoplasts by Electroporation

An aliquot containing 2×10^6 protoplasts was centrifuged at 100 *g* for 90 s in a swinging bucket rotor. The pellet was gently resuspended in 1 ml electroporation buffer (0.5 M mannitol, 5 mM MES and 70 mM KCl, pH 5.6) and mixed with 10–50 μg supercoiled plasmid DNA which had been purified on a CsCl gradient.

After 10 min on ice, the protoplasts were resuspended and transferred to a 1-ml semi-micro disposable cuvet (VWR, Batavia, IL). Electroporation was performed by using a discharge device (Fig. 5), made in our workshop, similar to the one described by Fromm et al. (1985). An electrode (P/N 471, BTX, San Diego, CA) and

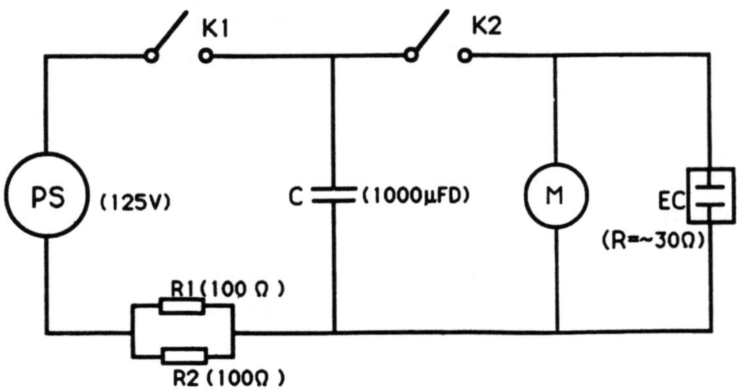

Fig. 5. Circuit of electroporation unit. *PS* power supply; *C* capacitor; *EC* electroporation chamber; *M* monitor (BTX Optimizer); R_1, R_2 resistors; K_1, K_2 keys for controlling charging and discharging of the capacitor. The discharge device is a slight modification of the apparatus described by Fromm et al. (1985)

a BTX Optimizer to monitor the process were included in the circuit. Normally a voltage of 125 V was used for charging the 1000 μF capacitor and a duration of 20–25 ms was needed. The transfected protoplasts were placed on ice for 10 min, transferred to a 15-ml culture tube, and then washed once with MMC by centrifugation at 100 *g* for 90 s. The pellet was gently resuspended in 8-ml incubation medium (0.5 M mannitol, 10 mM $CaCl_2$, 0.2 mM KH_2PO_4, 1 mM KNO_3, 1 mM $MgSO_4$, 1 μm KI, 0.01 μM $CuSO_4$, and 1 μg/ml 2,4-D, pH 5.6) and incubated at room temperature under light for 20–30 h. The concentration of living cells was then determined using a hemocytometer. An aliquot containing 2×10^5 intact protoplasts was collected in a 1.5 ml Eppendorf tube and centrifuged at speed 12 (maximum speed) in a Beckman 12 microfuge for 2 min. The supernatant was removed and the pellet stored at − 20 °C until the reporter gene assays were performed.

3.5 CAT Assays

CAT assays were performed using a modification of Gorman et al. (1982). Twenty five μl of 0.25 M Tris-Cl, pH 7.8 was added to each 1.5 Eppendorf tube containing a pellet of 2×10^5 protoplasts. The tube was vortexed at full speed for 1 min and then centrifuged in a Beckman microfuge at maximum speed for 10 min. After being transferred to a clean tube, the supernatant was mixed with 7.25 μl of substrate premix containing 0.25 μl of 0.2 μCi/μl ^{14}C-chloramphenicol (Amersham International, UK), 2 μl of 10 mM acetyl coenzyme A (Sigma chemical Co., St. Louis, MO), and 5 μl of 0.25 M Tris-Cl, pH 7.8. The samples were incubated at 37 °C for 2 h and then 250 μl of ethyl acetate was added to each sample. The tube was then vortexed at full speed for 20 s and centrifuged in the Beckman microfuge at maximum speed for 5 min. The organic phase (upper) was transferred to another clean tube and put under vacuum for 2–3 h until the ethyl acetate was completely evaporated. The dried material was redissolved in 15 μl ethyl acetate and spotted on a thin layer chromotography (TLC) silica gel plate (EM Science, Cherry Hill, NJ); the tube was rinsed with another 15 μl of ethyl acetate and spotted at the same place. The plate was then put in a TLC chamber containing chloroform/methanol (95:5, v/v) at room temperature for about 20 min until the solvent front was approaching the predrawn line 10 cm from the starting point. The plate was then allowed to dry and autoradiographed on Kodak X-Omat AR5 X ray film. For quantitations, each radioactive spot which contained either the original or the acetylated forms of chloramphenicol was located, cut out and immersed with 1 ml ethyl acetate in a scintillation vial. Five ml of ScintiVerse II (Fisher Sci., Pittsburgh, PA) was then added and agitated well with the spot. The radioactivity of the samples was determined in a Beckman LS 9000 liquid scintillation counter. The conversion rate of chloramphenicol into acetylated forms (1- and 3-acetates) by each protoplast sample was calculated.

3.6 GUS Assays

GUS assays were performed based on a modification of Jefferson et al. (1986). A pellet of 2×10^5 transfected protoplasts was homogenized with 0.45 ml of GUS

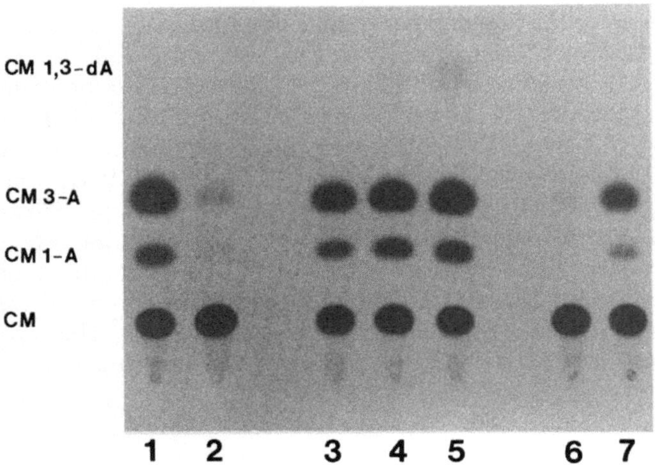

Fig. 6. Autoradiogram of chloramphenicol acetyltransferase (CAT) assay using protoplasts of *Nicotiana edwardsonii*. *Lane 1* contains the CAT standard; *lane 2* negative control; *lanes 3, 4 and 5* CAT gene fused downstream of the promoter for the major RNA transcript of peanut chlorotic streak mosaic virus, figwort mosaic virus and cauliflower mosaic virus, respectively; *lane 6* pFMV32CAT, a construct containing the CAT gene fused at the 5′ end to a small open reading frame of the 5′ leader of the major RNA transcript of FMV (see Gowda et al. 1989a); *lane 7* pFMV32CAT co-electroporated with pFMVRVI, a plasmid containing gene VI of FMV, showing post-transcriptional transactivation by gene VI. *CM* chloramphenicol; *CM 1-A* chloramphenicol 1-acetate; *CM 3-A* chloramphenicol 3-acetate; *CM 1,3-dA* chloramphenicol 1,3-diacetate

extraction buffer (50 mM sodium phosphate, pH 7.0, 10 mM β-mecaptoethanol, 10 mM EDTA-Na_2, 0.1% sodium lauryl sarcosine, 0.1% Triton X-100, 0.02% Na-azide, and 100 μg/ml bovine serum albumin) by vortexing at full speed for 1 min. The tubes were centrifuged at maximum speed for 10 min in a Beckman 12 microfuge. The supernatant was transferred to a clean Eppendorf tube and mixed with 50 μl of 10 mM p-nitrophenyl β-D-glucuronide. The samples were then incubated at 37 °C for 4–20 h depending upon the level of GUS activity, until the greenish-yellow color was stable. The enzymatic reaction was then stopped by adding 200 μl of 2.5 M 2-amino-2 methyl-1,3-propanediol. The absorbance of the samples at 415 nm was determined in a Beckman 35 spectrophotometer.

4 Conclusion

Production of transgenic plants of *Nicotiana edwardsonii* can be obtained by reducing the amount of antibiotic used for positive selection for neomycin phosphotransferase activity to the lowest level at which inhibition of callus or shoot production is first observed in noninoculated, nonresistant leaf squares. Careful selection of the plasmid vector and strain of *Agrobacterium tumefaciens* may improve yield of transformed shoots. Although use of the wild type (oncogenic) strain, A281, was highly effective in transforming Burley 21 tobacco and *Datura innoxia* (Goldberg et al. 1991), use of the disarmed strain C58C1:pGV3850 produced less callus and a larger number of shoots during transformation of *N. edwardsonii.* In addition, use of the strain C58Cl:pGV3850 to transform *N. edwardsonii* permitted a higher selection level of kanamycin to be used (100 mg/l) resulting in a much higher (90%) number of plants obtained actually being transformed compared to approximately 10–30% of those when strain A281 and a selection level of 0–50 mg/l kanamycin was used.

The suspension cell protoplast system is ideal for gene expression studies in that it eliminates the problems sometimes encountered with leaf protoplasts such as inconsistency of quality, difficulty in obtaining large quantities of protoplasts at one time, and the need for constant production of plants. A suspension of *N. edwardsonii* yields 10–15 million protoplasts per 100 ml and can be maintained for long periods without losing its capacity to produce useful protoplasts. Our particular suspensions have been maintained for over 3 years. CAT and GUS assays are easy to perform using *N. edwardsonii* protoplasts, as little background is observed from the protoplasts themselves. Using the reporter genes CAT and GUS and restructured DNAs, we have located some of the promoter elements on the genomes of figwort mosaic virus and the peanut chlorotic streak virus (Wu et al. 1988; Gowda et al. 1989a). We also discovered that gene VI of figwort mosaic virus played a key role in post-transcriptional transactivation (Gowda et al. 1989b; Scholthof et al. 1992). An example can be seen in Fig. 6. In addition, the requirement of gene VII in *cis* for the expression of downstream genes on the major transcript of figwort mosaic virus has been demonstrated using the protoplast/reporter gene system (Gowda et al. 1991).

Acknowledgments. We wish to thank Herman Scholthof for his help in producing the figures and Amy Wilson for her assistance in the laboratory and greenhouse. We also thank Dr. Arthur G. Hunt for providing us with the pKYLX7 plasmid. This work was supported by US. Department of Agriculture Grants 86-CRCR-1-2261 and 90-37262-5307.

References

An G, Watson BD, Stachel S, Gordon MP, Nester EW (1985) New cloning vehicles for transformation of higher plants, EMBO J 4:277–284

Christie SR (1969) *Nicotiana* hybrid developed as a host for plant viruses. Plant Dis Rep 53(12):939–941

Christie SR, Hall DW (1979) A new hybrid species of *Nicotiana* (Solanaceae). Baileya 20(4):133–136

Covey SN, Hull R (1981) Transcription of cauliflower mosaic virus DNA: Detection of transcripts, properties and location of the gene encoding the virus inclusion body protein. Virology 111:463–474

Daubert SD, Schoelz JE, Debao L, Shepherd RJ (1984) Expression of disease symptoms in cauliflower mosaic virus genomic hybrids. J Mol Appl Genet 2:537–547

Ditta G, Stanfield, S, Corbin D, Helinski DR (1980) Broad host range DNA cloning system for Gram-negative bacteria: construction of a gene bank of *Rhizobium meliloti.* Proc Natl Acad Sci USA 77:7347–7351

Flick CE, Evans DA, Sharp WR (1983) Organogenesis. In: Evans DA, Sharp WR, Ammirato PV, Yamada Y (eds) Handbook of plant cell culture vol 1, Macmillan, New York, pp 13–81

Fromm M, Taylor LP, Walbot V (1985) Expression of genes transferred into monocot and dicot plant cells by electroporation. Proc Natl Acad Sci USA 82:5824–5828

Gamborg OL, Miller RA, Ojima K (1968) Nutrient requirements of suspension cultures of soybean root cells. Exp Cell Res 50:151–158
Goldberg K-B, Kiernan J, Gowda S, Schoelz JE, Shepherd RJ (1988) Effects of genes I and VI of two caulimoviruses in transgenic plants. Phytopathology 78:1517 (Abstr)
Goldberg K-B, Kiernan J, Schoelz JE, Shepherd RJ (1990) Transgenic host response to gene VI of two caulimoviruses. In: Pirone TP, Shaw JG (eds) Viral genes and plant pathogenesis. Springer, Berlin Heidelberg New York, pp 58–66
Goldberg K-B, Kiernan J, Shepherd RJ (1991) A disease syndrome associated with expression of gene VI of caulimoviruses may be a nonhost reaction. Mol Plant-Microbe Interact 4:182–189
Gorman CM, Moffatt LF, Howard BH (1982) Recombinant genomes which express chloramphenicol acetyltransferase in mammalian cells. Mol Cell Biol 2:1044–1051
Gowda S, Wu FC, Shepherd RJ (1989a) Identification of promoter sequences for the major RNA transcripts of figwort mosaic virus and peanut chlorotic streak virus. J Cell Biochem 13-D (Suppl): 301 (Abstr)
Gowda S, Wu FC, Scholthof HB, Shepherd RJ (1989b) Gene VI of figwort mosaic virus (caulimovirus group) functions in posttranscriptional expression of genes on the full-length RNA transcript. Proc Natl Acad Sci USA 86:9203–9207
Gowda S, Scholthof HB, Wu FC, Shepherd RJ (1991) Requirement of gene VII in *cis* for the expression of downstream genes on the major transcript of figwort mosaic virus. Virology 185:867–871
Harrison BD, Finch JT, Gibbs AJ, Hollings M, Shepherd RJ, Valenta V, Wetter C (1971) Sixteen groups of plant viruses. Virology 45:356–363
Hood EE, Jen G, Kayes L, Kramer J, Fraley RT, Chilton M-D (1984) Restriction endonuclease map of pTi Bo542, a potential Ti plasmid vector for genetic engineering of plants. Bio/Technol 2:702–709
Horsch RB, Fry JE, Hoffmann NL, Eichholtz D, Rogers SG, Fraley RT (1985) A simple and general method for transferring genes into plants. Science 227:1229–1231
Hull R, Covey SN (1983) Does cauliflower mosaic virus replicate by reverse transcription? Trends Biochem Sci 8:119–121
Jefferson RA, Burgess SM, Hirsh D (1986) β-Glucuronidase from *Escherichia coli* as a gene-fusion marker. Proc Natl Acad Sci USA 83:8447–8451
Kiernan JM, Goldberg K-B, Young MJ, Schoelz JE, Shepherd RJ (1989) Transformation and regeneration of *Nicotiana edwardsonii.* Plant Sci 64:67–78
Linsmaier EM, Skoog F (1965) Organic growth factor requirements of tobacco tissue cultures. Physiol Plant 18:100–127
Linstead PJ, Hills GJ, Plaskitt KA, Wilson IG, Harker CL, Maule AJ (1988) The subcellular location of the gene I product of cauliflower mosaic virus is consistent with a function associated with virus spread. J Gen Virol 69:1809–1818
Malmburg R, Messing J, Sussex I (1985) Molecular biology of plants. Cold Spring Harbor NY: Cold Spring Harbor Laboratory, pp 106–107
Murashige T, Skoog F (1962) A revised medium for rapid growth and bioassays with tobacco tissue cultures. Physiol Plant 15:473–497
Ranch JP, Giles KL (1980) Factors affecting growth and aggregate dissociation in batch suspension cultures of *Datura innoxia* (Miller). Ann Bot 46:667–683
Richins RD, Scholthof HB, Shepherd RJ (1987) Sequence of figwort mosaic virus DNA (caulimovirus group). Nucleic Acids Res 15:8451–8466
Schardl CL, Byrd AD, Benzion G, Altschuler MA, Hildebrand DF, Hunt AG (1987) Design and construction of a versatile system for the expression of foreign genes in plants. Gene 61:1–11
Schoelz JE, Shepherd RJ (1988) Host range control of cauliflower mosaic virus. Virology 162:30–37
Schoelz JE, Shepherd RJ, Richins RD (1986) Properties of an unusual strain of cauliflower mosaic virus. Phytopathology 76(4):451–454
Scholthof HB, Gowda S, Wu FC, Shepherd RJ (1992) The full-length transcript of a caulimovirus is a polycistronic mRNA whose genes are *trans* activated by the product of gene VI. J. Virology 66(5):3131–3139
Shepherd RJ, Richins RD, Duffus JE, Handley MK (1987) Figwort mosaic virus: properties of the virus and its adaption to a new host. Phytopathology 77:1668–1673
Takatsuji H, Hirochika H, Fukushi T, Ikeda JE (1986) Expression of cauliflower mosaic virus reverse transcriptase in yeast. Nature 319:240–243
Volovitch M, Modjtahedi N, Yot P, Brun G (1984) RNA dependent DNA polymerase activity in cauliflower mosaic virus infected plant leaves. EMBO J 3:309–314

Wu FC, Gowda S, Scholthof HB, Kiernan JM, Shepherd RJ (1988) Comparative analysis of caulimovirus promoters in protoplasts. Phytopathology 78(12):1517 (abst)

Xiong C, Muller S, Lebeurier G, Hirth L (1982) Identification by immunoprecipitation of cauliflower mosaic virus *in vitro* major translation product with a specific serum against viroplasm protein. EMBO J 1:971–976

Zambryski P, Joos H, Gentello C, Leemans J, Van Montagu M, Schell J (1983) Ti plasmid vector for the introduction of DNA into plant cells without alteration of their normal regeneration capacity. EMBO J 2:2143–2150

II.17 Transformation in *Salvia chinensis*

ZHI-HONG XU and LI-JUN YANG[1]

1 Introduction

Salvia chinensis Benth (Labiatae), with the Chinese name Shijianchuan, is one of traditional Chinese medicinal plants in the South China. It is an annual herb (Fig. 1A), 20–70 cm in height, flowering during July–August, and growing mainly in South China. The plant contains steroid, triterpenoid, and procatechuic aldehyde. Stachyose was detected in its roots. The plants at the flowering stage are collected and dried for medical use. It is generally used as a heat-clearing and detoxifying drug for promoting blood circulation, regulating vital energy, and alleviating pain (Institute of Research and Development of Medicinal Plant Resource et al. 1988).

Our studies have shown that the leaf and stem explants of *S. chinensis* have a very high capacity for plant regeneration in culture, and transgenic plants can be readily produced by using *Agrobacterium tumefaciens*-mediated transformation (Yang and Xu 1989; Yang et al. 1993).

2 Plant Regeneration System

Sterile plants of *Salvia chinensis* can be propagated by subculture of shoot cuttings in MS basic medium without hormone (MSO). Leaf explants (5 × 5 mm^2) and stem segments (0.3–0.5 mm long) from the cultured sterile plants formed a large number of buds and shoots on the MS media with various combinations of cytokinin and auxins except 2,4-D (see Table 1). Among the combinations of different cytokinin and auxins tested, the treatments of 2 mg/l BA with 0.2 mg/l NAA or both 0.2 and 2 mg/l IAA, and 0.5 mg/l BA with 0.2 mg/l NAA gave the best results (Fig. 1B). 2,4-D stimulated callus formation and its subsequent growth, when it was added in the medium with cytokinin, and completely inhibited bud formation at 2 mg/l. Many small buds formed when the medium was supplemented with 0.2 mg/l 2,4-D and 2 mg/l BA. The regenerated shoots grew up and then formed roots in the culture flask. According to the above results, the MS medium with 2 mg/l BA and 0.2 mg/l NAA (MSBN) was selected for the transformation experiments.

[1]Shanghai Institute of Plant Physiology, Academia Sinica, Shanghai 200032, China

Biotechnology in Agriculture and Forestry, Vol. 22
Plant Protoplasts and Genetic Engineering III (ed. by Y.P.S. Bajaj)

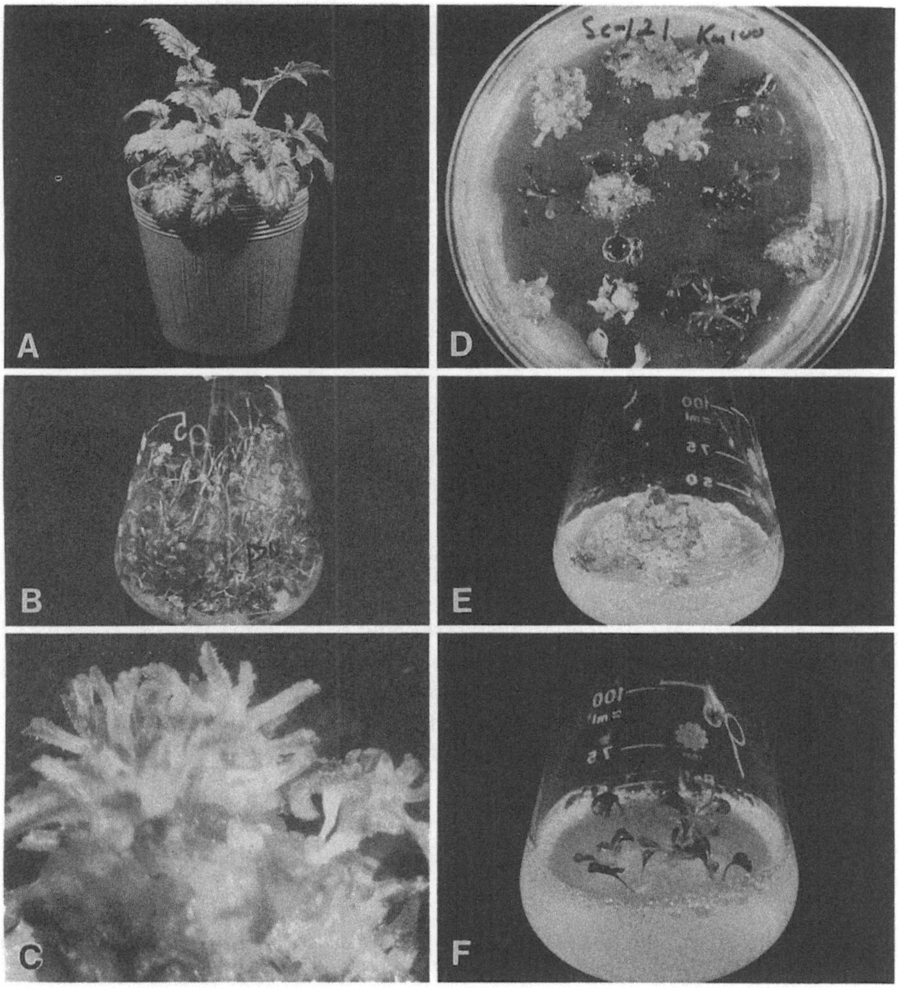

Fig. 1A–F. Tissue culture and transformation of *Salvia chinensis*. **A** *Salvia chinensis* plants in pot. **B** Multishoots and plants formed from leaf explants cultured on the MSBN medium. **C** A number of buds formed from leaf explants transformed with pBZ6111 and cultured on the hormone-free MS medium (MSO). **D** Shoots transformed with PBI121 grew normally on kanamycin-containing medium, while nontransformed ones became white and died. **E** Teratoma-like tissue formed from leaf explants transformed with pBI121 (Note that *black spots* are small buds). **F** Transgenic plants obtained from pBI121-transformed leaf plants

3 Transgenic Plants and Shoots Produced by *Agrobacterium*-Mediated Transformation

3.1 Bacterium Strains and Infection

Agrobacterium tumefaciens strains, LBA4404 with binary vectors (pAL4404 and pBI121) (courtesy of Dr. R. Jefferson) and C58C1 with reconstructed Ti plasmid

Table 1. Percentage of bud and callus formation in the culture of leaf and stem explants of *Salvia chinensis* (After 30 days of culture, means of two experiments, 58–60 explants for each treatment)

Plant growth regulator (mg/l)	Leaf		Stem	
	Bud	Callus	Bud	Callus
BA 2/2,4-D 2	0	100	0	100
BA 2/2,4-D 0.2	70	30	75	100
BA 2/IAA 2	100	0	100	0
BA 2/IAA 0.2	100	0	100	0
BA 2/NAA 2	98	5	82	10
BA 2/NAA 0.2	100	0	100	0
BA 0.5/NAA 2	80	10	63	47
BA 0.5/NAA 0.2	100	0	100	0
Zeatin 2/NAA 0.2	90	0	100	0
Kinetin 2/NAA 0.2	70	40	100	0

(pBZ6111) (from Dr. Q. Zhu of our laboratory) were used in our experiments. The plasmid pBI121 carries the genes for NPTII (neomycin phosphotransferase II) and GUS (β-glucuronidase) and pBZ6111 contains the genes for nopaline synthetase, CAT (chloramphenicol acetyltransferase), and isopentenyl transferase, an enzyme catalyzing the first step of cytokinin synthesis (gene 4). Leaf explants from the sterile plants were inoculated with the bacteria for 2 min, then cultured on the MS agar medium with 2 mg/l BA and 0.2 mg/l NAA (MSBN). Three days later, they were transferred on to MSBN medium with 300 μg/ml cefotaxime for killing bacteria.

3.2 Selection of Transformants

The results from the experiments on the optimal selection concentrations of kanamycin (Km) or chloramphenicol (Cm) (0–200 μg/l) showed that adding 10 μg/ml Km or 50 μg/ml Cm apparently inhibited the growth of the leaf explants cultured, while 25 μg/ml Km or 75 μg/ml Cm caused the explants to brown and die. Thus, the regenerated buds or shoots could be selected for putative transformants on the MSO medium containing 100 μg/ml Km or Cm, respectively, for the strains with pBI121 and pBZ6111.

After the infection of the leaf explants with C58C1 (pBZ6111), numbers of bud clusters formed on the hormone-free MS medium (Fig. 1C), and these transformed buds or shoots proved Cm-resistant on the medium with 100 μg/ml Cm. Nopaline synthetase activity was assayed following the method of Otten and Schilperoort (1978) (modified by Xu et al. 1987). Nopaline was detected in both shoots and small buds (Fig. 2B–E), but nopaline content in the bud clusters was much higher than that in the shoots. In the tissue from the normal salvia plant (control), no nopaline was detected (Fig. 2A). The transformed shoots could not root on the MS medium with 100 μg/ml Cm $\pm$ 0.1 mg/l NAA, indicating that endogenous cytokinin may increase greatly, which inhibits root formation, as was observed in *Populus tomentosa* (Wang et al. 1990) and *Solanum nigrum* (Zhana et al. 1989).

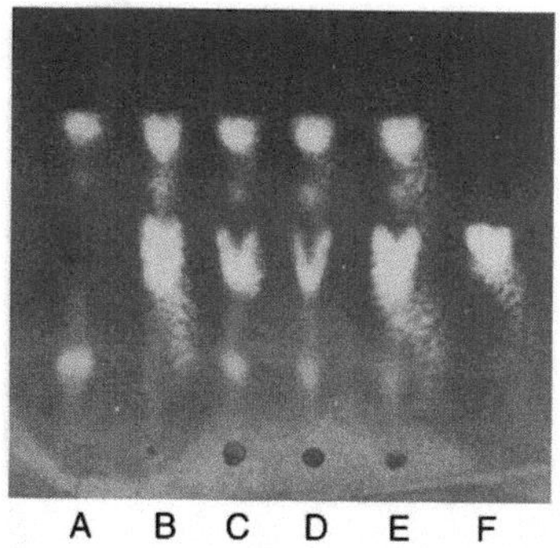

Fig. 2A–F. Nopaline detected in the buds regenerated from leaf explants transformed with pBZ6111. **A** Control (normal plant). **B–E** Different bud clusters transformed. **F** Standard nopaline sample

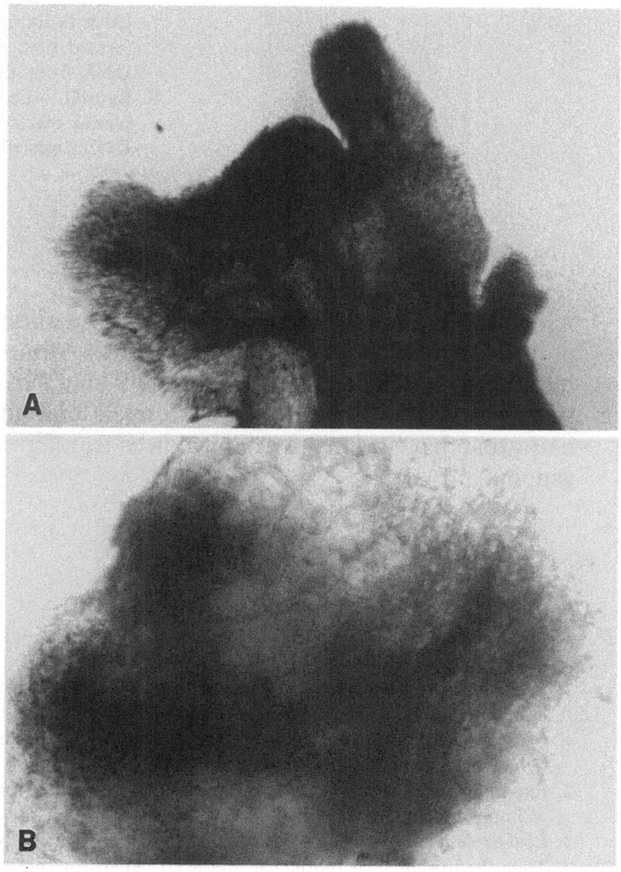

Fig. 3A, B. GUS activity detected in buds (**A**) and immature tissue (**B**) with X-Gluc

In the case of LBA4404(pAL4404/pBI121), both transformed shoots and teratoma-like buds were obtained on the MS selection medium with 100 μg/ml Km. Then the shoots could normally form roots. GUS activity was detected by the method of Jefferson (1987) and stably expressed in subculture. The buds, young leaves, and vascular tissues usually were stained dark blue by using X-Gluc (Fig. 3).

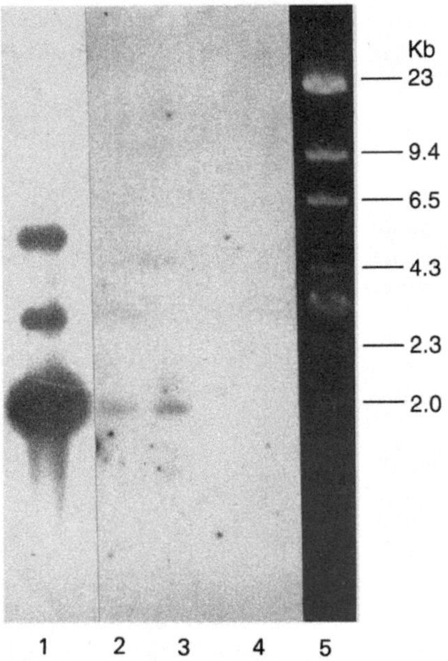

Fig. 4. Southern blot analysis of gus gene in *Salvia chinensis* transformed with pBI121. (1) pBI221 digested with BamH1 + EcoR1. (2) DNA from the transformed plants, digested with BamH1 + EcoR1. (3) DNA from the teratoma-like tissue, digested with BamH1 + EcoR1. (4) DNA from notransformed plants, digested with BamH1 + EcoR1. (5) λ-phage DNA digested with HindIII

According to the GUS activity, the transformation frequency was about 2% (two of 100 cultured explants formed the transformed plants or shoots). Southern blotting also demonstrated that the 1.9 kb DNA fragment, including the GUS gene, was present in both the transformed teratoma-like buds and plants (Fig. 4), indicating that the foreign gene had been transferred and integrated into the *Salvia* genome.

4 **Protocol** (After Yang et al. 1993)

1. *Agrobacterium tumefaciens* strains with disarmed or reconstructed Ti plasmid are inoculated into LB liquid medium and grown overnight (for 18 h) on orbital shaker at 28 °C. Then, the bacterium culture is centrifuged, and the supernatant is discarded. The bacterial pellet is resuspended in a certain volume of LB medium (adjusted to O.D. 0.2 at 600 nm).
2. Expanding leaves are taken from sterile plants cultured on MSO agar medium, and cut into 5 $\times 5$ mm^2 explants.
3. Leaf explants are incubated with *A. tumefaciens* by placing them into bacterial suspension for 2 min. Then the explants are blotted on sterilized paper to remove the liquid medium.
4. The blotted leaf explants are cultured on the MS agar medium supplemented with 2 mg/l BA and 0.2 mg/l NAA (MSBN) for 3 days and then transferred onto MSBN medium containing 300 μg/l cefotaxime to prevent bacterial growth.
5. The regenerated buds or shoots are selected on MS medium with 100 μg/ml kanamycin (Km) or chloramphenicol (Cm), depending on the bacterial strains used. The Km- or Cm-resistant buds or shoots are assayed according to need

5 Conclusions

Plants were readily regenerated from the cultured leaf explants or stem segments of *Salvia chinensis*, providing an efficient micropropagation method for this medicinal herb. *Agrobacterium tumefaciens* could be used for the transformation of this species, and the transgenic plants were obtained from the leaf explants infected by *Agrobacterium*.

Acknowledgment. This work was supported by the National Science Foundation.

References

Institute of Research and Development of Medicinal Plant Resources et al. (1988) Shijianchuan (*Salvia chinensis*). Chinese herbology, 2nd edn, vol 4. People's Health Press, Beijing, pp 247–250

Jefferson R (1987) Assaying chimeric genes in plants: the GUS gene fusion system. Plant Mol Biol Rep 5:387–405

Otten LA, Schilperoort RA (1978) A rapid microscale method for detection of lysopine and nopaline dehydrogenase activities. Biochem Biophys Acta 527:497–500

Wang SP, Xu ZH, Wei ZM (1990) Genetic transformation of leaf explants of *Populus tomentosa*. Acta Bot Sin 32:172–177

Xu Y, Jia JF, Zheng GC (1987) A simple and efficient method for the detection of opine synthase activities in plant tissues. Hereditas (Beijing) 9:41–43

Yang LJ, Xu ZH (1989) Production of transgenic plants of *Salvia chinensis* using *Agrobacterium*-mediated DNA transfer. Annu Rep National Laboratory of Plant Mol Genet, Shanghai Inst of Plant Physiol (1988/1989), pp 27–28

Yang LJ, Wang SP, Zhang Y, Shu XM, Xu ZH (1993) Plant regeneration and genetic transformation in tissue culture of *Salvia chinensis*. Acta Biol Exp Sin 26 (in press)

Zhana, Xu ZH, Wei ZM, Wang SP, Li TR (1989) Expression of foreign genes in transgenic plants of *Solanum nigrum*. Annu Rep National Laboratory of Plant Mol Genet, Shanghai Inst of Plant Physiol (1988/1989), pp 32–33

II.18 Transformation in *Scopolia*

Y.L. ZHANG[1]

1 Introduction

Plants of the genus *Scopolia* are perennial herbs of the family Solanaceae, and are distributed in Europe, China, Korea, and Japan. *Scopolia* plants contain tropane alkaloid and are used extensively for medicinal purposes.

Many Solanaceae plants are commercial sources of tropane alkaloid, in which the major alkaloids are scopolamine and hyoscyamine. In traditional medicine, tropane alkaloids are used as spasmolytics, anesthetics, and analgesics. Recently, they have played an important role in the study of the microcirculatory system, and are arousing increasing interest.

In China, 8 genera and 24 species of tropane alkaloid plants are used for medicinal purposes. Species such as *S. lurida* (Fig. 1), *S. tangutica*, *S. acutangulus*, etc. are an important source of tropane alkaloid (Xiao 1985).

Hairy roots can be subcultured and propagated on hormone-free basal medium. The hairy root culture system can be utilized as an experimental system to

Fig. 1. Plant of *Scopolia lurida*

[1] Institute of Medicinal Plant Development, Chinese Academy of Medical Sciences, Hai Dian District, Dong Bei Wang, Beijing 100094, China

Biotechnology in Agriculture and Forestry, Vol. 22
Plant Protoplasts and Genetic Engineering III (ed. by Y.P.S. Bajaj)

study physiological and biochemical problems, especially secondary metabolism (Heetore et al. 1987), and to produce useful secondary metabolites.

Recently, many research workers have reported that hairy root cultures induced from many plants produce secondary metabolists, such as alkaloids (Kamada et al. 1986; Payne et al. 1987; Parr et al. 1988; Hamill et al. 1988; 1989), glycyrrhizin, saponin (Ko et al. 1989; Yoshikawa and Furuya 1987), and flavonoids in licorice (Zhang et al. 1990), etc.

The T-DNA of the Ri plasmid also confers on the transformed plant tissue the ability to produce modified amino acids (opines). The opines (mannopine, agropine, etc. are based on bacteria strains) are in turn utilized only by the inciting bacteria as carbon, nitrogen, and energy source (Sinkar et al. 1987). As a result, opine assay of hairy root indicates that a transference of the T-DNA of *A. rhizogenes* Ri plasmid into the cell genomic DNA of the plant cell can be demonstrated.

2 Cell Culture of *Scopolia*

The development of callus and the cell culture of plants with the aim of producing valuable secondary metabolites is being studied in many laboratories. However, the cultures usually produce contents lower than the original plant, sometimes they do not synthesize the secondary metabolites that are produced in the original plant. Although regulation of medium and culture conditions (nutrient, hormone level, light, temperature, medium, pH, and aeration) can influence the synthesis of products, none of these effects help to produce secondary metabolites useful for commercial production (Fowler 1981).

Konoshima et al. (1967) have shown that the basal medium, hormone, and additives could not induce the tropane alkaloid in callus cultures of *S. japonica* to reach the content level of the original plant.

Some reviews have shown that by differentiation or reorganization of cell cultures, the synthesis of secondary metabolites can be greatly increased. This phenomenon has been demonstrated by many researchers (Kurz and Constabel 1979; Garre et al. 1980; Staba and Chung 1981; Lindsey and Yeoman 1983).

Tabata et al. (1972) have shown that a close relationship exists between root organization and tropane alkaloid production in callus cultures of *S. japonica.* They suggested that organ formation of cultures is a good method to produce more alkaloids. However, alkaloid contents were lower than those in the original plant.

3 Ri Plasmid Transformation and Hairy Root Culture

Agrobacterium rhizogenes is a Gram-negative soil bacterium. It causes hairy root disease in many dicotyledonous plants. It has been demonstrated that the Ri plasmid present in *A. rhizogenes* transforms plant cells by introducing its T-DNA into the genomic DNA of plant cells, and results in hairy roots as a transformed phenotype growing on the site of infection (White et al. 1985).

In our laboratary, the transformation of *S. lurida* by the Ri plasmid of *A. rhizogenes* and hairy root cultures have been studied (Zhang 1988).

The sterilized plants of *S. lurida*, after being infected by *A. rhizogenes*, were induced to produce hairy roots. They grew well on either hormone-free MS agar medium or in suspension culture (Fig. 2). The cultured hairy roots were shown to yield alkaloids such as hyoscyamine and scopolamine (0.1 and 0.35% dry wt., respectively). Since agropine and mannopine were also detected in the hairy root cultures, the transference of the T-DNA of *A. rhizogenes* Ri plasmid into the cell genomic DNA of *S. lurida* is clearly demonstrated (Fig. 3).

A high production of hairy root clones in *S. japonica* was established by clone selection (Mano et al. 1986); total of 125 hairy root clones were selected. The contents of scopolamine and hyoscyamine, which are major tropane alkaloids of *S. japonica*, were analyzed after the hairy root clones were cultured for 6 weeks in H liquid medium. Both alkaloids were found in all hairy root clones tested. Marked differences in the alkaloid contents were found among the hairy root clones analyzed. Clone S1 contained the highest amount of scopolamine, the content being slightly higher than that of natural roots. Clone S22 contained the highest amount of hyoscyamine with a content about four times higher than that of natural roots. Clone S1 was the best scopolamine-producing clone; clone S22, the best hyoscyamine-producing clone, was established by screening a number of hairy root clones.

As shown in Fig. 4, the growth of both clones reached a stationary phase at around 4 weeks. Clone S1 produced both scopolamine and hyoscyamine and the

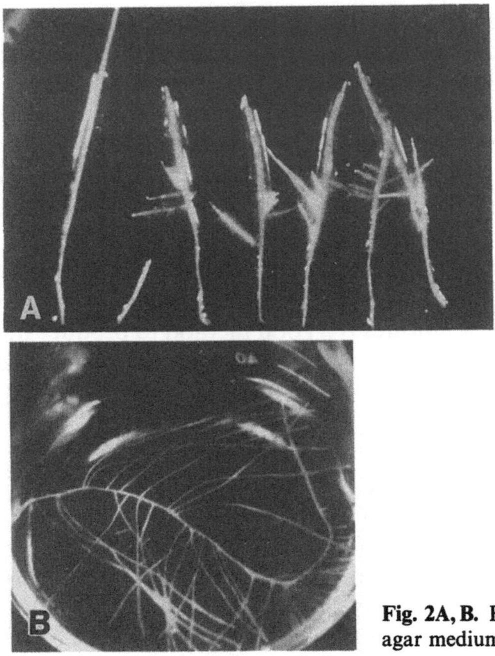

Fig. 2A, B. Hairy root culture of *S. lurida*. **A** Hairy roots on agar medium; **B** hairy roots in suspension culture

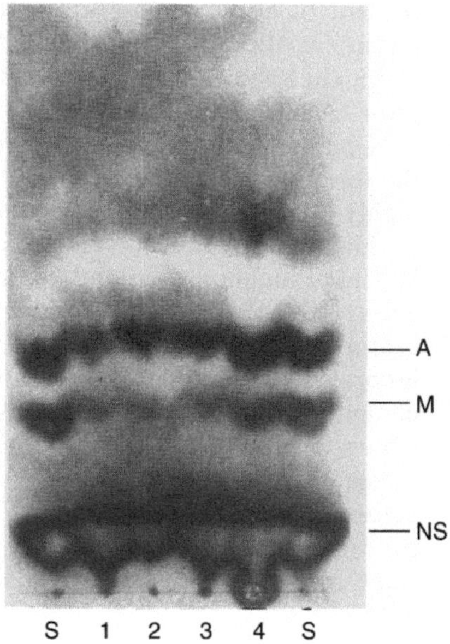

Fig. 3. Paper electrophoresis of agropine and mannopine in the extracts of the hairy roots. *A* Agropine; *M* mannopine; *NS* neutral sugar; *S* authentic sample, 1, 2, 3, 4

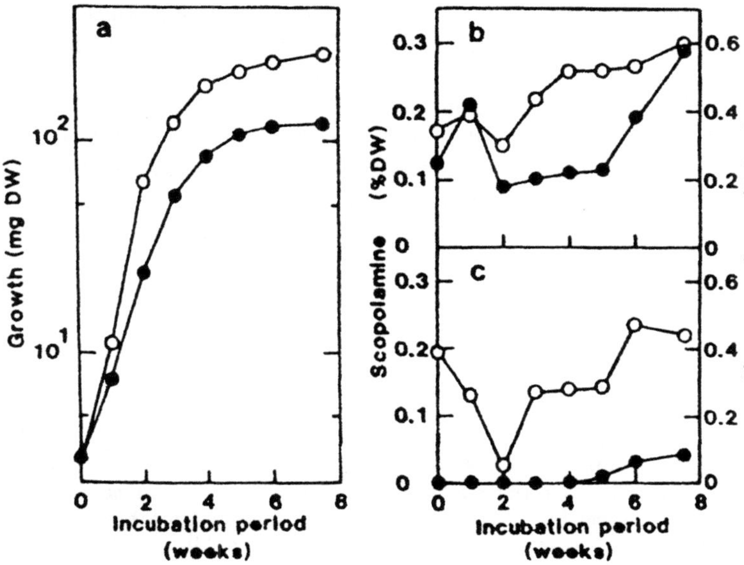

Fig. 4a, b. Time course of growth and alkaloid content of hairy root clones S1 and S22. Clones S1 and S22 were cultured in H liquid medium for 8 weeks at 25 °C in the dark **a** Growth of clones S1 (●) and S22 (○). **b** Alkaloid content of clone S1. **c** Alkaloid content of clone S22; (●) scopolamine; (○) hyoscyamine; DW, dry weight (Mano et al. 1986)

content of hyoscyamine was higher than that of scopolamine in every growth phase. On the other hand, clone S22 produced a fairly large amount of hyoscyamine, but showed a very low level of scopolamine even in the late logarithmic growth phase.

Comparison of the characteristics of *S. japonica* hairy root clones revealed that the selection of a number of the hairy roots can be a useful method to establish highly productive hairy root clones for tropane alkaloid production.

Hairy root cultures have a number of useful characteristics in the biotechnological production of secondary products, including faster growth than root or cell cultures, stable high-level production, and suitability for fermenter systems. For example, hairy root cultures of *Fagopyrum cymosum* were established in our laboratory. *F. cymosum* is a perennial herb of the Polygonaceae whose major useful content is dimeric procyanidin, used for pulmonary abscesses and cholecystitis in China. The hairy root cultures grew vigorously in MS hormone-free medium and increased 1861 times in fresh weight in 25 days, faster than in cell suspension cultures, which increased only 26.7 times in the same number of days. The dimeric procyanidin (about 4.5% dry wt.) was detected in hairy root cultures at nearly the same content level of the original plant (Zhang et al. 1992).

4 Summary and Conclusions

Scopolia is an important source of tropane alkaloid. Tissue and cell culture techniques employed to produce tropane alkaloid have been studied by many researchers; however, they failed to produce more alkaloids than the original plant, and could not even synthesize the alkaloids produced in the original plant.

Hairy root culture induced by the transformed Ri plasmid contained in *A. rhizogenes* into the plant cell is a useful culture system to produce secondary metabolites. Hairy root cultures of *Scopolia* grow fast and produce tropane alkaloid at nearly the same level as the original plant, sometimes even more than in the original plant by hairy root clone selection. It has been demonstrated that hairy root culture provides a new method to produce plant secondary products.

References

Chilton M, Tepfer D, Petit A, David C, Cassedelbart F, Tempé J (1982) *Agrobacterium rhizogenes* inserts T-DNA into the genomes of the host plant root cells. Nature 295:432–434

Fowler M (1981) plant cell biotechnology to produce desirable substances. Chem Ind (April) 4:229–233

Garre R, Luckner M, Vogel E, Tewes A, Nover L (1980) Growth, morphogenesis and cardenolide formation in long-term cultures *Digitalis lanata.* Planta Med 40:92–103

Hamill J, Robins R, Michael J (1989) Alkaloid production by transformed root cultures of *Cinchona ledgeriana.* Plant Med 55:354–357

Heetore E, Mary W, Julic J (1987) Secondary metabolites from root cultures. TIBTECH 5:354–357

Kamada H, Okamura N, Satake M, Harada H, Shimomura K (1986) Alkaloid production by hairy root culture in *Atropa belladonna,* Plant Cell Rep 5:239–242

Ko K, Noguchi H, Ebzuka Y, Sankawa U (1989) Oligoside production by hairy root cultures transformed by Ri plasmids. Chem Pharm Bull 37:245–248

Konoshima M, Tabata M, Hiraoka N, Myyake H (1967) Growth and alkaloid production of the callus cultures of *Scopolia japonica*. Jpn J Pharmacol 21:108–114

Kurz W, Constabel F (1979) Plant cell cultures, a potential source of pharmaceuticals. Adv Microbiol 25:209–240

Lindsey K, Yeoman M (1983) The relationship between growth rate, differentiation and alkaloid accumulation in cell cultures. J Exp Bot 34:1055–1065

Mano M, Nabeshima S, Matsui C, Ohkawa H (1986) Production of tropane alkaloids by hairy root cultures of *Scopolia japonica*. Agric Biol Chem 50:2715–2722

Parr A, Peerles A, Hamill J, Walton N, Robins P, Rhodes M (1988) Alkaloid production by transformed root cultures of *Catharanthus roseus*. Plant Cell Rep 7:309–312

Payne J, Hamill J, Robins R, Rhodes M (1987) Production of hyoscyamine by hairy root culture of *Datura stramonium*. Planta Med 53:474–478

Sinkar V, White F, Gordon M (1987) Molecular biology of Ri plasmid – a review. J Biosci 11(1–4) March: 47–57

Staba E, Chung A (1981) Quinine and quinidine production by cinchona leaf, root and unorganized cultures. Phytochemistry 20:2495–2498

Tabata M, Yamamoto H, Hiraoka M (1972) Organization and alkaloid production in tissue cultures of *Scopolia parriflora*. Phytochemistry 11:949–945

White F, Toylor B, Huffman G, Gordon M, Nester E (1985) Molecular and genetic analysis of the transferred DNA regions of the root-inducing plasmid of *Agrobacterium rhizogenes*. J Bacteriol 164:33–44

Xiao P (1985) The utilization of tropane alkaloid medicinal plant. Chin Tradit Herbal Drugs 16(6):19–21

Yoshikawa T, Furuya T (1987) Saponin production by cultures of *Panax ginseng* transformed with *Agrobacterium rhizogenes*. Plant Cell Rep 6:449–453

Zhang Y (1988) The transference of Ri plasmid from *Agrobacterium rhizogenes* to *Scopolia lurida* revealed by hairy root cultivation and alkaloid production. Acta Bot Sin 30:368–372

Zhang Y, Zhou X, Yang L, Lue Y (1990) On the hairy root culture of *Glycyrrhiza uralensis*. Chin Tradit Herbal Drugs 21(12):23–26

Zhang Y, Lu G, Zhou X, Zhao B, Li Y (1992) Study on hairy root cultures of Fagopyrum cymosum. Acta Bot Sin 34:603–608

II.19 Transformation in *Solanum melongena* L. (Eggplant)

M. LEONE[1], E. FILIPPONE[2], and P.F. LURQUIN[3]

1 Introduction

Eggplant (*Solanum melogena* L.) (2n = 24) belongs to the Solanaceae family. It is a biennial with a moderately extensive root system. Mature plants display a branched main stem with a woody base. Leaves are simple, alternate, and large. The violet or purple flowers, single or in cluster opposite the leaves, are hermaphroditic. Eggplant is mostly an autogamous species, the amount of cross-pollination ranging between 6 and 20% when plots are less than 50 m apart. Fruits at maturity vary in size (7.5 to 15 cm diameter), shape (long, ovate or pyriform), and color (dark purple to white). Descriptions for eggplant have been published by the International Board for Plant Genetic Resources, Rome.

Eggplant is a minor and usually underutilized crop in North America. Nevertheless, it is quite appetizingly prepared in areas of the world as culturally diverse as, for example, southern Italy and Thailand. Eggplant is, in fact, widely cultivated and consumed around the Mediterranean basin and in Asia. Unfortunately, this vegetable is susceptible to a whole host of pathogens (nematodes, fungi, bacteria, and Tobacco Mosaic Virus) and competing weeds, which can dramatically reduce production yields. Efforts to curtail this problem currently involve a battery of tissue culture techniques, an approach which one can nowadays consider classical. In this respect, eggplant displays properties not necessarily shared by many crop species. Indeed, the whole spectrum of plant cell manipulation, including protoplast culture and regeneration, production of haploids, and somatic embryogenesis, is available in eggplant (see references cited in Filippone and Lurquin 1989, and Rotino and Gleddie 1990).

Many applications of plant cell and tissue culture have been greatly expanded due to the use of selectable markers and gene tags, as, for example, in the isolation of somatic hybrids with potentially interesting agronomical characteristics. Additionally, a variety of cloned genes and molecular techniques now exists (i.e., the antisense technology), which could be used to protect eggplant against pests, viruses, and herbicides. All of these improvements are contingent upon the existence of a genetic transformation system in eggplant. This technology is now a

[1] Centro di Studio per il Miglioramento Genetico degli Ortaggi, CNR, 80055 Portici, Italy
[2] Dipartimento Scienze Agronomiche e genetica Vegetale Universita di Napoli, 80055 Portici, Italy
[3] Department of Genetics and Cell Biology, Washington State University, Pullman, WA 99164-4234, USA

Biotechnology in Agriculture and Forestry, Vol. 22
Plant Protoplasts and Genetic Engineering III (ed. by Y.P.S. Bajaj)

fait accompli and the purpose of this chapter is to review published transformation systems as well as to present our data on the production of transgenic somatic embryos.

2 General Transformation Strategies

Transgenes can be incorporated into plant cells via two basically different procedures, (1) the direct gene transfer technique in which purified DNA is forced into cell nuclei by chemical or physical effects (for review see Lurquin 1989), and (2) the cocultivation of plant cells or tissues with *Agrobacterium*, which relies on the ability of the latter to naturally transfer any DNA segment flanked by the T-DNA 25-bp repeat. Ti-plasmid-based vectors are either of the cointegrative type (in which case the transgene is inserted by recombination within a T-DNA itself present in the Ti plasmid) or belong to the binary class of vectors (where the T-DNA is harbored by a much smaller replicon, the *vir* regulon being carried by a helper pTi) (for review see Simpson and Herrera-Estrella 1989). It is of particular interest to note that transformation by cocultivation of leaf explants with *Agrobacterium* has been demonstrated previously in several solanaceous species (Horsch et al. 1985; Tavazza et al. 1988).

Direct gene transfer has not been reported in *Solanum melongena*; however, production of transgenic eggplant through cocultivation of leaf explants with *Agrobacterium tumefaciens* has been demonstrated (Guri and Sink 1988; Filippone and Lurquin 1989; Rotino and Gleddie 1990).

3 Eggplant Transformation by Cocultivation

The methodologies used in the three laboratories which have reported stable genetic transformation of *S. melongena* are discussed here.

3.1 Origin of Leaf Explants

In all three cases, successful transformation was observed with explants from young leaves obtained from plants grown in vitro. Rotino and Gleddie (1990) noted that greenhouse-grown plants yielded leaf fragments which died as a result of cocultivation with *A. tumefaciens*. In addition, Filippone and Lurquin (1989) also used explants originating from cotyledonary leaves. These proved to be about five times more efficient, at least as far as the number of transgenic calli produced was concerned. The same authors reported that no transformation events were observed after cocultivation of suspension cells with *Agrobacterium*.

3.2 *Agrobacterium* Strains, Vector Systems, and Selection

The chromosomal background of the bacterial strain(s?) used by Guri and Sink (1988) was not reported. Filippone and Lurquin (1989) observed transformation with hypervirulent strain A281 but were unsuccessful with LBA4404. Rotino and Gleddie (1990) used *A. tumefaciens* C58. All three studies effected successful selection of transgenic calli in the presence of kanamycin owing to the fact that the artificial T-DNAs used harbored a chimeric *nos-neo* gene. However, Guri and Sink (1988) observed transformation events only when the selectable marker was present in a cointegrated form in pMON200. Presence of the *neo* gene or the *hphIV* gene (conferring hygromycin resistance) in a binary vector produced no transformants. Nevertheless, both Filippone and Lurquin (1989) and Rotino and Gleddie (1990) demonstrated transformation with two types of binary vectors, pGA472 (An 1986) and pBCAT1 (a pBIN19 derivative, Bevan 1984), respectively. The reason for this discrepancy is unclear, since the chromosomal background of *A. tumefaciens* harboring the binary vectors was not given by Guri and Sink (1988). At any rate, it is certain that transformation was observed with two very different binary vectors harbored by completely independent *A. tumefaciens* strains (A281 or EHA105, see below, and C58). The fact remains that the susceptibility of eggplant leaf fragments to infection by various *A. tumefaciens* strains (and hence transformation efficiencies) has not been studied in detail.

3.3 Regeneration of Transgenic Calli

Eggplant regeneration can occur either through somatic embryogenesis or organogenesis, depending on the medium used (Matsuoka and Hinata 1979). All three transformation studies reported regeneration of transgenic plants via organogenesis, albeit in the presence of different hormone regimens. While Guri and Sink (1988) and Rotino and Gleddie (1990) used zeatin in their regeneration medium, Filippone and Lurquin (1989) attempted regeneration in the presence of 1 mg/l NAA, conditions known to favor embryogenesis in eggplant. Nevertheless, only organogenesis (starting with the formation of shoot primordia) could be observed after several months of culture on selective medium (200 mg/l kanamycin) (Fig. 1). The reason why embryogenesis did not occur under these conditions was not determined in this study. Either kanamycin (or its phosphorylated derivative) interfered with somatic embryogenesis or else hypervirulent (but not disabled) *A. tumefaciens* A281 (pTiBo542/pGA472) transformed some cells to both a kanamycin-resistant and crown gall phenotype. Such transgenic cells could have released enough phytohormones into the medium to prevent embryogenesis but not organogenesis. It will be shown below that transgenic somatic embryos can indeed be produced under appropriate circumstances.

3.4 Molecular and Genetic Analysis of Transgenic Eggplant

All three studies demonstrated the presence of the selectable gene (*neo*) by DNA hybridization. In addition, Guri and Sink (1988) demonstrated nopaline synthesis

Fig. 1. Regeneration of leaf primordia on transformed eggplant callus after 3 months of culture on selective medium (200 mg/l kanamycin)

in their transformants, an expected observation since the T-DNA of pMON200 contained an intact *nos* gene in addition to the chimeric *neo* selectable marker. Both Filippone and Lurquin (1989) and Rotino and Gleddie (1990) showed expression of neomycin phosphotransferase in kanamycin-resistant tissues. In addition, the latter authors showed expression of a *cat* reporter gene cotransformed with the *neo* selectable marker, both chimeric genes being present between T-DNA borders.

In several plant species, genetic analysis of selfed transgenic regenerants indicates that a selective marker such as *neo* often behaves as a hemizygous dominant Mendelian trait, thus yielding a classical 3:1 segregation ratio in the selfed R1 progeny. However, the possibility of multiple T-DNA insertions at different loci in a single genome cannot be discounted. Alternatively, T-DNA inserts may not be absolutely stable (Deroles and Gardner 1988a, b). Therefore, few conclusions can be drawn from such studies, unless one's interest is the analysis of T-DNA stability and/or multiplicity of integration at independent loci. Guri and Sink (1988) determined the resistance to kanamycin of 146 seedlings originating from two selfed transgenic regenerants. The 2.65:1 segregation ratio calculated from their pooled data is statistically compatible with *neo* behaving as a single dominant marker. However, a similar experiment performed by Rotino and Gleddie (1990), but this time involving 715 seedlings from five independent selfed

Table 1. Production of transgenic eggplant by cocultivation of leaf explants with *Agrobacterium tumefaciens*

Reference	Guri and Sink (1988)	Filippone and Lurquin (1989)	Rotino and Gleddie (1990)
Cultivar	Black beauty	Violetta lunga di Napoli	Picentia
Explant source	Young leaves	Young leaves Cotyledonary leaves	Young leaves
Agrobacterium strain	NS	A281	C58
Vector type	Cointegrate (pMON200)	Binary (pGA472)	Binary (pBCAT1)
Selection	Kanamycin (100 mg/l)	Kanamycin (200 mg/l)	Kanamycin (100 mg/l)
Hormones	Zeatin (2 mg/l)	NAA (1 mg/l)	Zeatin (1 mg/l)
Molecular techniques	DNA slot blots Nopaline synthase activity	DNA dot blots NPT II activity	Southern blots, CAT and NPT II activities
Progeny analysis (selfed regenerants)	106 resistant/ 40 sensitive (2.65:1)	ND	507 resistant/ 208 sensitive (2.44:1)

CAT, chloramphenicol acetyltransferase; NAA, naphthalene acetic acid; ND, not determined; NPT II, neomycin phosphotransferase; NS, not specified.

regenerants, yielded a 2.44:1 segregation ration, well outside the limit of acceptability for a single Mendelian trait as determined by chi square analysis. Interestingly, their results showed a strong bias towards kanamycin-sensitive seedlings. Whether this phenomenon was due to T-DNA instability, production of chimeric kanamycin-resistant regenerants, lower germination rates of transgenic seeds, or other causes remains to be seen.

The above results are summarized in Table 1 for comparative purposes.

4 Transgenic Somatic Embryos

Somatic embryogenesis considerably shortens the time it takes to obtain regenerants since shoot and root formation by organogenesis usually cannot be performed in a single step, as opposed to the former. In that respect, Scott and Draper (1987) first demonstrated the production by somatic embryogenesis of transgenic carrot plants from proembryogenic suspension cells cocultivated with *Agrobacterium.* Some other advantages of this approach are that somatic embryogenesis may reduce chances of formation of chimeric transformants and results in high transformation frequencies, at least in the case of carrot. This section describes conditions under which transgenic eggplant somatic embryos and regenerants were produced by us at high frequency.

In order to bypass potential problems associated with the use of fully armed *Agrobacterium* (see above), all experiments described below were conducted with avirulent *A. tumefaciens*. The two strains used were *A. tumefaciens* C58 (pGV2260) harboring a T-DNA-less Ti plasmid of the octopine type (Deblaere et al. 1985) and *A. tumefaciens* EHA105, a rifampicin-resistant/kanamycin-sensitive derivative of strain EHA101, itself a disarmed version of hypervirulent strain A281 (pTiBo542) (Hood et al. 1986). Both strains carried the binary transformation vector p35SGUSINT, a pBin19 derivative (Bevan 1984) harboring a plant-expressable *neo* gene as well as the *Escherichia coli gus* gene interrupted by a plant intron and under the control of the Cauliflower Mosaic Virus 35S promoter (Vancanneyt et al. 1990). Transgenic plant cells obtained with this vector will thus multiply in the presence of kanamycin and stain blue when treated with the chromogenic substrate X-Gluc (Jefferson 1987). It should be remembered that staining with X-Gluc is lethal and hence, staining for GUS and kanamycin selection cannot be performed in that order with the same tissue sample.

Cotyledonary leaf explants soaked in an *Agrobacterium* suspension as in Filippone and Lurquin (1989) and then cocultivated for 2 days on standard Murashige-Skoog medium (20 g/l sucrose, no hormones) with either one of the above disarmed *Agrobacterium* strains produced callus occasionally containing GUS^+ foci and few somatic embryos when further cultivated on MS medium supplemented with 20 g/l sucrose, 100 mg/l kanamycin and 1 mg/l NAA in order to induce somatic embryogenesis (Filippone and Lurquin 1989) (Table 2). Excluding the selective agent from the somatic embryogenesis medium did not

Table 2. Effect of conditioning medium on eggplant regeneration through somatic embryogenesis during cocultivation with *Agrobacterium.* Data scored after 1 month of culture on MS medium containing 100 mg/l kanamycin. Only the relevant differences between *Agrobacterium* strains (chromosomal background and disarmed pTi) are indicated. Both strains also harbored p35SGUSINT

		Explant number	
Agrobacterium strain	Cocultivation medium	Total	With GUS^+ somatic embryos
None	MS	40	0[a]
	Tav	30	0[a]
C58(pGV2260)	MS	30	2
	Tav	21	20
EHA105(pEHA105)	MS	30	2
	Tav	21	21

[a] Control (not cocultivated) explants regenerated in the absence of kanamycin yielded the following proportions of calli harboring somatic embryos: 38/40 (MS medium) and 23/30 (Tav medium). This difference is significant at $P < 0.5\%$. Thus, the use of Tav medium is justified only when the *Agrobacterium*-mediated transformation of proembryogenic cells is sought. There is no difference between the two *Agrobacterium* strains at $P < 0.5\%$ in their ability to produce transgenic embryos under the conditions used.

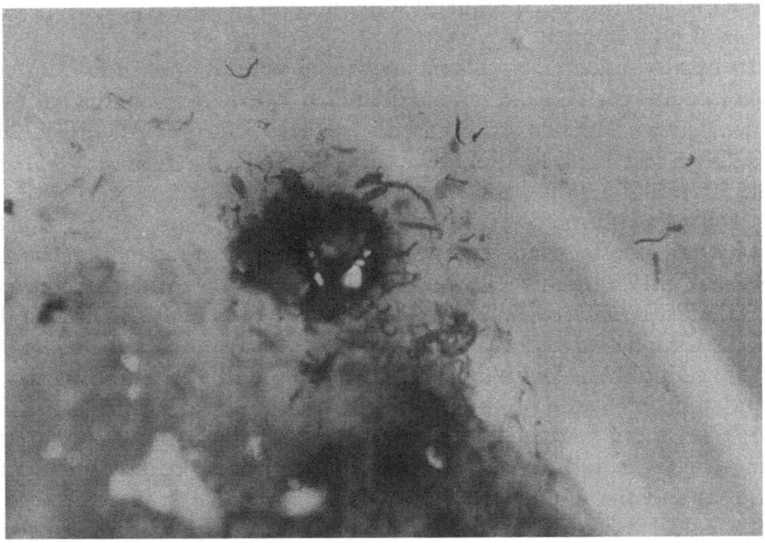

Fig. 2. Transgenic somatic embryo (late heart stage) and cells displaying GUS activity after 2 weeks of culture on selective medium (100 mg/l kanamycin). Callus tissue was sliced in half, and pieces were incubated at 37 °C overnight in the presence of 2 mM X-gluc in 50 mM Na phosphate buffer (pH 7.0) plus 0.1% Triton X-100 after vacuum infiltration

increase the number of somatic embryos either positive *or* negative for GUS staining (not shown). Interestingly, control, noncocultivated cotyledonary leaf explants displayed very high regeneration frequency through somatic embryogenesis on the same medium in the absence of the antibiotic (Table 2). Thus, cocultivation with *A. tumefaciens* or use of this type of cocultivation and regeneration medium in combination with a cocultivation step interfered with somatic embryogenesis.

When Tav medium (Tavazza et al. 1988) containing 5 mg/l 2,4-D and 0.25 mg/l kinetin as well as actively growing potato cells (as feeder layer) was used instead of MS medium during the 48-h cocultivation step, the number of cotyledonary leaf explants showing GUS^+ and kanamycin-resistant somatic embryos at various levels of development was one order of magnitude larger than previously observed with MS cocultivation medium (Table 2).

In conclusion, the use of conditioning medium (Tav plus potato cells feeder layer) and disabled *A. tumefaciens* allows high frequency recovery through somatic embryogenesis of regenerants resistant to kanamycin and expressing *gus* (Fig. 2). This technique should thus make possible the study of transgene expression and regulation during the early stages of somatic embryogenesis and differentiation in eggplant.

5 Conclusions

Transgenic *Solanum melongena* plants can be regenerated from leaf explants cocultivated with *A. tumefaciens* carrying transgenes either located on a binary vector or cointegrated within a Ti plasmid. Thus, it should now be possible to develop eggplant varieties carrying transgenes conferring resistance to some insects and fungi, two major pests for this crop. Similarly, plants tolerant to some broad-spectrum herbicides, for which the relevant resistant or degradative/inactivating genes have been isolated, could be produced.

Also of great interest is the ability to transform eggplant somatic embryos. Indeed, the ability to "tag" embryogenic cells at high frequency should facilitate the study of gene expression in differentiating organs and the influence of extrinsic or intrinsic factors on plant development. As far as we know, eggplant is one of only a few plant species where transformation followed by somatic embryogenesis has been reported. Other examples are cotton (Firoozabady et al. 1987), carrot (Scott and Draper 1987; Syrkin-Wurtele and Bulka 1988), maize (Rhodes et al. 1988), walnut (McGranahan et al. 1988), citrus spp. (Hidaka et al. 1990), and rice (Raineri et al. 1990).

Acknowledgments. Partially supported by a grant from Washington State University International Program Development Office. Part of this work was performed during PFL's Professional Leave at the University of Naples. Publication No. 81 of the Centro di Studio per il Miglioramento Genetico degli Ortaggi, CNR Portici, Italy.

References

An G (1986) Development of plant promoter expression vectors and their use for analysis of differential activity of nopaline synthase promoter in transformed tobacco cells. Plant Physiol 81:86–91

Bevan M (1984) Binary *Agrobacterium* vectors for plant transformation. Nucleic Acids Res 12:8711–8721

Deblaere R, Bytebier B, De Greve H, Debroeck F, Schell J, Van Montagu M, Leemans J (1985) Efficient octopine Ti plasmid-derived vectors for *Agrobacterium tumefaciens*-mediated gene transfer to plants. Nucleic Acids Res 13:4777–4788

Deroles SC, Gardner RC (1988a) Expression and inheritance of kanamycin resistance in a large number of transgenic petunias generated by *Agrobacterium*-mediated transformation. Plant Mol Biol 11:355–364

Deroles SC, Gardner RC (1988b) Analysis of the T-DNA structure in a large number of transgenic petunias generated by *Agrobacterium*-mediated transformation. Plant Mol Biol 11:365–377

Filippone E, Lurquin PF (1989) Stable transformation of eggplant (*Solanum melongena* L.) by cocultivation of tissues with *Agrobacterium tumefaciens* carrying a binary plasmid vector. Plant Cell Rep 8:370–373

Firoozabady E, DeBoer DL, Merlo DJ, Halk EL, Amerson NL, Rashka KE, Murray EE (1987) Transformation of cotton (*Gossypium hirsutumm* L.) by *Agrobacterium tumefaciens* and regeneration of transgenic plants. Plant Mol Biol 10:105–116

Guri A, Sink KC (1988) *Agrobacterium* transformation of eggplant. J Plant Physiol 133:52–55

Hidaka T, Omura M, Ugaki M, Tomiyama M, Kato A, Oshima M, Motoyoshi K (1990) *Agrobacterium*-mediated transformation and regeneration of *Citrus* spp. from suspension cells. Jpn J Breed 40:199–207

Hood E, Helmer GL, Fraley RT, Chilton M-D (1986) The hypervirulence of *Agrobacterium tumefaciens* A281 is encoded in a region of pTiBo542 outside of T-DNA. J Bacteriol 168:1291–1301

Horsch RB, Fry JE, Hoffmann NL, Eichholtz D, Rogers SG, Fraley RT (1985) A simple and general method for transferring genes into plants. Science 227:1229–1231

Jefferson RA (1987) Assaying chimeric genes in plants: the GUS gene fusion system. Plant Mol Biol Rep 5:387–405

Lurquin PF (1989) Uptake and integration of exogenous DNA in plants. In: Bajaj YPS (ed) Biotechnology in agriculture and forestry, vol 9. Plant protoplasts and genetic engineering II. Springer, Berlin Heidelberg New York, pp 54–74

Matsuoka H, Hinata K (1979) NAA-induced organogenesis and embryogenesis in hypocotyl callus of *Solanum melongena* L. J Exp Bot 30:363–370

McGranahan GH, Leslie CA, Uratan SL, Martin LA, Dandekar AM (1988) *Agrobacterium*-mediated transformation of walnut somatic embryos and regeneration of transgenic plants. Bio/Technol 6:800–804

Raineri DM, Bottino P, Gordon MP, Nester EW (1990) *Agrobacterium*-mediated transformation of rice (*Oryza sativa* L.) Bio/Technol 8:33–41

Rhodes CA, Pierce DA, Mettler, IJ, Mascarenhas D, Detmer JJ (1988) Genetically transformed maize plants from protoplasts. Science 240:204–210

Rotino GL, Gleddie S (1990) Transformation of eggplant (*Solanum melongena* L.) using a binary *Agrobacterium tumefaciens* vector. Plant Cell Rep 9:26–29

Scott RJ, Draper J (1987) Transformation of carrot tissues derived from proembryogenic suspension cells: a useful model system for gene expression studies in plants. Plant Mol Biol 8:265–274

Simpson J, Herrera-Estrella L (1989) DNA recombinants and transformation of agricultural crops. In: Bajaj YPS (ed) Biotechnology in agriculture and forestry, vol 9. Plant protoplasts and genetic engineering II. Springer, Berlin Heidelberg New York, pp 75–98

Syrkin-Wurtele E, Bulka K (1989) A simple, efficient method for the *Agrobacterium*-mediated transformation of carrot callus cells. Plant Sci 61:253–262

Tavazza R, Tavazza M, Ordas RJ, Ancora G, Benvenuto E (1988) Genetic transformation of potato (*Solanum tuberosum*): an efficient method to obtain transgenic plants. Plant Sci 59:175–181

Vancanneyt G, Schmidt R, O'Connor-Sanchez A, Willmitzer R, Rocha-Sosa M (1990) Construction of an intron marker gene: splicing of the intron in transgenic plants and its use in monitoring early events in *Agrobacterium*-mediated plant transformation. Mol Gen Genet 220:245–250

Subject Index

FSC
www.fsc.org
MIX
Papier aus verantwortungsvollen Quellen
Paper from responsible sources
FSC® C105338

If you have any concerns about our products,
you can contact us on
ProductSafety@springernature.com

In case Publisher is established outside the EU,
the EU authorized representative is:
Springer Nature Customer Service Center GmbH
Europaplatz 3, 69115 Heidelberg, Germany

Printed by Libri Plureos GmbH
in Hamburg, Germany